Power Transformer Online Monitoring Using Electromagnetic Waves

Power Transformer Online Monitoring Using Electromagnetic Waves

Gevork B. Gharehpetian
Amirkabir University of Technology (AUT), Tehran, Iran

Hossein Karami
High Voltage Studies Research Department,
Niroo Research Institute (NRI), Tehran, Iran

ACADEMIC PRESS
An imprint of Elsevier

ELSEVIER

Academic Press is an imprint of Elsevier
125 London Wall, London EC2Y 5AS, United Kingdom
525 B Street, Suite 1650, San Diego, CA 92101, United States
50 Hampshire Street, 5th Floor, Cambridge, MA 02139, United States
The Boulevard, Langford Lane, Kidlington, Oxford OX5 1GB, United Kingdom

Notices

Knowledge and best practice in this field are constantly changing. As new research and experience broaden our understanding, changes in research methods, professional practices, or medical treatment may become necessary.

Practitioners and researchers must always rely on their own experience and knowledge in evaluating and using any information, methods, compounds, or experiments described herein. In using such information or methods they should be mindful of their own safety and the safety of others, including parties for whom they have a professional responsibility.

To the fullest extent of the law, neither the Publisher nor the authors, contributors, or editors, assume any liability for any injury and/or damage to persons or property as a matter of products liability, negligence or otherwise, or from any use or operation of any methods, products, instructions, or ideas contained in the material herein.

ISBN: 978-0-12-822801-2

For information on all Academic Press publications visit our website at https://www.elsevier.com/books-and-journals

Publisher: Charlotte Cockle
Acquisitions Editor: Rachel Pomery
Editorial Project Manager: Andrea R. Dulberger
Production Project Manager: Kamesh R
Cover Designer: Miles Hitchen

Typeset by TNQ Technologies

Working together
to grow libraries in
developing countries

www.elsevier.com • www.bookaid.org

Contents

CHAPTER 5   Analyzing EMWs using wavelet transform............................**89**
Gevork B. Gharehpetian and Hossein Karami

CHAPTER 6   Frequency domain analysis of scattering parameters in transformers **115**
Gevork B. Gharehpetian, Hossein Karami and Seyed-Alireza Ahmadi

CHAPTER 9 Hyperbolic method...**247**

Gevork B. Gharehpetian, Hossein Karami and Seyed-Alireza Ahmadi

CHAPTER 10 Partial discharge monitoring using EMWs.....................**291**

Gevork B. Gharehpetian, Hossein Karami and Seyed-Alireza Ahmadi

Preface

Considering the new operating conditions of power systems, the need to achieve reliable power transformer protection and condition monitoring methods along with the economic solutions has been highlighted in recent years. Although the previous techniques have been used and applied to transformers, new approaches based on electromagnetic waves can add new values to the existing solutions. The current book presents and aggregates the authors' experiences with the application of electromagnetic wave-based solutions in different national and international research and industrial projects over several years.

The content of the book has been organized based on the needs of industry experts and researchers to reach new ideas and novel methods for online monitoring of power transformers. The experimental results presented in this book can attract a lot of attention from researchers to solve their problems with online monitoring systems using previous methods. Therefore, all researchers working in the electrical power industries, professors, and graduate students of electrical power engineering, who are working on their theses or projects, can use it.

It is indubitable that the results presented in this book cannot be achieved without the cooperation of many BSc and PhD students and experts such as Dr. Maryam Akhavan Hejazi and Dr. Yaser Norouzi. Great and special thanks to Dr. Seyed-Alireza Ahmadi for his cooperation in the book. His experiences and proficiency in power transformer protection and condition monitoring systems help us in different chapters of this book.

G. B. Gharehpetian and H. Karami

An introduction to power transformer monitoring

Gevork B. Gharehpetian[1], Hossein Karami[2] and Seyed-Alireza Ahmadi[3]

[1]*Amirkabir University of Technology (AUT), Electrical Engineering Department, Tehran, Iran;* [2]*Niroo Research Institute (NRI), High Voltage Studies Research Department, Tehran, Iran;* [3]*School of Electrical and Computer Engineering, College of Engineering, University of Tehran, Tehran, Iran*

1.1 Monitoring methods classification

1.1.1 Importance of transformer monitoring

Monitoring is the task of supervising the performance of different parts of equipment and detecting instantaneous condition to warn a defect occurrence or prevent propagation of defects. Transformers should be maintained and repaired due to their effective role in the stability and reliability of the power network. Thus the transformer monitoring systems should be used to quickly detect the defect and prevent its propagation in addition to the usual protection methods. Currently, there are many effective monitoring systems for power transformers, in which various sensors are used.

Most operators aim to understand the internal conditions of power transformers to manage the operation of the transformer and power system as well. The monitoring systems help to assess the state of transformers more accurately. The transformer significance in the network, its current condition, and economic issues are usually considered to determine the capabilities and facilities of a monitoring system for transformers.

Monitoring systems are grouped into two categories: online and offline, with many advantages, as follows (Blanc et al., 2008; Jarman et al., 2008; Tenbohlen et al., 2002):

- Increasing transformer reliability by minimizing unpredicted outages.
- Reducing the cost of undistributed energy.
- Conducting repairs based on the real condition of equipment and reducing maintenance costs.
- Optimally utilizing the transformer capacity.
- Increasing the service life of the transformer.
- Delaying the replacement cost.

1.1.2 Defects in power transformers

The power transformers include a tank, core, high-voltage (HV) and low-voltage (LV) coils, HV and LV bushings, insulation, and cooling tubes, etc. As mentioned before, the defects occurring in power transformers can be divided into three categories: electrical, thermal, and mechanical. Proper monitoring can detect and may prevent the spread of the first two, while the third one may persist for the rest

Power Transformer Online Monitoring Using Electromagnetic Waves. https://doi.org/10.1016/B978-0-12-822801-2.00001-X

of the transformer's life. Generally, the reason for such defects can be internal or external. Common external reasons for defects in transformers are lightning overvoltage, system overload, switching overvoltage, and system faults such as short circuits. In contrast, the internal reasons are due to insulation weakness, weakness in the winding clamp, increase in temperature, oxygen, moisture, insulating oil pollution, partial discharge, defect in design and construction, and resonance of internal winding (Abeywickrama et al., 2006; Rosentino et al., 2011). Based on the studies over a 5-year period, the reasons for 51% of transformer failures are as follows (Rosentino et al., 2011):

- Moisture, pollution, and aging, which reduce the insulation strength of transformers.
- Damage to the winding due to short circuit current (electromagnetic force).
- Damage to bushings due to loss of insulation strength.

It should be noted that there are other categories with more details in the literature, such as the one in (Coffeen, 2003), which divides the defects in transformers into those related to the windings, core, tap changer, tank and oil, terminals, and other equipment. The statistical results related to conventional defects in transformers should be evaluated to achieve a low-cost defect detection system. The defects associated with the windings account for about 19% of all transformer defects (Rosentino et al., 2011).

In another study, 188 transformers with different defect types have been examined (Abu-Siada et al., 2013). These transformers had various voltages and power levels between 88 and 756 kV and 20 and 800 MVA, respectively. Based on the results, it can be concluded that the main reason for defects in small (20−100 MVA), medium (100−400 MVA), and large (more than 400 MVA) transformers are the transformer aging, tap changer condition, and insulation incompatibility, respectively. The defects are classified by their occurrence as follows (Abu-Siada et al., 2013):

- Transformer aging, 30%
- Tap changer, 23%
- Core, 16%
- Lightning and switching voltages, 12%
- Short circuit, 8%
- Other defects, 11%

Based on the studies conducted on 63 transformers (Bagheri et al., 2013) in the Tehran Regional Electricity Company (TREC), the defects related to the tap changer and winding in the subtransmission and transmission sections are in the first place, which confirms the results of (Abu-Siada et al., 2013).

Aging in transformers puts the power networks at risk because the effects of a transformer outage can be catastrophic. In addition, the continuity of energy transfer has become more and more significant in smart and deregulated environment of power networks.

The common detection tests for each defect in the different parts of the transformer are described in standards (IEC Standard 60076-5, 2006). Based on the standards, comparing the results in a test with those in the previous ones on a similar device plays a significant role and affects the overall conclusion.

1.1.3 Time/condition-based monitoring

There are three main monitoring approaches in the empirical comparison: time-based, type-based, and structure-based (Christian & Feser, 2004). In addition, a model-based comparison has been proposed

(Rahimpour et al., 2003). The time-based comparison is regarded as the best approach in the scattering parameter method and the most common and accurate one in the frequency response analysis (FRA) method (Christian & Feser, 2004). This approach compares the measured parameter with the latest measurement results and those of a normal transformer. Continuous monitoring of the transformer winding by measuring a goal parameter (selected feature) provides the necessary data for the time-based comparison (Karami et al., 2019). Accurate information about the measurement setup, test steps, etc., is required to obtain repeatable and comparable results (Karami, Gharehpetian, Norouzi, et al., 2020). Such a method is applicable to all types of transformers. Type-based comparisons can be utilized for transformers of the same type when previous measurement data are unavailable. The results related to the new measurements can be compared with those of another transformer. Such an approach is applicable when the construction characteristics of both transformers are considered the same. The similarity of the core and winding is maximized when both transformers are designed and manufactured in the same factory. Accordingly, the test method and the related conditions should be the same.

The structure-based comparison can only be applied to a three-phase transformer with the same geometry. In this approach, one of the phases is considered as the reference to detect the defect of another phase by comparison of measurement results. However, such a comparison does not have high accuracy and is only used while the other two approaches cannot be applied. The model-based comparison uses a model related to a transformer that can determine goal/selected measuring parameters. The parameters obtained from modeling are compared with those achieved from measurements to validate the model. Then, the model can be used for monitoring aims.

1.1.4 Online/offline strategy

To fix the defective transformer, it should be out of service for a long time. Based on the data collected during 1997 on the reliability analysis of transformers above 220 kV in China, the unplanned outage time due to winding defects compared to other ones was equal to 79.49%, 72.31%, and 98.92% in 220, 330, and 500 kV transformers, respectively (Bagheri et al., 2013). Therefore fast detection of winding defects in the transformer can prevent many unwanted outages (Mahmoodi et al., 2020). Meanwhile, most defects in the transformer winding are deformations, which can be regarded as a starting point for creating other defects. For instance, transient waves due to switching and lightning may lead to a mechanical defect in the winding. Thus detecting, troubleshooting, and locating internal defects, especially in the early stages, play a significant role in extending the life and improving the performance of transformers (Karami et al., 2016).

There are two main maintenance methods: condition-based and periodic maintenance. Condition-based maintenance systems require regular monitoring, which can be conducted offline or online. On-time alerts for defect occurrence can keep the continuity of the electrical energy supply and prevent serious damage to the transformer. The transformer should be disconnected from the power supply or service in offline methods, while online monitoring steps are performed parallel to its service and operation, therefore the transformer should not be disconnected from the network. Generally, the use of online monitoring systems has a lot of advantages, some of which are as follows (Joshi & Kulkarni, 2008, 2010; Naiqiu et al., 2002):

• Detecting defects in the early stages and preventing transformer future outage
• Improving transformer reliability by minimizing unwanted outages

- Reducing the cost of undistributed energy
- Applying repairs based on the conditions and reducing maintenance costs compared to periodic ones
- Analyzing insulation aging
- Improving personnel safety and environmental protection
- Providing valuable information to analyze the main reason for defect occurrence
- Optimally utilizing transformer capacity specially in smart grid environment
- Increasing the transformer life and delaying the replacement cost
- Realizing testing repeatability.

Periodic maintenance, which is considered as time-based maintenance, is usually carried out based on the transformer life and history of defects. So far, most electric utilities focus on condition-based maintenance, which has various advantages such as improving system reliability, reducing maintenance costs, and decreasing human errors due to the reduced number of maintenance operations. However, it may fail to detect a defect that occurs in the intervals between two successive maintenances. Therefore online measurements seem to be a better solution for transformer monitoring and can reduce the duration of possible repairs and increase the reliability of the electric network due to the continuous operation and decreasing outage costs of the transformer (Karami, Azadifar, Mostajabi, et al., 2020).

1.1.5 Time/frequency domain analysis

As described before, different features, such as chemical or electrical ones, are used to monitor the condition of power transformers. In most methods, an electric or magnetic signal is recorded and analyzed to detect defects. The transformer condition can be obtained not only by analyzing a signal in the time domain but also by investigating its important information in the frequency domain. Therefore there are two main approaches for using the transformer signals in the monitoring methods, i.e., use of them in frequency and time domains.

One of the well-known methods for using the frequency domain of signals in transformers is FRA (Behkam et al., 2021; Tarimoradi et al., 2021). In this method, the mechanical defects of a winding can be detected by sending a wideband frequency signal, usually a pulse, to one end of the winding and receiving it from another. The Fourier transform of the recorded signal is analyzed and by comparing it with that of the sound condition of the winding and applying some computational algorithms, the defect could be detected.

A well-known method in time domain analysis is the time difference of arrival for partial discharge (PD) localization (Karami et al., 2012, 2013). A PD in the power transformer emits electromagnetic waves (EMWs) inside the tank, so these signals are recorded by installing antennas in the transformer tank (Karami et al., 2018; Tabarsa et al., 2019). As will be described in Chapter 10, the time difference of the received signals for the four antennas can be used to localize the PD position.

The useful information describing the condition of a transformer can be extracted through various methods in the time or frequency domains. One researcher may use a recorded signal in the frequency domain, while another may extract information from that signal in the time domain. This shows the importance of the used method to extract information. For example, in (Golsorkhi et al., 2012), the researchers detected the radial deformation of acwinding by sending an electromagnetic pulse toward the winding and analyzing the time domain of the received signal, while in (Hejazi et al., 2011), the radial deformation has been detected by analyzing the signal in the frequency domain using wavelet transform. Therefore it is expected that the combination of frequency and time domains will attract more attention in the near future, to achieve more information from transformer conditions.

1.1.6 Life management

Transformers are among the largest and most expensive equipment in the network. It is well known that the useful life of a transformer can be increased by assessing its condition to determine the necessary measures for its optimal repair and maintenance. The aging of a transformer is the change of physical and chemical properties of its components due to environmental parameters such as pressures and stresses resulting from operating conditions, including thermal, mechanical, and electrical factors. When a transformer is not properly repaired or maintained, a lot of investment is lost and the reliability of the transformer and power system reduces. Thus each of the transformers, especially the old ones, should be maintained and monitored, and their remaining lives should be estimated.

For transformer life management, accurate views and information about its condition are required. This could be obtained with the help of different tests, operation histories, and records of transformers, along with taking corrective measures to improve their condition. It is impossible to perform all the tests on a transformer to achieve life management objectives. By contrast, its condition can determine the need to perform a special test. The utility engineers seek to find a solution to increase equipment reliability. The knowledge of the transformer life management can answer such questions.

It is worth noting that the remaining life of the transformer cannot be determined since the amount and intensity of imposed stresses on the transformer in the future are unknown. However, the useful life, efficiency, reliability, productivity, etc., can be increased for the transformers utilizing life management methods.

To perform life management measures on power transformers, knowing the type, method, and time of collecting information plays a significant role. From an engineering perspective, the information related to transformer life management includes the following ones:

- Data on the design, construction, testing, and operation of the transformer in the network.
- History of events.
- Information and history related to maintenance, and reconstruction, or replacement.
- Information related to monitoring and detecting the defects in transformers and related equipment considering environmental and economic constraints.

1.2 Monitoring of mechanical defects

Available methods for detecting winding deformation and displacement include short circuit testing, frequency response or transfer function analysis, low-pressure shock test, an ultrasonic approach, and electromagnetic waves, which are distinguished in terms of accuracy, amount of obtained information, ease of implementation, and the requirement of transformer disconnection. In some methods, there is no need to interrupt the transformer operation to make measurements and detect the defect when it is in service. Therefore the defect can be detected before spreading. The mechanical defects and some monitoring methods are briefly reviewed in the next.

1.2.1 Winding defects

Mechanical defects, including radial deformation and axial displacement, may be occurred due to improper transportation and the electrodynamic force due to high short-circuit currents. Winding deformation or displacement can increase the voltage gradient in the deformed parts, resulting in damaging or impairing the insulation and creating defects due to an increase in the temperature around the above-mentioned parts, which raises the insulation decay rate (García et al., 2005). In addition, an increase in temperature in oil-filled transformers can alter the properties of the oil and may lead to a PD, resulting in producing inappropriate gases, which can be detected by analyzing the oil or operation of the Buchholz relay. The insulation strength of the windings is lost depending on the severity of the mechanical defect, leading to a possible electrical short circuit and operation of protection relays. Thus timely detection of the type, location, and severity of the winding defect can indicate the transformer condition and prevent its unwanted outage. Accordingly, any minor deformation in the winding structure should be detected as soon as possible, along with making the necessary repairs.

1.2.2 Short circuit impedance method

This method has been developed based on IEC 60,076 (IEC Standard 60076-5, 2006), along with performing a large number of practical tests (Santhi et al., 2005). Based on the method, the short circuit reactance of a power transformer changes when the winding is deformed, or its physical dimensions alter. Thus the effect of electrodynamic forces on the windings can appropriately be estimated by comparing the two values related to short circuit reactance before and after its occurrence. Generally, a change of more than 2% in the short circuit reactance can indicate a deformation in the windings (Xu & Li, 1998).

The short circuit reactance measurement can simultaneously be performed during transformer operation (Xu & Li, 1998). Such a method does not provide any information about the location and type of deformation in the windings (Christian & Feser, 2004).

1.2.3 Vibration method

Vibration analysis is another method for detecting the internal defects in the transformer. In this method, the components inside the transformer tank are moved repetitively, and the transformer condition is assessed by analyzing the changes in the vibration response. It should be noted that the mechanical properties of the transformer affect the vibration response. The online implementation of this method has been proposed in (Alpatov, 2004; Hu et al., 2011; Shi-bin & Guo, 2004; Std 62-1995, 1995), indicating that the vibration in a transformer tank depends on the square of the voltage and current signals. Therefore the main harmonic of the vibration in the winding is twice the main frequency of the power system. The relations for tank vibration in terms of the square of voltage and current, along with some defined coefficients, are indicated in (Alpatov, 2004; Hu et al., 2011; Shi-bin & Guo, 2004; Std 62-1995, 1995).

As indicated, the vibration in the transformer tank can be regarded as an online method for detecting winding deformation. However, external factors such as loosening the clamps affect the test results, which sometimes cannot be interpreted.

1.2.4 Ultrasonic method

This method was introduced in 2002 (Naiqiu et al., 2002). The ultrasonic waves are sent by an ultrasonic transducer installed outside the transformer tank and received again by the same transducer after reflection from transformer oil and windings. The distance between the transformer winding and the tank can be determined in different places, considering the velocity of waves and round-trip time.

The accuracy and simplicity in implementation and inference, as well as the possibility of detecting the defect in the transformer winding in the operation mode, are among the advantages of this method, while the measurement errors due to temperature changes and oil circulation inside the tank are among its disadvantages. However, some methods have been indicated to reduce the effect of the mentioned errors (Naiqiu et al., 2002).

1.2.5 Current deviation coefficient method

In this method, a high-frequency low-voltage signal is applied to the power system line. Then the high-frequency currents in both the line and ground outputs are continuously measured by accurate current probes and digital filtering methods (Feser et al., 2000; Malewski et al., 1992). Capacitive reactance changes due to winding deformation lead to changes in the high-frequency currents of the terminal. The ratio of changes is calculated. The current deviation coefficient (CDC), which is used as a justifiable relation, is calculated as follows:

$$CDC = \log_{10}\left(\frac{I_{1H} - I'_{1H}}{I_{2H} - I'_{2H}}\right) \tag{1.1}$$

where, I_{1H} and I_{2H} indicate the high-frequency currents of reference or normal mode of output terminals, and I'_{1H} and I'_{2H} represent the high-frequency currents measured at the same terminals.

The CDC is regarded as an online method, which has some disadvantages, including the effects of noise on the output response and its low sensitivity to winding displacement and deformation. In addition, the CDC cannot determine the location of the defect.

1.2.6 Frequency response analysis method

This method is based on the two-port network theory. Considering the transformer as a linear, complex, and passive network (Fig. 1.1), for one input voltage signal, several outputs can be defined. Each of the defined output signals creates a transfer function, in which some of them are as follows (Christian et al., 1999):

$$TF_{u,U}(f) = \frac{U(f)}{u(f)} \tag{1.2}$$

$$TF_{u,i}(f) = \frac{i(f)}{u(f)} \tag{1.3}$$

$$TF_{u,I}(f) = \frac{I(f)}{u(f)} \tag{1.4}$$

FIGURE 1.1

Transformer as linear two-port network.

Fourier transform (FT) of the impulse response in a linear system is called the frequency response or the system transfer function. Theoretically, a transfer function can describe a network completely. The frequency response can be measured directly by a network analyzer in the frequency domain or by applying the fast FT on the time domain measurements (Tarimoradi & Gharehpetian, 2017). In other words, there are two common FRA types: the frequency sweep method (FRA-S) and the impulse method (FRA-I). In the first method, a voltage with a variable frequency is applied to the winding. The response is recorded at the other end of the winding or terminal. However, in the second method, an impulse voltage is applied to the HV winding with the same method of recording the response (Wang et al., 2004).

In the case of an internal defect, the output responses to the given inputs differ from the one obtained in the healthy state and cause the defect to be detected. Thus the user should utilize the most sensitive signal to detect the defect. The transfer function is considered as a comparative method, not an absolute one. The actual measurement results are compared with the previous ones (transformer fingerprint). There is probably no defect inside the transformer when no significant alteration is observed in the fingerprint. The values of inductances and capacitances in the transformer change in the case of deformation and displacement in the transformer windings. Therefore the changes in the transfer functions of the transformer winding can indicate the type and location of the defect (Leibfried & Feser, 1999; Rahimpour et al., 2003).

The computer simulations can be used to achieve the intended information and avoid costly practical tests. To this aim, the effects of winding defects on the transfer function can be evaluated by the simulated transformer model. In addition, various factors affecting the FRA test results, such as a change in the position of power transformer tap changer, ambient temperature change, shunt impedance value effect, HV bushing effect, transformer neutral point connection effect, effect of length of measuring wires and connections, and wire displacement effect should be regarded (Wang et al., 2004).

1.2.7 Electromagnetic waves-based methods

The method which can be utilized online plays a significant role among those specialized in detecting the displacement of the transformer winding. The present book aims to discuss the electromagnetic waves-based method for defect detection, which can be applied alongside the previous ones, such as the transfer function method. Utilizing EMWs to detect winding defects was proposed (Hejazi et al., 2007, 2008) for the first time.

The electromagnetic approach is based on sending signals and analyzing reflected ones in the time or frequency domain (Karami, Gharehpetian, Hejazi, et al., 2020). To this aim, the EMWs are transmitted toward the transformer winding by an antenna, and the reflected waves are received by the same antenna, resulting in storing in a database for normal and defective modes. Then, the change in the winding can be detected by comparing the normal and defective modes (Rahbarimagham et al., 2015, 2019).

The change in amplitude and phase of the return waves can only be considered a function of deformation in transformer windings. This is because the behavior of wave propagation in the transformer environment can be considered constant and predicted because a stationary system is used in the transformer. Different methods based on EMWs will be discussed in the next chapters. In Chapter 2, after describing winding mechanical defects, the advantages of EMWs-based methods compared with other monitoring ones will be explained for these defects. Chapter 3 briefly describes ultra- wideband systems and explains some specifications and applications. The purpose of Chapter 4 is to model the channel using the minimum number of parameters to show the channel behavior in simulations correctly. Different models will be considered for the channel transfer function to detect defects. The applicability of using the wavelet transform to discriminate mechanical (axial and radial) defects in addition to determining their extents through classification methods will be studied in Chapter 5.

Chapter 6 presents the use of scattering parameters to study the feasibility of defect detection and their extents, in addition to axial displacement and radial deformation discrimination, at early stages using EMWs through classification methods such as Bayesian methods, neural networks, k-nearest neighbors, Parzen window, and support vector machines. The CST environment will be used in Chapter 7 to simulate the traveling of electromagnetic signals in transformers and detect radial and axial faults by applying classification methods on indices extracted from time domain signals. Then, the results of the simulation and experimental studies will be compared and the performance of the proposed approach will be discussed. In Chapter 8, the syntactic aperture radar imaging method to detect the defect's occurrence, location, and extent without any need for defected situation data will be described,sss and a solution for interactions with PD signals will be proposed. Chapter 9 investigates static ultra-wideband antennas and the subtraction of input signals for three main monitoring stages, including detecting, locating, and determining radial deformation and axial displacement extents. Finally, PD monitoring using EMWs will be described in Chapter 10.

References

Abeywickrama, K. N. B., Serdyuk, Y. V., & Gubanski, S. M. (2006). Exploring possibilities for characterization of power transformer insulation by frequency response analysis (FRA). *IEEE Transactions on Power Delivery, 21*(3), 1375−1382.

Abu-Siada, A., Hashemnia, N., Islam, S., & Masoum, M. A. (2013). Understanding power transformer frequency response analysis signatures. *IEEE Electrical Insulation Magazine, 29*(3), 48–56.

Alpatov, M. (2004). Online detection of winding deformation. In *Conference record of the 2004 IEEE international symposium on electrical insulation* (pp. 113–116).

Bagheri, M., Naderi, M. S., Blackburn, T., & Phung, T. (2013). Frequency response analysis and short-circuit impedance measurement in detection of winding deformation within power transformers. *IEEE Electrical Insulation Magazine, 29*(3), 33–40.

Behkam, R., Karami, H., Naderi, M. S., & Gharehpetian, G. B. (2021). Generalized regression neural network application for fault type detection in distribution transformer windings considering statistical indices. *COMPEL-The International Journal for Computation and Mathematics in Electrical and Electronic Engineering, 41*(1), 381–409.

Blanc, R., Buffiere, G., Taisne, J. P., Tanguy, A., Guvinic, P., Long, P., Moutin, E., & Devaux, F. (2008). *Transformer refurbishment policy at RTE conditioned by the residual lifetime assessment.* A2-204.

Christian, J., & Feser, K. (2004). Procedures for detecting winding displacements in power transformers by the transfer function method. *IEEE Transactions on Power Delivery, 19*(1), 214–220.

Christian, J., Feser, K., Sundermann, U., & Leibfried, T. (1999). Diagnostics of power transformers by using the transfer function method. In , *Vol. 1. 1999 eleventh international symposium on high voltage engineering* (pp. 37–40).

Coffeen, L. T. (2003). *System and method for online impulse frequency response analysis.* Google Patents.

Feser, K., Christian, J., Neumann, C., Sundermann, U., Leibfried, T., Kachler, A., & Loppacher, M. (2000). The transfer function method for detection of winding displacements on power transformers after transport, short circuit or 30 years of service. *CIGRE, 12*, 33–04.

García, B., Burgos, J. C., & Alonso, Á. (2005). Winding deformations detection in power transformers by tank vibrations monitoring. *Electric Power Systems Research, 74*(1), 129–138.

Golsorkhi, M. S., Hejazi, M. S. A., Gharehpetian, G. B., & Dehmollaian, M. (2012). A feasibility study on the application of radar imaging for the detection of transformer winding radial deformation. *IEEE Transactions on Power Delivery, 27*(4), 2113–2121.

Hejazi, M. A., Gharehpetian, G. B., & Mohammadi, A. (2007). Characterization of online monitoring of transformer winding axial displacement using electromagnetic waves. In *15th international symposium on high voltage engineering.*

Hejazi, M. A., Choopani, M., Dabir, M., & Gharehpetian, G. B. (2008). Effect of antenna position of transformer winding axial displacement measurement using electromagnetic waves. In *2nd IEEE international conference on power and energy* (pp. 1–3).

Hejazi, M. S. A., Ebrahimi, J., Gharehpetian, G. B., Mohammadi, M., Faraji-Dana, R., & Moradi, G. (2011). Application of ultra-wideband sensors for online monitoring of transformer winding radial deformations—A feasibility study. *IEEE Sensors Journal, 12*(6), 1649–1659.

Hu, G., Zhang, L., Wu, X., Correia, D., & He, W. (2011). Detecting the capacity of distribution transformer based on an online method. In *2011 Asia-Pacific power and energy engineering conference* (pp. 1–4).

IEC Standard 60076-5, T. (2006). Part 5: Ability to withstand short circuit. In *IEC standard, 60076–5.*

Jarman, P., Tenbohlen, S., Judd, M. D., Olof-Stenestam, B., & Viereck, K. (2008). Recommendations for condition monitoring and condition assessment facilities for transformers. *Electra, 237*, 48–57.

Joshi, P. M., & Kulkarni, S. V. (2008). Transformer winding diagnostics using deformation coefficient. In *2008 IEEE power and energy society general meeting-conversion and delivery of electrical energy in the 21st century* (pp. 1–4).

Joshi, P. M., & Kulkarni, S. V. (2010). A novel approach for online deformation diagnostics of transformer windings. *IEEE PES General Meeting*, 1–6.

Karami, H., Hejazi, M. S. A., Naderi, M. S., Gharehpetian, G. B., & Mortazavian, S. (2012). Three-dimensional simulation of PD source allocation through TDOA method. In *The 4th conference on thermal power plants* (pp. 1–4).

Karami, H., Gharehpetian, G. B., & Hejazi, M. S. A. (2013). Oil permittivity effect on PD source allocation through three-dimensional simulation. In *International conference on power system Tehran-Iran* (pp. 1–5).

Karami, H., Gharehpetian, G. B., Norouzi, Y., & Hejazi, M. A. (2016). GLRT-based mitigation of partial discharge effect on detection of radial deformation of transformer HV winding using SAR imaging method. *IEEE Sensors Journal, 16*(19), 7234–7241.

Karami, H., Gharehpetian, G. B., Norouzi, Y., & Hejazi, M. A. (2018). Experimental study on elimination of partial discharge effect on detection of radial deformation of high voltage transformer winding using electromagnetic waves. In *2018 IEEE international conference on environment and electrical engineering and 2018 IEEE industrial and commercial power systems europe (EEEIC/I&CPS europe)* (pp. 1–5).

Karami, H., Tabarsa, H., Gharehpetian, G. B., Norouzi, Y., & Hejazi, M. A. (2019). Feasibility study on simultaneous detection of partial discharge and axial displacement of HV transformer winding using electromagnetic waves. *IEEE Transactions on Industrial Informatics, 16*(1), 67–76.

Karami, H., Azadifar, M., Mostajabi, A., Rubinstein, M., Karami, H., Gharehpetian, G. B., & Rachidi, F. (2020a). Partial discharge localization using time reversal: application to power transformers. *Sensors, 20*(5), 1419.

Karami, H., Gharehpetian, G. B., Hejazi, M. A. A., & Norouzi, Y. (2020b). *Detection of radial deformations of transformers*. Google Patents.

Karami, H., Gharehpetian, G. B., Norouzi, Y., & Akhavan-Hejazi, M. (2020c). Simultaneous radial deformation and partial discharge detection of high-voltage winding of power transformer. *IET Electric Power Applications, 14*(3), 383–390.

Leibfried, T., & Feser, K. (1999). Monitoring of power transformers using the transfer function method. *IEEE Transactions on Power Delivery, 14*(4), 1333–1341.

Mahmoodi, M., Abadi, S. M. N., Karami, H., Hejazi, M. A., & Gharehpetian, G. B. (2020). Design and implementation of dielectric windows for detection of radial deformation of HV transformer winding using radar imaging. *IET Science, Measurement & Technology, 14*(4), 478–485.

Malewski, R., Gockenbach, E., Maier, R., & Fellmann, K. H. (1992). Five years of monitoring the impulse test of power transformers with digital recorders and the transfer function method. *International Conference on Large High Voltage Electric Systems, 1*, 12–201.

Naiqiu, S., Can, Z., Fang, H., Qisheng, L., & Lingwei, Z. (2002). Study on ultrasonic measurement device for transformer winding deformation. In , *Vol. 3. Proceedings. International Conference on Power System Technology* (pp. 1401–1404).

Rahbarimagham, H., Porzani, H. K., Hejazi, M. S. A., Naderi, M. S., & Gharehpetian, G. B. (2015). Determination of transformer winding radial deformation using UWB system and hyperboloid method. *IEEE Sensors Journal, 15*(8), 4194–4202.

Rahbarimagham, H., Karami, H., Esmaeili, S., & Gharehpetian, G. B. (2019). Determination of transformer HV winding axial displacement extent using hyperbolic method—a feasibility study. *IET Electric Power Applications, 13*(7), 1004–1013.

Rahimpour, E., Christian, J., Feser, K., & Mohseni, H. (2003). Transfer function method to diagnose axial displacement and radial deformation of transformer windings. *IEEE Transactions on Power Delivery, 18*(2), 493–505.

Rosentino, A. J. P., Saraiva, E., Delaiva, A. C., Guimarães, R., Lynce, M., Oliveira, J. C., Fernandes, D., Jr., & Neves, W. (2011). Modelling and analysis of electromechanical stress in transformers caused by short-circuits. *Renewable Enegry and Power Quality Journal ISSN, 1*(9), 717–722.

Santhi, S., Jayalalitha, S., & Jayashankar, V. (2005). *Online analysis of short circuit tests on windings. IPST-05 Montreal, Canada*.

Shi-bin, G., & Guo, W. (2004). *Study on online monitoring of windings deformation of power transformer.*

Std 62-1995, I. (1995). *IEEE guide for diagnostic field testing of electric power apparatus-part 1: Oil filled power transformers, regulators, and reactors.* Piscataway, NJ, USA: IEEE.

Tabarsa, H., Hejazi, M. A., & Gharehpetian, G. B. (2019). Detection of HV winding radial deformation and PD in power transformer using stepped-frequency hyperboloid method. *IEEE Transactions on Instrumentation and Measurement, 68*(8), 2934−2942. https://doi.org/10.1109/TIM.2018.2868491

Tarimoradi, H., & Gharehpetian, G. B. (2017). Novel calculation method of indices to improve classification of transformer winding fault type, location, and extent. *IEEE Transactions on Industrial Informatics, 13*(4), 1531−1540.

Tarimoradi, H., Karami, H., Gharehpetian, G. B., & Tenbohlen, S. (2021). *Sensitivity analysis of different components of transfer function for detection and classification of type, location and extent of transformer faults. Measurement* (p. 110292).

Tenbohlen, S., Stirl, T., Bastos, G., Baldauf, J., Mayer, P., Stach, M., Breitenbauch, B., & Huber, R. (2002). Experienced-based evaluation of economic benefits of on-line monitoring systems for power transformers. *Cigré Session,* 12−110, 2002.

Wang, M., Vandermaar, A. J., & Srivastava, K. D. (2004). Transformer winding movement monitoring in service-key factors affecting FRA measurements. *IEEE Electrical Insulation Magazine, 20*(5), 5−12.

Xu, D. K., & Li, Y. M. (1998). A simulating research on monitoring of winding deformation of power transformer by online measurement of short-circuit reactance. In , *Vol. 1. POWERCON'98. 1998 international conference on power system technology. Proceedings (cat. No. 98EX151)* (pp. 167−171).

Using electromagnetic waves for mechanical defects monitoring

Gevork B. Gharehpetian[1], Hossein Karami[2] and Seyed-Alireza Ahmadi[3]

[1]*Amirkabir University of Technology (AUT), Electrical Engineering Department, Tehran, Iran;* [2]*Niroo Research Institute (NRI), High Voltage Studies Research Department, Tehran, Iran;* [3]*School of Electrical and Computer Engineering, College of Engineering, University of Tehran, Tehran, Iran*

2.1 Transformer winding deformation types

As mentioned in the previous chapter, the defects related to the winding are the major type of defects in power transformers. The windings of power transformers, which are made of copper or aluminum conductors with round or rectangular cross-sections in layers or disks, are separated by insulation from each other and the transformer body and core. The defects related to transformer winding can be divided into two categories, including electrical and mechanical.

2.1.1 Winding electrical defects

The electrical defects occur when the transformer insulation is weak or lost for any reason and a short circuit is created. The fault between the transformer winding turns is considered the most common one, which may become a short circuit between the phases if its duration is long. A connection between one of the phases and the transformer tank is a rare case. Due to its higher voltage, such defects can occur more in high-voltage (HV) winding than in low-voltage (LV) ones.

Various types of insulation, such as paper, oil, and ceramics are utilized in different parts of the transformer. Any change, which reduces the electrical strength of the insulation leads to the partial discharge (PD) phenomenon. An increase in the temperature and transformer aging are regarded as significant factors in reducing its insulation lifetime. The PD in the transformer increases losses and decreases efficiency, and may lead to irreparable damage.

Some electrical defects are due to the spread of mechanical failure in the transformer winding. It means that after the occurrence of a mechanical defect, some electrical defects may be created in the transformer. In this situation, the transformer should be kept out of service in the early stage of defect occurrence until the maintenance program and protected from worse events such as explosion, due to spreading the defects (Mahmoodi et al., 2020).

2.1.2 Winding mechanical defects

The mechanical defects in the transformer winding can occur due to improper transporting or electrodynamic forces resulting from a short circuit near the transformer. Such defects are divided into two general categories.

The first category concerns the whole winding axial displacement of HV winding related to LV in the direction of the winding axis. This defect changes the electrical and electromagnetic conditions of the transformer. Therefore it can be detected by electric or electromagnetic-based approaches such as frequency response analysis (FRA) (Behkam et al., 2021; Tarimoradi et al., 2021) or the synthetic apparatus radar imaging method (Mortazavian et al., 2012).

In some studies, the disk space variations defect is categorized as axial displacement. Generally, the sound winding has equivalent spaces between disks. However, due to some forces in the winding, these spaces are changed and in some parts, the distance of two adjacent disks is increased and in some others are decreased, so the equivalency of these distances is not a valid assumption.

The second category is related to radial deformation, in which the distance between the winding turns and its axis changes at different locations and its circular cross-section deforms due to the force applied. It may be a bulgy or concave radial deformation.

2.2 Mechanical forces on windings

Winding mechanical deformations due to electromagnetic forces, which are mainly the result of short circuit currents, are among the most serious defects for transformers in service. Such forces are generally applied to the windings radially or axially. Designing the internal structure to withstand against these forces for a long time increases the transformer lifetime. However, the aforementioned forces can reach millions of pounds of power, resulting in pulling the windings up and down 50 or 60 times per second or pushing them toward each other or at a distance from each other (Blanc et al., 2008).

Winding deformation has different shapes, such as spiral tightening, conductor tilting, radial/hoop buckling, shortened or opened connection loops, loosening of clamp structure, axial displacement, core displacement, and collapse of winding end retainers (García et al., 2005). Such internal defects can be difficult to detect with conventional test methods.

Axial forces tend to compress the windings, while radial forces push the inner winding inward and the outer one outward. Fig. 2.1 shows the force directions, current flow, and magnetic field directions in the winding. According to the current flow direction, the force may be inward or outward.

In most cases, moving or changing the mechanical shape of the windings does not prevent energy transfer and does not cause a transformer outage. But there is a risk that by passing the time, the mechanical impact on the winding insulation causes insulation failure and short circuit inside the transformer as a result of subsequent overvoltages. Therefore it is very important to identify and diagnose the displacements in the transformer. For this task, there are various methods in the literature. Each of these methods has some drawbacks, which limits its practical application.

2.3 Drawbacks of previous methods

Various methods have been proposed to monitor the transformer mechanical defects, some of which are mentioned below:

- Short circuit impedance (IEC Standard 60076–5, 2006)
- Vibration analysis

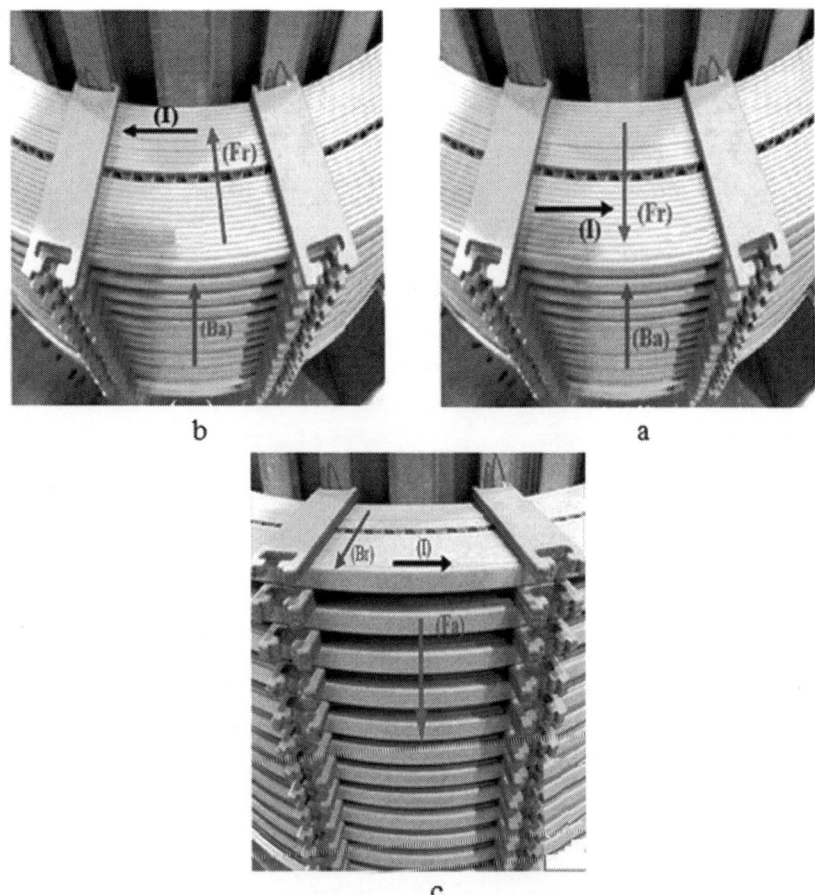

FIGURE 2.1

Forces on transformer windings, (A) External radial force, (B) Internal radial force (toward core), and (C) Axial force.

- Ultrasonic method
- Current deviation coefficient method
- Frequency response analysis
- EMW-based methods

These methods have widely been used in the industry for transformer monitoring. The details of these methods have been described in the previous chapter. Each method can be applied to a specific type of problem and has its own advantages and disadvantages. The disadvantages of their compared to electromagnetic waves (EMWs) method, are as follows:

- The short circuit test method requires short circuit reactance measurement and can simultaneously be performed during the transformer operation (on-line). However, such a method does not provide any information on the type and location of the defect (Christian & Feser, 2004).

- The ultrasonic wave method cannot be used due to its high sensitivity to temperature, noise, and wave propagation medium (oil).
- To monitor the transformer using the FRA method (Leibfried & Feser, 1996, 1999) by injecting a signal, the transformer should be disconnected from the circuit. This method utilizes transient overvoltages created by switching operations or lightning strikes and needs consideration of other influencing factors such as surge arresters, various power network structures, and the like. In addition, the monitoring schedule cannot be realized by the user when the transient mode does not occur (Leibfried & Feser, 1994). Also, the above-mentioned method is in the research phase and has not yet been used for any transformers.

2.4 Partial discharge and EMWs

PD is actually the main source of insulation failure in power transformers. The presence of PD with the passage of time will destroy the insulation between the winding turns of the transformer and finally results in the complete failure of the insulation and transformer outage. PD is actually small sparks inside an insulating material located under a strong electric field. PD occurs when the electrical strength of the electrical insulation is not strong enough in a part of the insulating material due to a nonuniform electric field. The existence of a PD in the transformer is obvious and inevitable. Even in new transformers, discharges of several picocoulombs can be measured, and the standard determines the amount of discharge allowed for different equipment. However, an increase in the amount of PD in a transformer can be a good sign for an upcoming insulation failure. To this end, various methods are used in the power industry to monitor PD in a transformer, one of which is the use of EMWs emitted from the PD source location (Karami, Azadifar, Mostajabi, et al., 2020; Karami et al., 2012).

Many researchers have conducted simulation and experimental studies to locate PD using EMWs (Azadifar et al., 2020; Karami, Gharehpetian, & Hejazi, 2013, pp. 1—5; Karami, Hejazi, & Gharehpetian, 2013). However, most of them only focused on the emitted signal from PD and the effects of this signal on other monitoring systems, and the compatibility of merging monitoring systems has not been investigated. It must be mentioned that the emitted waves from PD can disturb the functionality of other monitoring methods. For example, in (Karami et al., 2016) the effect of emitted PD waves on radial deformation detection has been investigated. Also, similar works have been done to show the effect of PD on mechanical defect detection (Karami et al., 2018, 2019).

The first step is eliminating the PD effect on other monitoring systems and the next is studying the possibility of merging monitoring systems to detect defects at a lower cost. In (Karami, Gharehpetian, Hejazi, et al., 2020), the generalized likelihood ratio test method has been applied to distinguish the effect of PD signals from the monitoring system, and in (Karami, Gharehpetian, Norouzi, et al., 2020), a monitoring system for detection of both mechanical defects and PD has been introduced and designed. This procedure should be investigated for other monitoring systems to study the interaction of defects on detection systems and the possibility of merging.

2.5 Advantages of EMW-based approach

The EMW-based methods are in the early stages of entering the power industry and it seems that they will be used on a large scale in the near future. This is due to the great benefits that they can provide to engineers. The advantages of the electromagnetic method over other ones in transformer winding monitoring can be summarized as follows:

- This method does not require making the transformer out of service during monitoring. The test is performed in long intervals in off-line methods, despite the high probability of failure at such intervals. The electromagnetic method can continuously be utilized to detect defects during the operation of transformers when it is implemented practically, which results in increasing the availability of the transformer and its reliability.
- Economically, the on-line method prevents the unplanned outage of transformers. Thus the defect can be detected in its initial stages before spreading, and the transformer can be repaired or replaced with a sound one with the help of the aforementioned method. One hour outage of a transformer with an average load of 100 MW costs $ 40,000 when the outage is estimated at 40 cents per kWh.
- The EMW-based method can detect two types of defects, i.e., mechanical defects and PD. This can reduce the installation cost of a monitoring system.

Apart from the mentioned advantages, which prove the new benefits gained by using the EMW-based methods in the industry, they can bring new opportunities in developing new protection and monitoring methods. This is due to the fact that the new EMW-based monitoring system can be used to extract other important features from the transformer status, which is important in smart grids.

2.6 EMW-based monitoring methods and comparison approaches

The EMW-based methods can be divided into two general categories, frequency and time domain analysis. The scattering parameter method, estimating the EMWs propagation channel, and the wavelet method are named in the first category, while applying amplitude/phase of the time signal to introduce the index and using it in classification algorithms, radar imaging, and hyperbolic methods are indicated in the second one. Both of these categories will be discussed in the coming chapters in detail.

Imaging and hyperbolic methods do not require a classification approach. The operator can only compare and detect the defect by utilizing and recreating the image or signal received in the normal mode at any time. In contrast, other methods require a database with different defects and calculated signals or indices to compare the new signal or index of the transformer winding with the previous data and classify the defect using classification algorithms. Some of the methods applied for classification include the decision tree method, k-nearest neighbors, neural network, partition window, and support vector machine, which will be discussed later in the next chapters.

Also, the EMW-based methods can be divided into two other categories, UHF and UWB, considering the used frequency of signals. The first one is proper for simultaneously detecting PD and mechanical defects. However, the UWB technology uses an extremely narrow pulse bandwidth, which creates a large bandwidth, resulting in separating the reflections of the environment in the receiver from each other. Therefore the changes in environmental components can be assessed more separately, such as the feature utilized for the electromagnetic method in monitoring the mechanical changes of the transformer winding.

Detecting the occurrence and separation of radial deformation and axial displacement defects in the transformer winding, estimating the volume of radial deformation and the amount of axial displacement, along with detecting the radial defect location, are evaluated in the next chapters. The effect of the transformer tank and oil on the results is also examined. Also, the methods for detecting axial displacement and radial deformation in transformer windings utilizing the EMW-based approach will be presented in detail in the next chapters.

References

Azadifar, M., Karami, H., Wang, Z., Rubinstein, M., Rachidi, F., Karami, H., Ghasemi, A., & Gharehpetian, G. B. (2020). Partial discharge localization using electromagnetic time reversal: A performance analysis. *IEEE Access, 8*, 147507−147515.

Behkam, R., Karami, H., Naderi, M. S., & Gharehpetian, G. B. (2021). Generalized regression neural network application for fault type detection in distribution transformer windings considering statistical indices. In *COMPEL-the international journal for computation and mathematics in electrical and electronic engineering*.

Blanc, R., Buffiere, G., Taisne, J. P., Tanguy, A., Guvinic, P., Long, P., Moutin, E., & Devaux, F. (2008). Transformer refurbishment policy at RTE conditioned by the residual lifetime assessment. In *A2-204*.

Christian, J., & Feser, K. (2004). Procedures for detecting winding displacements in power transformers by the transfer function method. *IEEE Transactions on Power Delivery, 19*(1), 214−220.

García, B., Burgos, J. C., & Alonso, Á. M. (2005). Transformer tank vibration modeling as a method of detecting winding deformations-part I: Theoretical foundation. *IEEE Transactions on Power Delivery, 21*(1), 157−163.

IEC Standard 60076-5, T. (2006). Part 5: Ability to withstand short circuit. In *IEC Standard, 60076−5*.

Karami, H., Hejazi, M. S. A., Naderi, M. S., Gharehpetian, G. B., & Mortazavian, S. (2012). Three-dimensional simulation of PD source allocation through TDOA method. In *The 4th conference on thermal power plants* (pp. 1−4).

Karami, H., Gharehpetian, G. B., & Hejazi, M. S. A. (2013a). Oil permittivity effect on PD source allocation through three-dimensional simulation. In *Int'l. Power system conf., tehran-Iran*.

Karami, H., Hejazi, M. S. A., & Gharehpetian, G. B. (2013b). Simulation of transformer oil effect on PD source allocation. In *4th conference on partial discharge in electrical apparatus (PDC'13)* (pp. 26−27).

Karami, H., Gharehpetian, G. B., Norouzi, Y., & Hejazi, M. A. (2016). GLRT-based mitigation of partial discharge effect on detection of radial deformation of transformer HV winding using SAR imaging method. *IEEE Sensors Journal, 16*(19), 7234−7241.

Karami, H., Gharehpetian, G. B., Norouzi, Y., & Hejazi, M. A. (2018). Experimental study on elimination of partial discharge effect on detection of radial deformation of high voltage transformer winding using electromagnetic waves. In *2018 IEEE international conference on environment and electrical engineering and 2018 IEEE industrial and commercial power systems europe (EEEIC/I&CPS europe)* (pp. 1−5).

Karami, H., Tabarsa, H., Gharehpetian, G. B., Norouzi, Y., & Hejazi, M. A. (2019). Feasibility study on simultaneous detection of partial discharge and axial displacement of HV transformer winding using electromagnetic waves. *IEEE Transactions on Industrial Informatics, 16*(1), 67−76.

Karami, H., Azadifar, M., Mostajabi, A., Rubinstein, M., Karami, H., Gharehpetian, G. B., & Rachidi, F. (2020). Partial discharge localization using time reversal: Application to power transformers. *Sensors, 20*(5), 1419.

Karami, H., Gharehpetian, G. B., Hejazi, M. A. A., & Norouzi, Y. (2020). *Detection of radial deformations of transformers. Google Patents.*

Karami, H., Gharehpetian, G. B., Norouzi, Y., & Akhavan-Hejazi, M. (2020). Simultaneous radial deformation and partial discharge detection of high-voltage winding of power transformer. *IET Electric Power Applications, 14*(3), 383−390.

Leibfried, T., & Feser, K. (1994). On-line monitoring of transformers by means of the transfer function method. In *Proceedings of 1994 IEEE international symposium on electrical insulation* (pp. 111−114).

Leibfried, T., & Feser, K. (1996). Off-line-and on-line-monitoring of power transformers using the transfer function method. In*, Vol 1. Conference record of the 1996 IEEE international symposium on electrical insulation* (pp. 34−37).

Leibfried, T., & Feser, K. (1999). Monitoring of power transformers using the transfer function method. *IEEE Transactions on Power Delivery, 14*(4), 1333−1341.

Mahmoodi, M., Abadi, S. M. N., Karami, H., Hejazi, M. A., & Gharehpetian, G. B. (2020). Design and implementation of dielectric windows for detection of radial deformation of HV transformer winding using radar imaging. *IET Science, Measurement & Technology, 14*(4), 478−485.

Mortazavian, S., Gharehpetian, G. B., Hejazi, M. A., Golsorkhi, M. S., & Karami, H. (2012). A simultaneous method for detection of radial deformation and axial displacement in transformer winding using UWB SAR imaging. In *The 4th conference on thermal power plants* (pp. 1−6).

Tarimoradi, H., Karami, H., Gharehpetian, G. B., & Tenbohlen, S. (2021). Sensitivity analysis of different components of transfer function for detection and classification of type, location and extent of transformer faults. *Measurement, 187*, 110292.

Introduction to ultra wide band (UWB) systems

Gevork B. Gharehpetian[1] and Hossein Karami[2]

[1]*Amirkabir University of Technology (AUT), Electrical Engineering Department, Tehran, Iran;* [2]*Niroo Research Institute (NRI), High Voltage Studies Research Department, Tehran, Iran*

3.1 Concepts of UWB systems

In recent years, ultra wideband systems have been in the spotlight. Although UWB is a new technology in wireless communication, it has been applied for a long time. UWB systems, similar to broadband systems, were initially used for military and radar applications. Until February 2002, the Federal Commission of Communication (FCC) declared the commercial use of this technology in its first report on UWB systems for indoor and outdoor environments subject to certain spectral restrictions (Federal Communications Commission, 2002, pp. 02–48). UWB systems use pulses with high time accuracy and low amplitude, extended in the frequency domain.

According to established standards, UWB systems are those systems in which the used relative bandwidth is more than a quarter, or the bandwidth is more than 500 MHz. The relative bandwidth is defined as follows:

$$2\frac{(f_H - f_L)}{(f_H + f_L)} \tag{3.1}$$

where, f_H as the high cut-off frequency is $-10\ dB$ and f_L as the low cut-off frequency is $-10\ dB$. The reasons for using $-10\ dB$ instead of $-20\ dB$ are the low level of power of UWB systems and also the closeness of their power to the noise level, which makes it very difficult to measure the cut-off frequency (Federal Communications Commission, 2002, pp. 02–48). The bandwidth allocated to UWB systems is 7.5 GHz, which is much more than the bandwidth of similar conventional systems. This wide bandwidth means that UWB technology can transfer data at speeds of several Gbps. When the FCC announced its decision to use UWB technology, many dissenting views were expressed. To end these criticisms, the FCC imposed a limit on the maximum power available to UWB systems. The FCC allowed UWB systems to operate in the frequency range of 3.1–10.6 GHz, if the spectral constraint was applied (Roy et al., 2004).

The spectral density function of a measured UWB system should not exceed the mentioned frequency range of -41.45 dBm/MHz. If a UWB system uses all the available bandwidth, its radiant power will be around 0.5 mW. Obviously, this level of available power is very low, which makes UWB technology applicable for high-speed data transmission and short distances indoors. Applications such as high-speed wireless personal area networks with high data rates (above 100 MHz) and short-range (1−10 m) require such systems, which must comply with IEEE 802.15.3a standards (Kamoun et al., 2009; Qiu et al., 2005).

One of the most important features of UWB is the ability to transmit data at very high speeds over short distances due to the very low available power for each bit. Nowadays, most wireless communication methods are based on sinusoidal waves, but the first communication systems were based on pulses (Ghavami et al., 2007). Impulse radio-based communication systems use very short pulses to transmit information, resulting in an extremely wide frequency range. This method is known as the pulse modulation technique in radio applications because the modulated information is generated by pulse position modulation.

Impulse radio-time-modulated signal is seen like a baseband transfer without a carrier. The lack of carrier frequency is an essential property that distinguishes impulsive radio from narrowband applications as well as from the direct sequence spread spectrum and multi-carrier, which is a broadband technique (Zhou et al., 2001).

There are three important factors to consider when a UWB system must be designed:

- What is the design purpose?
- What are the system resources?
- What are the limitations of the system?

3.2 Power spectrum density of ultra wideband systems

Power spectral density (PSD) for a UWB system is generally considered to be very low. PSD is defined as the average power that a signal carries in 1 Hz of bandwidth (Ghavami et al., 2007).

$$PSD = \frac{P}{B} \tag{3.2}$$

where, PSD is in W/Hz, P is the transmitted power in W, and B is the signal bandwidth in Hz.

Narrowband wireless communication systems have low bandwidth, resulting in high power spectrum densities. From another point of view, since frequency and time are inversely related to each other, it can be seen that in narrowband sinusoidal systems with small bandwidth, the time period T is long. But for UWB systems with small T-pulse pulses, the bandwidth is very wide. Table 3.1 presents the specifications of conventional wireless systems (Ghavami et al., 2007).

Table 3.1 Specifications of conventional wireless systems.

System type	Power spectral density [W/MHz]	Transmission power [W]	Bandwidth [Hz]	Classification
UWB	0.013	1 mW	7.5 GHz	Ultra wideband
802.11a	0.05	1 W	20 MHz	Wideband
2G cellular	1.2	10 mW	8.33 kHz	Narrowband
Television	16,700	100 kW	6 MHz	Narrowband
Radio	666,600	50 kW	75 kHz	Narrowband

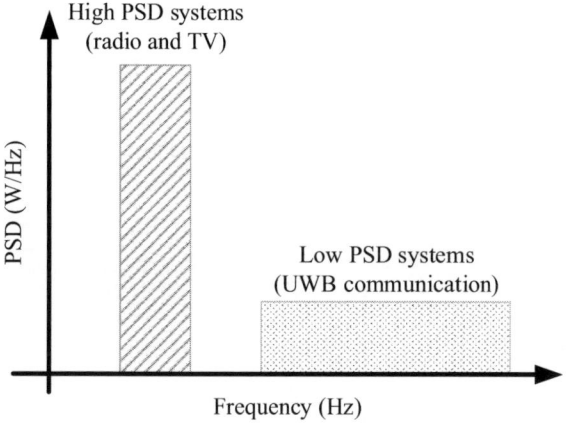

FIGURE 3.1

Systems via high and low PSD.

The energy used to send a signal in wireless systems should be as low as possible. Suppose we have a constant amount of energy. In that case, we can send a high density of energy over a small bandwidth or a low energy density over a large bandwidth, especially for consumer electronics devices. For UWB systems, energy is propagated over a wide bandwidth, which usually has a low power spectrum density (Ghavami et al., 2007). Fig. 3.1 compares the PSD of the UWB system with that of radio and television systems.

One of the advantages of low power spectrum density is the low probability of detection, which is useful for military applications. It is also widely applicable in information security for communication systems.

3.3 Specifications of ultra wideband systems
3.3.1 Signal pulse

UWB systems emit very narrow pulses with a width of about nanoseconds (Martin, 2003). By sending such narrow pulses, the power transmitted by these systems is greatly reduced. In general, UWB pulses are very short pulses with low ascent and descent times, which have a very wide spectrum and low energy contents.

The shape of the transmitted pulse in any wireless system is so important because it must be processed to extract the necessary information after receiving. Sine waves are commonly used in wireless systems. One of the reasons for being such applicable is that multiple transmitters can be used while the receiver can select the desired signal.

The effect of transmitter and receiver antennas on signal waves is very important in UWB systems, while it is less important in systems with a signal carrier. The impact of the transmitter antenna on the transmitted signal is a derivation in the time domain. Thus the transmitted pulse will not have a direct current (DC) component, and the result of the pulse integral over the pulse duration will be zero. Although a single pulse waveform seems to be suitable for UWB systems, other waveforms are also used in these systems. Today, nonsinusoidal waves are also used in UWB systems. In general, the purpose of using different signals is to have a smooth spectrum in the bandwidth area of the transmitted signal and avoid the DC component. These nonsinusoidal waves include Gaussian, Riley, Cubic, Laplacian, and modified Hermit waves (Karami et al., 2016, 2019; Martin, 2003).

3.3.2 Signal penetration

One of the important advantages of a UWB communications system is the ability of pulses to pass through doors, walls, partitions, and objects in the company or home. We know that the frequency (f) and the wavelength (λ) are related to the speed of light by the following equation.

$$\lambda[m] = \frac{c \ [m/s]}{f \ [Hz]} \tag{3.3}$$

In other words, the higher the frequency, the shorter the wavelength and the longer the wavelength, the lower the frequency will be. In conventional narrowband communication waves, the frequency is low and as a result, the wavelength is high, and they can pass easily through doors, walls, and windows. In other words, some of the energy of high-frequency waves, due to their shorter wavelength, is reflected from doors and walls.

In the end, UWB pulses consist of a wide range of frequencies. One of the main features of the UWB communication system compared to the wireless local area network system is the ability to pass through walls. Of course, it is noteworthy that this property is related to low-frequency components.

3.3.3 Bandwidth

One of the most important benefits of UWB systems is the efficiency in maintaining and managing the frequency. Because the power of UWB pulses is very low, the frequency range of transmitted signals is very close to noise. For this reason, UWB systems are capable of operating in the presence of current systems, and the bandwidth allocated to them has not considerable impact on them. This phenomenon eliminates the need for dedicated bandwidth for UWB systems and uses the bandwidth allocated to existing systems for their performance. Perhaps the most obvious privilege of UWB is its high transfer rate. Higher information rates enable new applications and devices that have not been implemented before. Transferring data via high-speed as 100 Mbps is created by this technology, and higher speeds are also available over short distances. The very large bandwidth occupied by UWB has made this possible (Ghavami et al., 2007).

3.3.4 Transmission speed

Another advantage of UWB systems is the ability of these systems to cope with disturbance. Since UWB systems have large bandwidth, the energy of the transmitted signal is expanded in this bandwidth. In such a situation, narrowband interferences have little effect on system performance. In other words, separately sending the narrow pulses used in UWB systems, which generate a very large bandwidth, results in clear reflected pulses in the receiver. This is important in any wireless communication system, whether pulsed or sinusoidal, i.e., no interference of signals is a key parameter in reducing the error of a communication system. Also, the wide bandwidth used in these systems, does not make it possible to use broadband disruptors due to power consumption limitations.

The main advantage of UWB can be derived from Shannon's formula (Schneider, 2006):

$$C = B \log\left(1 + \frac{S}{N}\right) \tag{3.4}$$

where, B is the bandwidth in Hz, C is the maximum channel capacity in bps, S and N are the signal and noise power in W.

As can be seen in this relationship, channel capacity increases linearly with bandwidth, and logarithmically with signal-to-noise. As a result, little power propagation is required to achieve high speeds when the bandwidth is high.

The most important issue in allocating bandwidth to a wireless system is the high cost of purchasing bandwidth. This issue is the reason for using low-power systems, without band registration and with high bandwidth, which eventually ended up with UWB systems.

3.3.5 Cost of devices

Another advantage of UWB systems is the simplicity of their transmitters and receivers. Transmitters and receivers of communication systems have complex components such as an oscillator, mixer, modulator, and filter. But since UWB systems are not limited to the frequency band constraint, their transmitter consists of a pulse generator, filter, and modulator, and we do not need to use a mixer in their construction (Li, 2006). In essence, the ability to directly modulate the pulse to the antenna simplifies the transmitter structure. This results in the removal of many of the components required by conventional sinusoidal transmitters and receivers.

3.3.6 Power consumption

Via the ideal design of electronic chips, the consumption power of UWB will be very low. For chip-related technologies, with optimal circuit design and proper use of signal processing techniques, the UWB chip will be small with low consumption. Currently, the power consumption of UWB chips is less than 25 mW (Jose et al., 2005).

3.3.7 Multipath problems

The multipath phenomenon is investigated for indoor UWB wireless systems. Due to the very short time-width of the pulses, the multipath effects will be less problematic if these pulses are processed in the time domain (Mortazavian et al., 2012). The multipath phenomenon occurs when an electromagnetic signal is delivered to the receiver from different directions (Mortazavian et al., 2015). The cause of this phenomenon is the absorption, reflection, scattering, and deflection of electromagnetic energy by objects between the receiver and transmitter paths (Rahbarimagham et al., 2015). If these objects do not exist to reflect or absorb energy, this effect will not exist, and the energy will be released only due to the properties of the antenna. But it is clear that there are always objects between the receiver and transmitter in the real world that cause absorption, reflection, deflection, and scattering, which also leads to the presence of several paths. Due to the difference in path lengths, the pulses reach the receiver at different times, proportional to the path length.

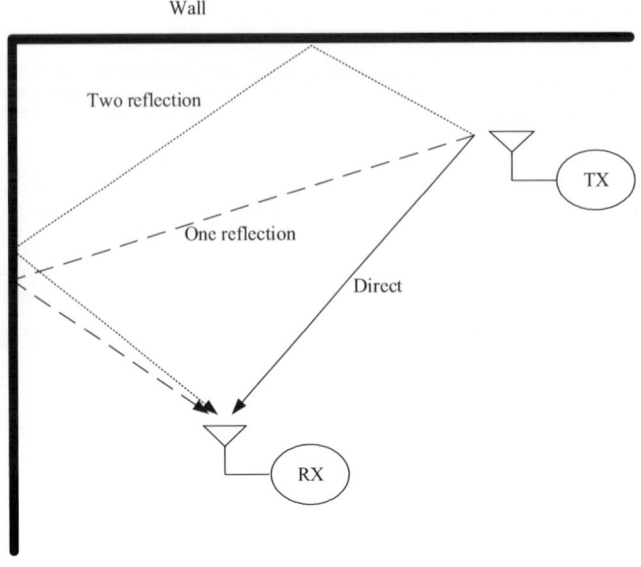

FIGURE 3.2

Reflection of the transmitted pulse in an indoor environment.

If the pulses do not overlap, this problem can be easily solved by filtering. Therefore to prevent the pulses from overlapping, a certain distance is required to separate the pulses related to several paths, and the required distance decreases with decreasing pulse width. Fig. 3.2 shows how the transmitted pulse propagates in the environment between the transmitter and the receiver. The transmitter and receiver are inside a room. The transmitted pulse is reflected from different objects, and the receiver records copies of the pulse with different delays and attenuated magnitudes (Ghavami et al., 2007). The power of the signal is decreased by increasing the number of reflections.

3.3.8 Modulation methods

The modulations used to send these pulses are well-known modulations, which are (Ghavami et al., 2007):

- Pulse position modulation
- Pulse amplitude modulation
- Bipolar signaling
- On-off keying

3.3.9 Ultra wideband antennas

Antennas are considered as an essential part of wireless communication systems. Antennas are used to send and receive signals from one device to another. An antenna is generally a metal device for radiating, transmitting, and receiving electromagnetic waves. The effect of transmitter and receiver

Table 3.2 Comparison between parameters of UWB and narrow-bandwidth radars.

Parameter	Signal	
	Narrow bandwidth	**UWB**
Radiation	From all antenna holes (one wave)	Only the middle and edges of the antenna (several waves)
The shape of the radiation field in time	Current derivative (signal)	Repetition of the current pattern (signal)
Radiation field amplitude	Dependent on polar coordinates	Dependent on polar coordinates and time
Radiation field shape in space (antenna pattern)	Dependent on polar coordinates	Depending on the polar coordinates, time, and current waveform (signal)
Lateral radiation	Lateral side	Uniform reduction

antennas on the signal is in the form of a derivative, which means that the signal in the receiver is a high-order derivative of the signal sent by the transmitter. This signal is derived at least once in the transmitter and once in the receiver antenna (Ghavami et al., 2007).

The radiation of short-wavelength signals with wide bandwidth is entirely different compared to the signals with narrow-bandwidth. Table 3.2 presents a comparison between the parameters related to UWB radars and narrow-bandwidth radars (Ghavami et al., 2007).

3.3.10 Signal length

Reducing the signal length on UWB radars gives us the following capabilities (Immoreev, 2002):

- Correcting the measurement accuracy at the detected target distance.
- The possibility of identifying the type of target, because the received signal carries not only information about different components of the target but also its general information.
- The effect of factors such as rain, fog, and particulates on UWB radars is low. This is because the signal scatter rate in these factors is much lower than the target itself.
- Increasing the target observation stability, because the return signal from the target itself and other obstacles reach the receiver at different time intervals, and their separability increases.
- Increased likelihood of target detection
- Slim pattern for antennas
- Improved radar immunity to narrow bandwidth and noise signals
- Reducing dead zone for radar
- Difficulty in detecting radar

3.4 Ultra wideband system applications
3.4.1 Imaging systems

These systems include all ground-penetrating radars (Reisenzahn et al., 2006), through-wall imaging (Aftanas, 2009; Huang et al., 2009), surveillance, and medical systems (Shang et al., 2008). These systems use the ability to penetrate pulses to hard objects. The FCC specifies the bandwidth for each of these systems, which we refer to:

- Ground-penetrating radars: These systems operate in the frequency range of 3.1–10.6 GHz and must be operated when they are tangential to or close to the ground. These systems are applicable for detecting objects buried in the ground.
- Through-wall imaging: The bandwidth allocated to these systems is similar to ground-penetrating radars. These tools are used to detect objects inside thick concrete walls.
- Behind-wall imaging: The bandwidth allocated to these systems is in the frequency range of 1.99–10.6 GHz. The movement and location of objects behind the wall should be detected by these means.
- Medical systems: The bandwidth allocated to these systems is between 3.1 and 10.6 GHz. Imaging body parts and movements is the task of these systems.

3.4.2 Radar systems

These radars are used to detect objects approaching vehicles and ultimately prevent them from colliding. The bandwidth intended for these systems is 22–29 GHz.

3.4.3 Communication systems

The UWB communication systems include a wide range of tools and systems, which can be referred to as a high-speed home or business network. According to existing laws, such systems must be restricted to indoor (Federal Communications Commission, 2002, pp. 02–48).

3.5 Conclusion

In this chapter, a brief introduction of UWB systems and their applications have been prepared. In addition, their specifications, such as bandwidth and penetration, are explained. Its application for transformer monitoring will be described in the next chapters.

References

Aftanas, M. (2009). *Through wall imaging with UWB radar system*. Department of Electronics and Multimedia Communications, Technical University of Kosice.

Federal Communications Commission. (2002). *Revision of part 15 of the commission's rules regarding ultrawideband transmission systems. First Report and Order*. FCC.

Ghavami, M., Michael, L., & Kohno, R. (2007). *Ultra wideband signals and systems in communication engineering*. John Wiley & Sons.

Huang, Q., Qu, L., Wu, B., & Fang, G. (2009). UWB through-wall imaging based on compressive sensing. *IEEE Transactions on Geoscience and Remote Sensing, 48*(3), 1408–1415.

Immoreev, I. J. (2002). *Main possibilities and main features of ultra-wideband (UWB) radars*.

Jose, S., Lee, H.-J., Ha, D., & Choi, S. S. (2005). A low-power CMOS power amplifier for ultra wideband (UWB) applications. In *2005 IEEE international symposium on circuits and systems* (pp. 5111–5114).

Kamoun, M., Mazet, L., De Courville, M., & Duhamel, P. (2009). A multicode approach for high data rate UWB system design. *IEEE Transactions on Communications, 57*(2), 553–561.

Karami, H., Gharehpetian, G. B., Norouzi, Y., & Hejazi, M. A. (2016). GLRT-based mitigation of partial discharge effect on detection of radial deformation of transformer HV winding using SAR imaging method. *IEEE Sensors Journal, 16*(19), 7234–7241.

Karami, H., Tabarsa, H., Gharehpetian, G. B., Norouzi, Y., & Hejazi, M. A. (2019). Feasibility study on simultaneous detection of partial discharge and axial displacement of HV transformer winding using electromagnetic waves. *IEEE Transactions on Industrial Informatics, 16*(1), 67–76.

Li, K. (2006). Experimental study on UWB pulse generation using UWB bandpass filters. In *2006 IEEE international conference on ultra-wideband* (pp. 103–108).

Martin, R. F. (2003). Ultra-wideband (UWB) rules and design compliance issues. In *2003 IEEE symposium on electromagnetic compatibility. Symposium record (cat. No. 03CH37446)* (Vol. 1, pp. 91–96).

Mortazavian, S., Gharehpetian, G. B., Hejazi, M. A., Golsorkhi, M. S., & Karami, H. (2012). A simultaneous method for detection of radial deformation and axial displacement in transformer winding using UWB SAR imaging. In *The 4th conference on thermal power plants* (pp. 1–6).

Mortazavian, S., Shabestary, M. M., Mohamed, Y. A.-R. I., & Gharehpetian, G. B. (2015). Experimental studies on monitoring and metering of radial deformations on transformer HV winding using image processing and UWB transceivers. *IEEE Transactions on Industrial Informatics, 11*(6), 1334–1345.

Qiu, R. C., Liu, H., & Shen, X. (2005). Ultra-wideband for multiple access communications. *IEEE Communications Magazine, 43*(2), 80–87.

Rahbarimagham, H., Porzani, H. K., Hejazi, M. S. A., Naderi, M. S., & Gharehpetian, G. B. (2015). Determination of transformer winding radial deformation using UWB system and hyperboloid method. *IEEE Sensors Journal, 15*(8), 4194–4202.

Reisenzahn, A., Buchegger, T., Scherrer, D., Matzinger, S., Hantscher, S., & Diskus, C. G. (2006). A ground penetrating UWB radar system. In *2006 3rd international conference on ultrawideband and ultrashort impulse signals* (pp. 116–118).

Roy, S., Foerster, J. R., Somayazulu, V. S., & Leeper, D. G. (2004). Ultrawideband radio design. The promise of high-speed, short-range wireless connectivity. *Proceedings of the IEEE, 92*(2), 295–311.

Schneider, T. D. (2006). Claude Shannon: Biologist [information theory used in biology]. *IEEE Engineering in Medicine and Biology Magazine, 25*(1), 30–33.

Shang, Y., Wang, D., & Birru, D. (2008). Performance analysis of WiMedia UWB system for medical applications with human blockage. In *2008 proceedings of 17th international conference on computer communications and networks* (pp. 1–5).

Zhou, S., Giannakis, G. B., & Swami, A. (2001). Comparison of digital multi-carrier with direct sequence spread spectrum in the presence of multipath. In *2001 IEEE international conference on acoustics, speech, and signal processing. Proceedings (cat. No. 01CH37221)* (Vol. 4, pp. 2225–2228).

UWB wave emission channel transfer function estimation

4

Gevork B. Gharehpetian[1] and Hossein Karami[2]

[1]*Electrical Engineering Department, Amirkabir University of Technology (AUT), Tehran, Iran;* [2]*High Voltage Studies Research Department, Niroo Research Institute (NRI), Tehran, Iran*

4.1 Loss of propagation path

Path loss refers to the attenuation of electromagnetic waves along the path from transmitter to receiver. This attenuation can occur for a variety of reasons. It is usually expressed in decibels (dB). Some of the attenuation reasons are as follows (Ghavami et al., 2007):

- Loss of free space
- Refraction
- Reflection
- Diffraction
- Clutter
- Coupling losses
- Absorption

4.1.1 Free space loss

Normally, most of the wave energy is lost around the transmitter environment. As the distance from the propagation source increases, the wave becomes like a flame from the torch, which means that some of the wave energy goes to spaces of the environment. Only a small amount of the primary wave energy is recorded by the receiver. Free-space loss means that the wave energy per unit area decreases with increasing distance. Therefore, after the wave reaches the receiver, a small fraction of the transmitted wave energy is attained by the receiver.

Here is a brief description of the general relationship between wave propagation in free space with respect to frequency dependence as well as the structure of antennas. The average power attained by the receiving antenna can be written as follows (Ghavami et al., 2007):

$$P_R(f) = \frac{P_T(f)G_T(f)G_R(f)c^2}{(4\pi d)^2 f^2} \tag{4.1}$$

In Eq. (4.1), $P_T(f)$ is the average power of the transmitter, c is the light speed, and $G_T(f)$ and $G_R(f)$ are the frequency responses of the transmitter and receiver antennas, respectively.

Obviously, Eq. (4.1) depends on the antenna frequency response and the problem arises by increasing bandwidth. It is better to create $G_T(f)P_T(f)$ in the range of bandwidth that is appropriate. Hence, in the first step, we approximate the frequency response for convergence. Then, the average power of the ultra wideband (UWB) wave taken at the receiver, P_{Rav}, can be determined using Eq. (4.2). The UWB wave has the bandwidth around the center frequency, f_c, from f_c-$W/2$ to $f_c + W/2$, power spectral density (PSD) of P_{av}/W, and frequency response of the receiver antenna with a fixed gain of G_R (Foerster, 2002).

$$P_{Rav} = \int_{f_c-W/2}^{f_c+W/2} P_R(f)\, df = \frac{P_{av}G_R c^2}{W(4\pi d)^2}\left[\frac{1}{f_c - W/2} - \frac{1}{f_c + W/2}\right]$$

(4.2)

$$= \frac{P_{av}G_R c^2}{W(4\pi d)^2 f_c^2}\left[\frac{1}{1 - (W/2f_c)^2}\right]$$

By defining P_{av}^{NB} as Eq. (4.3), P_{Rav} can be rewritten as given in Eq. (4.4).

$$P_{av}^{NB} = \frac{P_{av}G_R c^2}{w(4\pi d)^2 f_c^2}$$

(4.3)

$$P_{Rav} = P_{av}^{NB}\left[\frac{1}{1 - (w/2f_c)^2}\right]$$

(4.4)

In view of the receiver antenna response, the analysis can be presented by Eq. (4.5), where A_R is the effective area of the antenna.

$$G_R(f) = \frac{4\pi A_R f^2}{c^2}$$

(4.5)

It can be seen that there is more gain for higher frequencies. Therefore, the result of the above analysis in the average power of the receiver can be determined by using Eq. (4.6), where $G_R(f_c)$ is the antenna gain at the f_c of the transmitted signal (Ghavami et al., 2007):

$$P_{Rav} = \frac{P_{av}4\pi A_R}{(4\pi d)^2} = \left[\frac{P_{av}c^2}{(4\pi d)^2 f_c^2}\right]\left[\frac{4\pi A_R f_c^2}{c^2}\right] = P_{av}^{NB}G_R(f_c)$$

(4.6)

As a result, the narrow-band model can be obtained by approximating the path loss of the UWB system considering the above hypotheses.

4.1.2 Refraction

Generally, the redirection of a wave passing through a boundary between two different materials and coatings is called refraction. For two different environments, the refractive index, n, and the degree of refraction, θ, can be approximated by Snell's law as follows (Parazzoli et al., 2003):

$$n_1 \sin(\theta_1) = n_2 \sin(\theta_2)$$

(4.7)

If the refractive index of the first material is greater than the second one, the angle of the beam path will be larger than normal, and vice versa.

4.1.3 Reflection

A rapid change in the wave path at the boundary of two different environments, which returns part of the wave to the first environment is called reflection. The reflection may be mirror-like or diffuse according to the interface nature. As mentioned, when a wave hits a dielectric surface or a conducting body, some of the wave energy is reflected and some of it enters the material, which is dependent on the frequency. The dielectric properties of many materials have special behaviors at different frequency ranges. These properties determine the reflection coefficients.

4.1.4 Diffraction

Off-path wave propagation is called diffraction. All waves tend to propagate off the edge of the track, and more diffraction occurs around corners and curvatures. The diffraction phenomenon is also frequency-dependent.

4.1.5 Clutter

The ripple wave model is called clutter due to the rough surface or the boundary between the two materials. The general mechanism of the clutter is not clear, so lots of ideas of narrow-band mode are compared to each other. Understanding the electromagnetic wave interactions between a very short pulse and the insulation are essential factors in designing a UWB system that can guide clutter cancellation and specify the amount of efficiency.

4.1.6 Absorption

In the transmission of electrical, electromagnetic, or acoustic signals, absorption means converting energy into other forms. Absorption is one of the reasons for signal attenuation.

4.1.7 Coupling loss

Coupling losses occur when wave energy is transferred from one material to another. Coupling losses are the difference between the theoretical antenna gain and the gain obtained in the real performance. These losses are related to the scatter angle and antenna bandwidth.

4.2 Signal waveform deformation during sending, propagating and receiving

Narrow-bandwidth signals that are sinusoidal or quasi-sinusoidal have unique properties. The summation, derivation, and integral of these signals do not change the waveform. The received signal has the same shape as the original and transmitted signal, and may only have changes in amplitude and delay in time (Karami et al., 2020a, 2020b). In the following, the signal changes are investigated during the transmission, propagation, and reception steps.

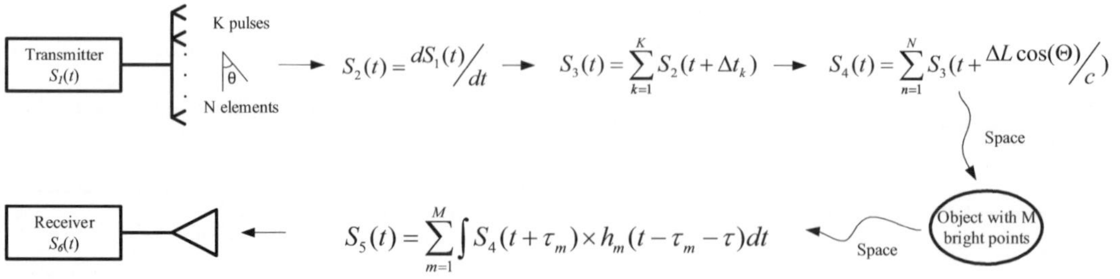

FIGURE 4.1

Signal waveform changes in path of transmitter to receiver.

Suppose S_1 is the transmitted UWB pulse with the pulse length of $c.\tau$, where c is the speed of light and τ is the pulse time duration. As shown in Fig. 4.1, the first change in the UWB signal (S_2) shape is when a signal is transmitted. The electromagnetic field changes are relative to the derivative (first-order or higher) of the antenna current. The second change in the signal waveform (S_3) occurs when the antenna is excited and the pulse travels to the transmitter. In this case, the elements of the antenna emit pulses of electromagnetic waves sequentially. As a result, a pulse signal is converted to a sequence of K pulses divided into Δt time intervals.

The third change that occurs in the waveform (S_4) is due to the delays of the fields propagated by various N antenna elements with the length of L in space. One of the antenna elements pulse propagated via angle θ, has the time delay as $(d/_c) \cos \theta$, relative to the pulse propagated by neighboring elements. The combination of these pulses will lead to various waveform shapes with different lengths and angles at different distances from the source. The waveform of the transmitted signal is deformed in the account of reflecting by the target considering the scattering phenomenon, resulting in the next change in the signal (S_5).

In Fig. 4.1, it is considered that the target has M bright points which return pulses. The number of pulses, time delays τ_m, and pulses power depend on the target shape and its different elements response h_m. This sequence of pulses is called "target image". The whole image shows the distribution of all scattered energy by the collision with the target. This image is formed through $t_0 = 2\frac{L}{c}$ time intervals.

The fifth change in signal occurs at the time of receiving (S_6). The reason is exactly the same as the signal transmission time. The time-domain transmission view of the signal is related to the strength of the electromagnetic field with respect to the distance of the various antenna elements from the target.

4.3 Models of transfer function between transmitter and receiver

The propagation environment through which a signal passes from transmitter to receivers is called a "channel". UWB channel modeling aims to show the channel with the minimum parameters to model the channel behavior in computer simulations correctly. Different models can be considered for the channel transfer function (Ikonen & Najim, 2001) in which some of them are described in the next sections.

4.3.1 Autoregressive moving average[1] model

Generally, in the discrete time domain, a sampled random process can be considered as the output of a time-invariant linear system. The Z-transform of this model can be written as follows (Juang, 1994):

$$H(z) = \frac{a_0 + a_1 z^{-1} + a_2 z^{-2} \cdots + a_m z^{-m}}{1 + b_1 z^{-1} + b_2 z^{-2} \cdots + b_n z^{-n}} = \frac{A(z^{-1})}{B(z^{-1})} \tag{4.8}$$

According to the above transfer function, the input and output values in the time domain can be expressed as follows:

$$y_t + b_1 y_{t-1} + b_2 y_{t-2} \cdots + b_n y_{t-n} = a_0 x_t + a_1 x_{t-1} + a_2 x_{t-2} \cdots + a_m x_{t-m} \tag{4.9}$$

The above model is called the autoregressive moving average (ARMA) model. The ARMA model has two specific states, AR and MA. In the AR model, $A(z^{-1}) = 1$ and in the MA model, $B(z^{-1}) = 1$.

4.3.2 Transfer function models considering dynamics of error

The system error can be considered as the output of a linear time-invariant system and its input is white noise. Therefore, each sampled random process can be considered in the ARMA model. $\frac{A(z^{-1})}{B(z^{-1})}$ is the system transfer function, $\tilde{y}_t$ is the error-free output that is not accessible and y_t is the actual output. The general structure is shown in Fig. 4.2.

The following equation can express this structure:

$$y_t = \frac{A(z^{-1})}{B(z^{-1})} x_t + \frac{C(z^{-1})}{D(z^{-1})} v_t \tag{4.10}$$

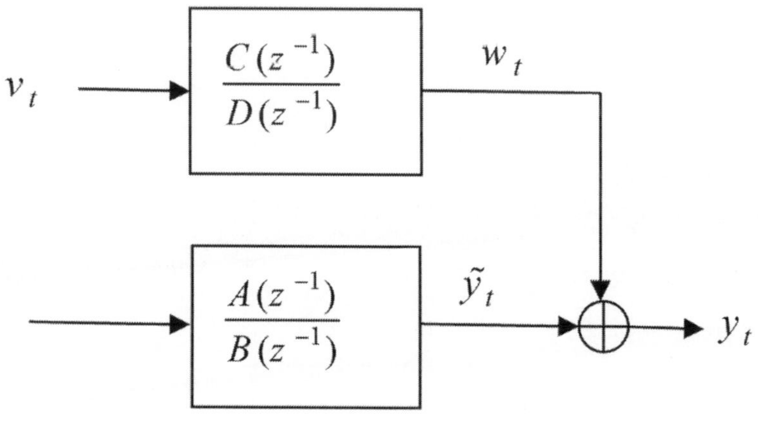

FIGURE 4.2

ARMA model including errors as a sampled random process.

[1]Autoregressive moving average.

In general, the parameters of four polynomials $A(z^{-1})$, $B(z^{-1})$, $C(z^{-1})$, and $D(z^{-1})$ should be estimated. If Eq. (4.10) is considered in the following form, the autoregressive moving average exogenous (ARMAX)[2] model is obtained.

$$B(z^{-1})y_t = A(z^{-1})x_t + C(z^{-1})v_t \tag{4.11}$$

This model states that the model of system and noise have zero and poles, and the poles intended for noise modeling, are the same as the system poles.

The autoregressive exogenous (ARX)[3] model can be obtained from Eq. (4.11), as follows:

$$B(z^{-1})y_t = A(z^{-1})x_t + v_t \tag{4.12}$$

This model states that the system model has zero and poles, but the noise model has only some poles. Also, the poles for noise modeling are the same as the poles of the system.

Another transfer function models considering error dynamics are the output error (OE)[4] and Box-Jenkins (BJ)[5] models, which can be formulated as Eqs. (4.13) and (4.14), respectively. In the OE model, the system noise is considered to be white noise.

$$y_t = \frac{A(z^{-1})}{B(z^{-1})}x_t + v_t \tag{4.13}$$

$$y_t = \frac{A(z^{-1})}{B(z^{-1})}x_t + \frac{C(z^{-1})}{D(z^{-1})}v_t \tag{4.14}$$

The BJ model is the most general case among the abovementioned models and models the system and noise considering zeros and poles for both of them separately.

4.3.3 State space model

The discrete equations of the system in state space model are expressed as follows (Aström & Wittenmark, 2013):

$$\begin{aligned} q_{k+1} &= A\,q_k + B\,x_k \\ y_k &= C\,q_k + D\,x_k \end{aligned} \tag{4.15}$$

In this model, the system is expressed by the matrices A, B, C, and D (Aström & Wittenmark, 2013; Juang, 1994). In most cases, D is considered equal to zero. A particular system has countless realizations that all will have the same response to the same input. Minimum realization means that a model with less state space dimensions is modeled compared to other realizations in a way that has the same answer as the main answer of the system.

---

[2] Autoregressive moving average exogenous.
[3] Autoregressive exogenous.
[4] Output error.
[5] Box—Jenkins.

4.4 Estimating transfer function between transmitter and receiver

Transfer function estimation approaches include online and offline or en-bloc approaches. In the en-bloc approach, estimation is carried out offline. At first, the experimental study is performed, and the input and output are sampled. At the end of the test, the necessary calculations are performed, and the system parameters are estimated. In the online approaches, the system parameters are calculated using the previous input and output sampling. The parameters are modified by receiving a new sample of the input and output. In other words, the estimation of the parameters is repeated by achieving new input/output values.

4.4.1 En-bloc estimation approach

The most straightforward parameter estimation approach is the en-bloc approach, but the condition for using this fast and straightforward approach is that the error signal of the residual noise should be white (Zheng, 1999). The main methods of the en-bloc estimation approach are the classical and the least-squares error methods.

The classical methods are based on the theory of linear systems that they have not changed over time. These methods are based on step response, impulse response, and frequency response of systems. The classical methods have many problems and limitations. Therefore, they are less applicable. One of the critical problems with these methods is their vulnerability to noise.

The least-squares method is one of the most practical and comprehensive methods of estimation. It can be boldly said that many estimation methods have been introduced as a kind of extension of this method (Giunta et al., 1997). The en-bloc least squares estimation methods include the ordinary and weighted least squares methods. In the following section, a complete description of this method will be presented.

4.4.1.1 Least squares error estimation method

This method was first invented by Gauss (1795) to calculate the orbit of a planet with astronomical observations in hand. This method assumes that any information measurement is accompanied by error and the transfer function used for modeling differs from reality. Under these conditions, at any moment, such as t, there is an error in the model as e_t between the estimated output and the real output of the model.

In the least-squares method, the goal is to estimate the factors so that the minimized sum of squares error, $\sum e_t^2$, has been achieved. The first step in implementing this transfer method is as follows:

$$y_t = u_t^T \cdot \theta + e_t \tag{4.16}$$

In this equation, θ is the unknown vector, u_t^T is the information vector at moment t, e_t is the modeling error, and y_t is the system output. If there is no error and the equation is $y_t = \underline{u}_t^T \underline{\theta}$, for the total number of unknowns equal to K, the parameters can easily be calculated by measuring K samples of the input and output. But since the error is generally not zero, N samples are taken from the input and output. Assume N samples are available at moments $t = 1$, $t = 2$, and, $t = N$. For different moments we can write:

$$
\begin{cases}
t = 1 : y_1 = u_1^T . \theta + e_1 \\
t = 2 : y_2 = u_2^T . \theta + e_2 \\
\vdots \\
t = N : y_N = u_N^T . \theta + e_N
\end{cases}
\tag{4.17}
$$

The above formulations can be expressed in the form of vectors, as follows:

$$
\begin{bmatrix} y_1 \\ y_1 \\ \vdots \\ y_N \end{bmatrix} = \begin{bmatrix} u_1^T \\ u_2^T \\ \vdots \\ u_N^T \end{bmatrix} . \theta + \begin{bmatrix} e_1 \\ e_2 \\ \vdots \\ e_N \end{bmatrix}
\tag{4.18}
$$

$$
Y_{(N \times 1)} = U_{(N \times P)} . \theta_{(P \times 1)} + e_{(N \times 1)}
\tag{4.19}
$$

In the least-squares method, the goal is to find the best estimate of unknown parameters, i.e., θ, to minimize $S = \sum\limits_{t=1}^{N} e_t^2$. On the other hand, according to the definition of e vector, S can be written as follows:

$$
S = e^T . e
\tag{4.20}
$$

or,

$$
S = \sum e_t^2 = e^T . e = \left(Y - U\theta \right)^T \left(Y - U\theta \right) = Y^T Y - \theta^T U^T Y - Y^T U\theta + \theta^T U^T U\theta
\tag{4.21}
$$

To find θ, we derive S respect to θ and set it to zero as given in Eq. (4.22).

$$
\frac{\partial S}{\theta} = 0 \Rightarrow -U^T Y - U^T Y + \left(U^T U + U^T U \right)\theta = 0
\tag{4.22}
$$

Therefore, we should have the following equations:

$$
-2U^T Y + 2U^T U\theta = 0
\tag{4.23}
$$

$$
U^T Y = U^T U\theta
\tag{4.24}
$$

$$
\theta = \left[U^T U \right]^{-1} U^T Y
\tag{4.25}
$$

The above equation is the main formulation in the least-squares method. Since the matrix U and the vector Y are known, it is possible to calculate θ if $\left[U^T U \right]$ is invertible and $\det \left[U^T U \right]$ is not zero.

In two cases, the above determinant may be zero:

- The degree considered for the system is higher than its actual degree, and
- The input does not excite the system sufficiently.

It should be noted that the system input should be summed with feedback so that the system input is changed to a combination of input and previous output. If the structure of the system differential equation is assumed to be Eq. (4.26), then we can write the system equation as Eq. (4.27).

$$y_t + a_1 y_{t-1} + a_2 y_{t-2} + \ldots + a_n y_{t-n} = b_0 x_t + b_1 x_{t-1} + \ldots + b_m x_{t-m} \tag{4.26}$$

$$y_t = - a_1 y_{t-1} - a_2 y_{t-2} - \ldots - a_n y_{t-n} = b_0 x_t + b_1 x_{t-1} + \ldots + b_m x_{t-m} \tag{4.27}$$

In other words, we have:

$$y = \begin{bmatrix} - y_{t-1} & - y_{t-2} \ldots & - y_{t-n} & x_t & x_{t-1} & \ldots & x_{t-m} \end{bmatrix} \times \begin{bmatrix} a_1 \\ a_2 \\ . \\ . \\ a_n \\ b_0 \\ b_1 \\ . \\ . \\ b_m \end{bmatrix} \tag{4.28}$$

Now, if we have the definition of $\underline{u}_t^T$ and $\underline{\theta}^T$ as Eqs. (4.29) and (4.30), we will have Eq. (4.31).

$$\underline{u}_t^T \overset{\Delta}{=} \begin{bmatrix} - y_t & 1 & - y_{t-2} \ldots & y_{t-n} & x_t & x_{t-1} & \ldots & x_{t-m} \end{bmatrix} \tag{4.29}$$

$$\underline{\theta}^T \overset{\Delta}{=} \begin{bmatrix} a_1 & a_2 & \ldots & a_n & b_0 & b_1 & \ldots & b_m \end{bmatrix} \tag{4.30}$$

$$y_t = \underline{u}_t^T \cdot \underline{\theta} \tag{4.31}$$

No error is considered in the abovementioned equations. The following equation is obtained by considering the error, which is the preferred model.

$$y_t = \underline{u}_t^T \cdot \underline{\theta} + e_t \tag{4.32}$$

Generally, the transfer function of a discrete system for the path between transmitter and receiver is as follows:

$$H(z) = \frac{b_0 + b_1 z^{-1} + b_2 z^{-2} \cdots + b_m z^{-m}}{1 + a_1 z^{-1} + a_2 z^{-2} \cdots + a_n z^{-n}} \tag{4.33}$$

According to Eq. (4.33), the input and output values in the time domain can be expressed as follows:

$$y_t + a_1 y_{t-1} + a_2 y_{t-2} \cdots + a_n y_{t-n} = b_0 x_t + b_1 x_{t-1} + b_2 x_{t-2} \cdots + b_m x_{t-m} \tag{4.34}$$

Therefore, the output at the moment t can be written as Eq. (4.35) according to the input and output values at previous times:

$$y_t = b_0 x_t + b_1 x_{t-1} + b_2 x_{t-2} \cdots + b_m x_{t-m} - a_1 y_{t-1} - a_2 y_{t-2} \cdots - a_n y_{t-n} \tag{4.35}$$

Therefore, the values of $y_0 - y_L$ can be expressed as follows:

$$\begin{cases} y_0 = b_0 x_0 \\ y_1 = b_0 x_1 + b_1 x_0 - a_1 y_0 \\ y_2 = b_0 x_2 + b_1 x_1 + b_2 x_0 - a_1 y_1 - a_2 y_0 \\ \vdots \\ y_{n+m} = b_0 x_{n+m} + b_1 x_{n+m-1} + \cdots + b_m x_n - a_1 y_{n+m-1} - a_2 y_{n+m-2} \cdots - a_n y_m \\ \vdots \\ y_L = b_0 x_L + b_1 x_{L-1} + \cdots + b_m x_{L-m} - a_1 y_{L-1} - a_2 y_{L-2} \cdots - a_n y_{L-m+1} \end{cases} \tag{4.36}$$

where, L is the number of input or output samples. The matrix form can be written as follows:

$$\begin{bmatrix} y_0 \\ y_1 \\ y_2 \\ \vdots \\ y_{n+m} \\ \vdots \\ y_L \end{bmatrix} = \begin{bmatrix} x_0 \; 0 \; 0 \; \ldots \; 0 \; 0 \; 0 \; 0 \; \ldots \; 0 \\ x_1 \; x_0 \; 0 \; \ldots \; 0 \; -y_0 \; 0 \; 0 \; \ldots \; 0 \\ x_2 \; x_1 \; x_0 \; \ldots \; 0 \; -y_1 \; y_0 \; 0 \; \ldots \; 0 \\ \vdots \\ x_{n+m} \; x_{n+m-1} \; \ldots \; x_n \; -y_{n+m-1} \; -y_{n+m-2} \; \cdots \; -y_m \\ \vdots \\ x_L \; x_{L-1} \; \ldots \; x_{L-m} \; -y_{L-1} \; -y_{L-2} \; \cdots \; -y_{L-n+1} \end{bmatrix} \cdot \begin{bmatrix} b_0 \\ b_1 \\ b_2 \\ \vdots \\ b_m \\ a_1 \\ a_2 \\ \vdots \\ a_n \end{bmatrix} \tag{4.37}$$

$$Y \qquad\qquad\qquad U \qquad\qquad\qquad \theta$$

4.4.1.2 Weighted least-squares method

In the least-squares method mentioned in the previous section, errors at different moments are of equal importance. But errors may have different importance at different moments. In the weighted least-squares method, different weights are given to errors at different moments. Therefore, S can be calculated as follows:

$$S = \sum_{t=1}^{N} w_t . e_t^2 = e^T . W . e \tag{4.38}$$

In general, the matrix W can be any matrix as long as it is positive. It is usually considered diagonal. Based on this, the best vector of parameters, in this case, is obtained as follows:

$$\underline{\theta} = \left[U^T W U \right]^{-1} U^T W Y \tag{4.39}$$

If the noise of the error estimation is white, the weighted least-squares method has the characteristics of an ideal estimator. In other words, if the noise of error estimation is white, the estimation will be without bias. Therefore, the condition for using these fast and straightforward en-bloc approaches is that the noise error should be white.

4.4.2 Adaptive estimation approach

This category of parameter estimation methods assumes that the input and output samples are used up to moment t and parameter estimation is performed online. The previous estimate is corrected based on the input and output information at the moment $t + 1$. In other words, there is an instantaneous connection between the system and the parameter estimator, and the parameters are updated using each new instance of the input and output information. Some types of adaptive methods for parameter estimation are as follows:

- Projection algorithm (PA)
- Orthogonalized projection algorithm (OPA)
- Recursive least squares (RLS)
- Recursive least squares with exponential data weighting (RLSEDW)
- Recursive least squares with selective data weighting (RLSSDW)
- Recursive least squares with covariance resetting (RLSCR)

In the recursive least-squares method, the values of the system parameters at time $t + 1$ are expressed as a function of the parameters at time t and the new information, x_{t+1} and y_{t+1} as given in Eq. (4.40):

$$\theta(t) = f(\theta(t-1), x_t, y_t) \tag{4.40}$$

Different adaptive methods use different functions for this purpose. In the next section, the formulation of these methods will be described.

4.4.2.1 Projection algorithm

In PA, the system parameters at moment t are expressed as follows:

$$\theta(t) = \theta(t-1) + \frac{a . u(t-1)}{c + u(t-1)^T u(t-1)} e(t) \tag{4.41}$$

where, a and c are two integers that usually considered equal to 0.1 and 1, respectively.

4.4.2.2 Recursive least squares with exponential data weighting

The formulations of this algorithm are as follows:

$$
\begin{cases}
\widehat{\theta}(t) = \widehat{\theta}(t-1) + \dfrac{a(t-1)P(t-2)\Phi(t-1)}{1 + a(t-1)\Phi(t-1)^T P(t-1)\Phi(t-1)} e(t) \\[4mm]
P(t-1) = P(t-2) - \dfrac{a(t-1)P(t-2)\Phi(t-1)\Phi(t-1)^T P(t-2)}{1 + a(t-1)\Phi(t-1)^T P(t-2)\Phi(t-1)}, P(-1) = P_0 > 0
\end{cases}
\tag{4.42}
$$

where, a is an integer which is selected as follows:

$$
a(t-1) =
\begin{cases}
k_1 & \Phi(t-1)^T P(t-2)\Phi(t-1) > \varepsilon \\
k_2 & \Phi(t-1)^T P(t-2)\Phi(t-1) < \varepsilon
\end{cases}
, \quad k_1 \& k_2 > 0
\tag{4.43}
$$

4.4.2.3 Recursive least squares with selective data weighting

The formulations of RLSSDW algorithm are as follows:

$$
\begin{cases}
\widehat{\theta}(t) = \widehat{\theta}(t-1) + \dfrac{P(t-2)\Phi(t-1)}{\alpha(t-1) + \Phi(t-1)^T P(t-1)\Phi(t-1)} e(t) \\[4mm]
P(t-1) = \dfrac{1}{\alpha(t-1)}\left(P(t-2) - \dfrac{P(t-2)\Phi(t-1)\Phi(t-1)^T P(t-2)}{\alpha(t-1) + \Phi(t-1)^T P(t-2)\Phi(t-1)} \right), P(-1) = P_0 > 0
\end{cases}
\tag{4.44}
$$

where, $\alpha(t)$ is defined as follows:

$$
\alpha(t) = \alpha_0 \times \alpha(t-1) + (1 - \alpha_0),
\begin{cases}
\alpha(0) = 0.95 \\
\alpha_0 = 0.99
\end{cases}
\tag{4.45}
$$

4.4.2.4 Recursive least squares with covariance resetting

The formulations of RLSCR are as follows:

$$
\begin{cases}
\widehat{\theta}(t) = \widehat{\theta}(t-1) + \dfrac{P(t-2)\Phi(t-1)}{1 + \Phi(t-1)^T P(t-1)\Phi(t-1)} e(t) \\[4mm]
P(t-1) =
\begin{cases}
P(t-2) - \dfrac{P(t-2)\Phi(t-1)\Phi(t-1)^T P(t-2)}{1 + \Phi(t-1)^T P(t-2)\Phi(t-1)} & \text{if } t \notin Z_s , \ Z_s = \{t_1, t_2, ...\} \\[4mm]
K_t I & \text{otherwise}
\end{cases} \\[4mm]
P(-1) = k_0 I
\end{cases}
\tag{4.46}
$$

4.5 Experimental setup

The setup for laboratory studies has different parts, including winding model, antennas and transceiver. In the following, the specification of each part will be described.

4.5.1 Transformer winding model

Making laboratory prototypes in the real dimensions of a transformer is very expensive and practically difficult. Indeed, due to the complexity of the transformer and its dimensions, designing a simpler and smaller single-phase model is reasonable in the research phase. Given that the size and dimensions of the transformer will not affect the results of our tests and data processing, two prototypes of the transformer winding model have been made, each with its own advantages and disadvantages (Hejazi et al., 2011). The most important point is the lightness and high material similarity to the real transformer winding for electromagnetic wave studies. Materials that can be easily shaped should be used.

For this reason, one of the best solutions is to use Plexiglas for the desired structure, which is covered with a layer of copper. Plexiglas can easily be shaped. On the other hand, it has the necessary strength. The specifications of two models of transformer windings will be expressed in the following section.

4.5.1.1 Transformer winding model no. 1

This model is one-third of a real transformer winding size in terms of winding diameter and number of disks. The core view of this model is shown in Fig. 4.3. The transformer winding model is made similar to the disk type of the real transformer winding. Spacer plates are located between every two adjacent disks. The specifications of each disk and spacer are given in Table 4.1.

FIGURE 4.3

View of winding model no. 1.

Table 4.1 Specifications of winding model no. 1.	
Parameter	**Value**
Diameter of the main disk	60 cm
Diameter of spacer disk	60 cm
Thickness of the main disk	2 cm
Thickness of spacer's disk	0.5 cm

To apply radial deformations to the model, sectors in the radial direction have been made on one of the Plexiglas disks. These sectors can be made in different dimensions and sizes, as presented in Table 4.2. These sectors can only be moved outwards. Radial deformations are modeled by moving sectors. An example of a sector outward shift is shown in Fig. 4.4. These sectors are named according to their external surface/area. All sectors can be moved from 2 mm to 10 cm. To change the location of the radial deformation along the vertical axis of the winding, the disk with the sectors can be replaced with a healthy disk at a desired height.

In order to model axial displacement, all disks and spacers are simultaneously displaced. Plastic disks with the desired thickness are placed under the model for different movements in the axis direction.

The advantage of this model is that due to the large perimeter of its disks, it is easy to separate different parts of it and model various deformations in a radial direction on it. The disadvantage is that it is large and bulky, which creates problems for the axial displacement and disk displacement with sectors. Another disadvantage is that the sectors cannot be moved inward. So, the concave deformation cannot be modeled.

4.5.1.2 Transformer winding model no. 2

One of the reasons for proposing a smaller model for the transformer winding is its ability to be simulated in the CST software, which will be used as a simulation tool in the following sections and chapters. The previous model could not easily be implemented in CST. As shown in Fig. 4.5, this model consists of 5 disks with an outer radius of 15 cm covered by a layer of copper. The disk thickness is 2 cm, and five spacers with the same radius and thickness of 0.5 cm are placed between every two adjacent disks.

Table 4.2 Area of sectors created on the simulator.		
Created groove	**Sector dimensions (cm * cm)**	**Movable area (cm²)**
s_1	5/0*2	1
s_2	1*2	2
s_3	2*2	4
s_4	4*2	8

FIGURE 4.4

Bulgy deformation modeling by sectors in model no. 1.

FIGURE 4.5

Transformer winding model no. 2.

Like the previous model, sectors with 0.5, 1, 2, and 4 cm width have been made on one of the disks to model the radial deformation. In this model, the sectors can be moved in the range of −2.5 cm− +2.5 cm.

According to Fig. 4.6, which is a top view of the transformer winding model, it is clear that the sectors are perpendicular to the disk diameter. Therefore, to model the radial deformation, the sectors are pulled straight out. In the previous model, when a sector was pulled out, air space was created between the moved sector and the disk, which caused unwanted and possible vibrations and, therefore, a reduction of measurement accuracy. But in this model, the mentioned problem has been decreased.

FIGURE 4.6

Top view of segments in winding model no. 2.

4.5.2 Transmitter and receiver antennas

One of the most suitable antennas for broadband applications, especially UWB applications, is the Vivaldi antenna shown in Fig. 4.7. This antenna is a tapered slot antenna, and one of its inherent features is good impedance bandwidth. Since in measuring the mechanical defects of the transformer winding, a narrow pattern is needed with high gain (to prevent the effects of the environment), Vivaldi antennas have been used. The operating frequency of the used antennas is between 3 and 10 GHz. However, the used transceiver has a frequency of 3.2–6.1 GHz and so, the antennas are used in this bandwidth. The structure and dimensions of the used antenna are as follows (Mehdipour et al., 2007).

The characteristic of the Vivaldi antenna pattern for 3 and 6 GHz frequencies is shown in Fig. 4.8. It should be noted that according to the defined angles on the figure, the desired pattern is obtained.

4.5.3 Transceiver

To transmit and receive UWB waves, PulsON 220 is used (PulsON 220 Specifications, 2010). The specifications of this device is listed in Table 4.3.

To get data from this device, BSR software is used. The procedure is that the antennas are first connected to the transceiver connected to the computer. Then BSR software is run for each transmitter and receiver antennas. Some related parameters, such as Tx (transmitting) interval and bin, window length, and others are adjusted, which will be defined in the next section. One of the antennas is set in

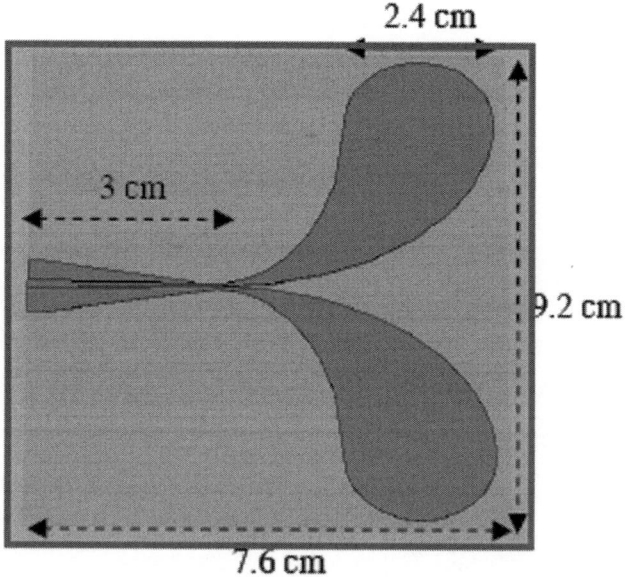

FIGURE 4.7

Dimensions of vivaldi antenna.

transmitter mode and the other in receiver mode, and the values of the parameters are adjusted. The critical point is that the values of the device parameters in the transmitter and receiver must be the same. Otherwise, there will be an error in the measurements. The BSR program stores any incoming data in a .txt file. After recording the data, it is time to analyze it. The definition of parameters and analysis of the recorded signals are discussed in the following section.

4.5.3.1 Selecting appropriate Tx interval

Tx interval is the time interval between sending two consecutive waves by the transmitter. From one point of view, this value can be considered as the sampling time of a continuous process. By each wave, the transformer winding is monitored. Therefore, it is clear that the shorter the Tx interval, the faster the monitoring of any model changes can be detected. However, our aim is not to catch the defect occurrence speed, so the Tx interval is unimportant. However, it should be noted that this value should be such that after sending a signal, sufficient attenuation of multipath signals in the environment has occurred to send the next signal. According to the above description and the device manual, the value of 50 ms is considered as the appropriate value in most tests.

4.5.3.2 Selecting appropriate bin

This parameter is the sampling time at the receiver. The received signals are sampled and stored at equal intervals as defined by the bin parameter. The minimum value of this parameter is 1 bin or 3.1789 ps. Selecting lower bin values leads to higher accuracy of recording signals and increases the amount of information. Therefore, a lower bin value equals more computation cost, and the time required to process the data. For example, if bin is set to 10, the distance between the received samples is 31.789 ps. If the BSR software setting for the received wave is set to the period of $[-10 \text{ ns}, 10 \text{ ns}]$, the number

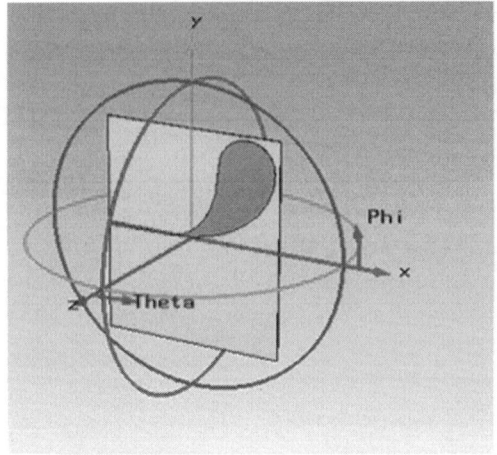

a

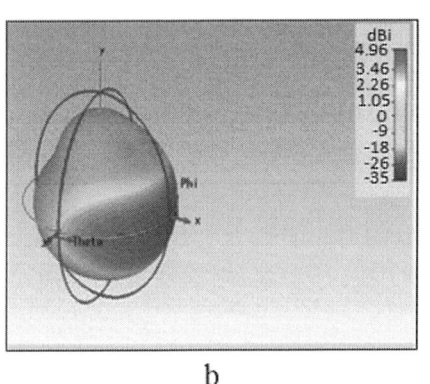

b

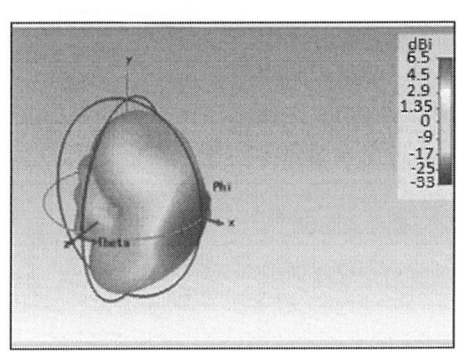

c

FIGURE 4.8

Field distribution around the antenna: (A) placed in three-dimensional coordinates, (B) at 3 GHz and, (C) at 6 GHz.

Table 4.3 Specifications of PulsON 220.	
Parameter	**Value**
PRF (pulse repetition frequency)	9.6 MHz
Center frequency (radiated)	4.7 GHz
Bandwidth (10 dB radiated)	3.2 GHz
EIRP	−12.8 dBm
Power consumption	5.7 watts
Dimension	16.5 cm*10.2 cm*5.1 cm

of samples is obtained by dividing the duration of this interval by the distance between two adjacent samples. For example, for bin = 10, we have 640 samples, equal to the division of 20 ns−31.789 ps. In most tests, the bin parameter is set to 10 or 31.789 ps.

4.5.3.3 Collecting data

After performing the measurements, to collect the data, all the stored files with the .txt extension are read through a program in MATLAB software, and each file is stored in a row of the data matrix. Therefore, a data matrix is a matrix in which each row represents an incoming wave. So, the number of rows is equal to the total number of received waves during the test. In other words, the time interval between two rows is equal to the Tx interval. If we consider a time axis in the direction of the column, the distance between the samples on this axis is the Tx interval. This axis is defined in the range of zero to the entire experiment time.

The size of the calculations depends on the size of the data matrix. The fewer rows and columns in this matrix, the smaller the computational volume. The number of columns in this matrix depends on the recorded time interval and the interval between samples, both of which are set during the experiments in the BSR program. For example, with the parameter bin = 10, if we have the time interval in terms of bin as [−3145, 3145], the samples will be recorded at the following times:

$$[-3145, -3135, -3125, \ldots, -15, -5, 5, 15, \ldots, 3125, 3135, 3145] \tag{4.47}$$

We will have a matrix with 640 columns. Therefore, the smaller the time interval and the larger the bin parameter, the smaller the number of columns in the matrix. Also, the number of matrix rows depends on two parameters: experiment duration and Tx interval. The longer the test, the greater the number of waveforms received, which results in more rows. The experiment time is divided into equal intervals according to the Tx interval parameter, and a sample is recorded in each interval. So, the number of rows and the computational volume decrease by reducing the experiment time and increasing the Tx interval.

4.5.3.4 Removing inappropriate data

Due to the device error and ambient noise, some received packets are entirely incorrect and should not be included in the data processing. To remove this type of data from the implemented algorithm, the following two filters are considered.

Noise elimination filter: According to various experiments in the desired distance range, the maximum amplitude of the waveform corresponding to the straight path after normalization will not be less than a predetermined value. Therefore, in a waveform where the maximum value is less than the predetermined value, only the noise is received, and this data must be deleted. An example of noise is shown in Fig. 4.9, where the predetermined value is 200.

Filter of eliminating data containing error: An error has occurred if we have large amplitude values comparable to the original signal amplitude value before receiving the first multipath signal, whose value will not be less than 200. This data will also be removed. Fig. 4.10 shows an example of these signals.

4.5.3.5 Elimination of time delays in received signals

In UWB systems, a single pulse cannot have sufficient information. The information must be modulated by a string of pulses, called a pulse train. When pulses are sent at regular intervals, called pulse repetition speed, they lead to peak power points at the center frequency of the spectrum. These

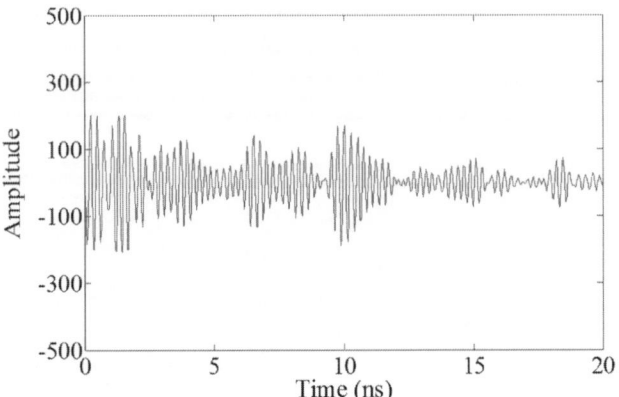

FIGURE 4.9

Received noise signal.

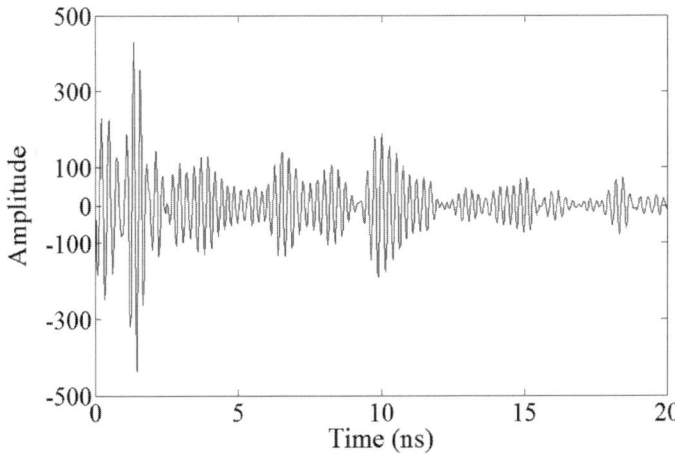

FIGURE 4.10

Received signal containing error.

peak points unfavorably limit the total transmitted power. One way to solve this problem is to delay the pulses. This delay is generated in accordance with the pseudo-noise sequence, which is also used in telecommunication systems such as wireless optical telecommunications.

In the PulsON 220 device, delaying the pulses means that the signals from the transmitter do not exactly have the same time interval. Therefore, the received signals do not have a fixed time of origin. In other words, if a fixed time origin is considered relative to the transmitter, the received waveforms will have delays to each other. Therefore, if we average the data in this case, it will not lead to the correct signal. Fig. 4.11 shows two examples of received signals. It is clear that two signals have some time shifts of samples to each other.

Therefore, these time shifts must be compensated. To do this, we define a time origin. Then, we find the corresponding time to the maximum amplitude of the received signal and transfer it to the time origin. In this way, the maximum value of all received signals is at the zero position. In other words, the

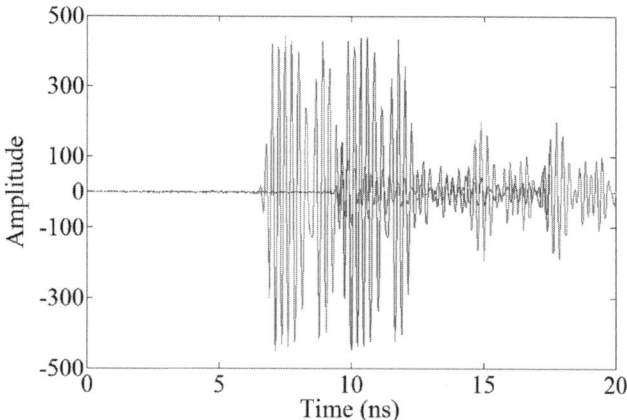

FIGURE 4.11

Two received signals before synchronization.

signals are synchronized. Fig. 4.12 shows two signals shown in Fig. 4.11 after synchronization. A waveform can be obtained that represents the state and noise effects in different samples by averaging all the synchronized signals.

4.6 Transfer function estimation based on experimental studies

The wave propagation environment is the transformer oil surrounded by a metal transformer tank in a real transformer. In this project, to simplify the propagation environment, it is initially assumed that the environment is air, and the transformer tank is not modeled. Then, the effect of the transformer tank is examined.

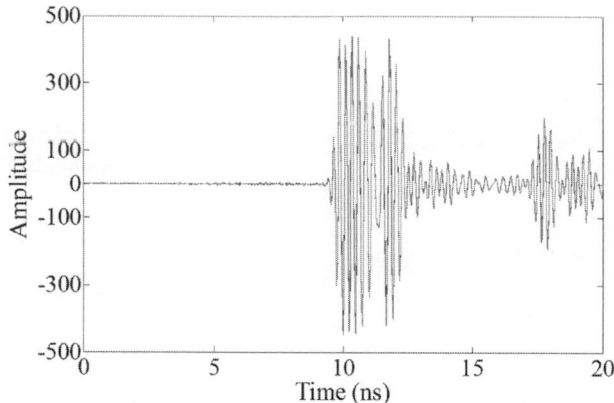

FIGURE 4.12

Two signals received at receiver after synchronization.

4.6.1 Analysis of received signal

UWB waves that are emitted in the environment, depending on the type of transmitted waveform and the frequency bandwidth in the channel, will be reflected from different parts of the environment, penetrated, or attenuated. Due to various metallic barriers inside the transformer, the waves reach the receiver from different directions/paths. In other words, different multipath signals are recorded in the receiver. These multipath signals will be further weakened when there is oil inside the transformer. On the other hand, the presence of oil causes the wave to dissipate in the environment. It is assumed that there is no other source of high-frequency electromagnetic waves inside the transformer. For example, when a partial discharge occurs, high-frequency waves are generated and can be received by the UWB receiver, which requires further investigation into the simultaneous effect of partial discharge signals and reflected signals.

The basis of the work is that first, as shown in Fig. 4.13, the UWB pulses are transmitted into the transformer winding model at time t. The transmitted pulses have a bandwidth of 3.2 GHz and a center frequency of 4.7 GHz. The distance between two antennas is d_1, the model distance from the center of the line connecting two antennas is d_2, and the distance between the transmitter and receiver is d_3. The distance of the antennas from the ground is d_4.

In the first test, there is no deformation on the transformer winding model and the winding model is in ideal condition. The experiment is performed at time t, and the received signals are stored. The changes are then applied to the transformer winding model and the measurements are repeated. There is no change in the transmitter and receiver positions and the transmitted pulses. The received signals are stored for different situations of winding faults. According to the time of arrival (TOA) method, the

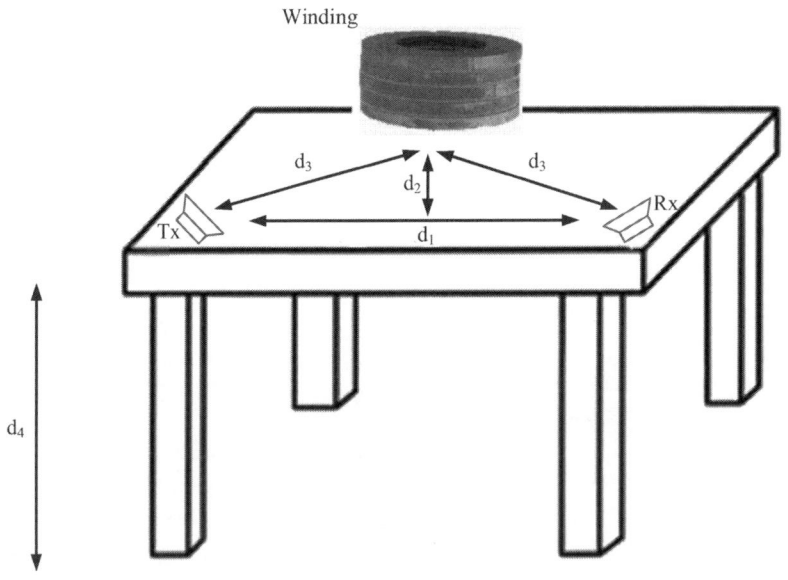

FIGURE 4.13

Setup of measurements and model of transformer winding.

speed of the wave and the time of receiving the reflections, the distance of the reflectors are determined (Kay, 1988). If there is previous information about the position of the reflectors, it can help to determine the reflected waves from the received signal.

According to the laboratory configuration shown in Fig. 4.13, via sending each UWB pulse from the transmitter, a wave which is the sum of the reflections from the entire environment will be recorded. Therefore, each time period of the received wave can represent the sum of multipaths received from the channel in that period. The purpose of the analysis is to identify and distinguish multipaths from each other. For this reason, the TOA method has been used. To determine which path and reflections of the received multipaths have been returned from, the best approach is to investigate the time domain of the signal due to the complexity of UWB channels and transformer environments. The reason for this is based on the fact that the positions of the winding model and antennas are fixed. To implement this method, the steps presented in the following sections are carried out.

4.6.1.1 Step 1: forming matrix from received signals

The time interval between two consecutive waveforms, T_s, has an average of 50 ms. The interval to be measured, T_{meas}, is about 10 s. Therefore, the number of waveforms recorded in the receiver is equal to $N_r = T_{meas}/T_s = 100$. The received signals are used to form a matrix. Each received wave for each UWB transmitted pulse is used to create a matrix row, and the next received wave is inserted in the next row. The experiment results are stored in the W_1 matrix in the first step, where there is no deformation on the transformer winding model. Received waves are used for situations where changes are made to the transformer winding model and applied to matrices W_2, W_3, and …. W_n.

4.6.1.2 Steps 2 and 3: eliminating inappropriate data and time delays in received signals

It is carefully observed in the received waveforms that the waves obtained in laboratory conditions have time delay relative to each other and also, some of them are unsuitable. The way of removing delays and inappropriate data is described in the previous sections.

4.6.1.3 Step 4: selecting time origin

In the received waveform, the first multipath signal recorded in the receiver is from the direct path between the transmitter and receiver antennas. We consider the moment of receiving this wave as the origin of time.

4.6.1.4 Step 5: calculating time window of target reflected signal

The first part of the received signal is related to the direct path, and the other parts belong to the multipaths, including the reflected signal from the transformer winding model. In experimental studies, the direct path is not important and should be omitted. To do this, the distances of the transformer winding model from the transmitter and receiver are calculated. According to that, the approximate arrival and end time of the reflected signal from the target is calculated. Then a window considering the start and end of the reflected signal from the target is applied, and the values outside the window are set to zero. This eliminates the direct path, which is the biggest problem for analysis, and other unwanted reflected signals from the environment. The window is calculated using the distance of the antennas and winding, and the wave propagation speed in the environment. If we assume that the speed of wave propagation is c, the time, which the wave needs to reach the receiver from a path of length d is:

$$t = \frac{d}{c} \tag{4.48}$$

For example, in Fig. 4.13, the summation of the distance of the transmitter to the model (Tx-Target) and the model to the receiver (Target-Rx) is $2d_3$. Given that the first multipath recorded by the receiver is related to the direct path between two antennas, so, the difference in arrival time of multipaths related to the Tx-Target-Rx can be obtained from the difference between the direct path and the path related to the Tx-Target-Rx. According to Fig. 4.13, Δd is obtained as follows:

$$\Delta d = 2d_3 - d_1 \tag{4.49}$$

The time difference is equal to:

$$\Delta t = \frac{2d_3 - d_1}{c} \tag{4.50}$$

So, the expected wave should be in the waveform with this time difference after the first multipath. An example of a received signal, its direct path component, and its target reflected signal is shown in Fig. 4.14.

4.6.1.5 Step 6: determining recursive multi-path matrix of transformer winding model

Due to the point that the location of the first multipath received from the winding model is known, it is possible to use the winding dimensions in terms of radius and height to determine the range of multiple recursive paths of the winding in the time axis. Therefore, the W_1 matrix can be converted to a smaller matrix, which reduces the computational volume. According to the angle of radiation, the range of the window can be determined for multipaths, and based on that, the dimensions of the matrix can be reduced. This window is considered in the time interval $T_1 - T_2$. The formed window is a multipath that returns from the winding. The window width is determined according to the dimensions of the transformer winding model and the distance between the antennas and the winding.

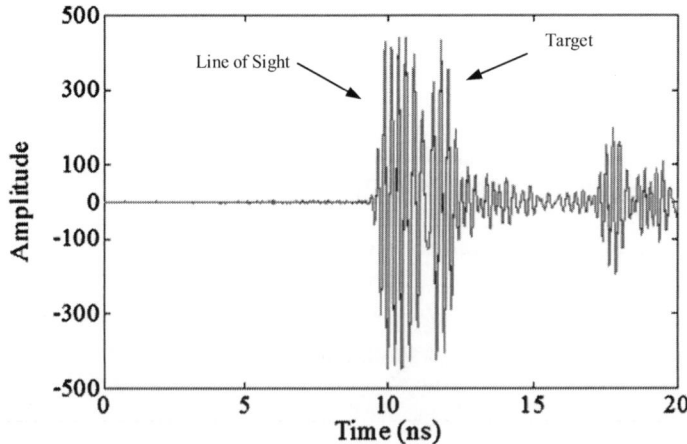

FIGURE 4.14

Example of waveform recorded by receiver and its direct path and target parts.

4.6.1.6 Step 7: form representative vector of each winding state

After selecting a common time origin for all rows of the matrices W_1, W_2, W_3, ... W_n, a representative vector can be calculated for each one of them. For this purpose, an average of all received signals in each winding state, e.g., W_1, is stored in the vector W'_1. In other words, the column of each of the matrices is averaged. The mean calculated for each column is stored in its corresponding place in a vector, and finally, this vector represents the waveform of the winding state. For each winding state, we have vector matrices as W'_1, W'_2, W'_3, and, W'_n.

4.6.1.7 Step 8: compare waveforms with reference state

In this part, the vector matrices obtained in the faulty cases are compared with the vector matrix of the reference transformer winding in the healthy state. The comparison of healthy and faulty states is performed using the transfer function. As the winding conditions change, so does the transfer function.

4.6.2 ARMA model estimation of transfer function

The signal in the performed tests is the UWB pulse sent by the transmitter. The transmitted pulse is of the Gaussian type, as shown in Fig. 4.15.

The next issue is the size of the UWB pulse range sent by the device. Suppose that the transmitted pulse attenuation rate is obtained from the distance between the transmitter and the receiver. In that case, the size of the transmitted pulse can be obtained concerning the size of the first multi-path signal received by the receiver. To do this, we place the transmitter and receiver antennas in a straight line facing each other. The transmitter at prespecified locations sends pulses between the transmitter and receiver. After performing the steps described in Section 4.6.1, the first multipaths are selected from the main signal, and the average amplitude is calculated to obtain the appropriate data. Therefore, at different transmitter locations and different distances between the transmitter and receiver, the average

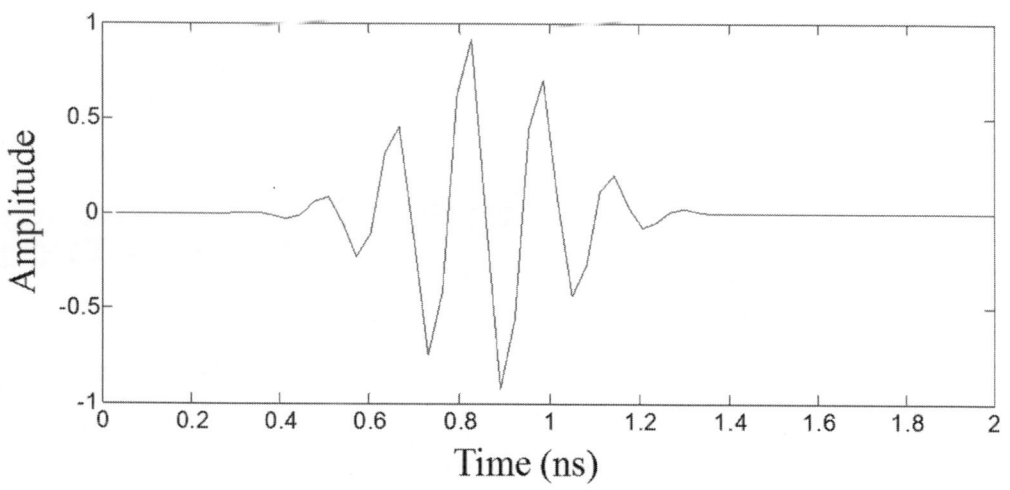

FIGURE 4.15

UWB pulse transmitted.

of the received signal is obtained. Suppose $x(t)$ is the transmitted signal, and the received signal can be expressed as follows:

$$y(t) = x(t) * h_1(t) * h_2(t) * h_a(t) \qquad (4.51)$$

where, $h_1(t)$ is the impulse response of the transmitter antenna and electronic system producing UWB pulse, $h_2(t)$ is related to the receiver antenna and electronic receiving circuit, $h_a(t)$ is the impulse response of the environment between the receiving and transmitting antennas. Different distances between the transmitter and receiver antennas lead to changes in h_a, while h_1 and h_2 are constant. The effect of the environment between the transmitter and receiver is to weaken the signal amplitude. Therefore, signal attenuation can be obtained in terms of the distance between the transmitter and receiver. Fig. 4.16 shows the average amplitude of the first multipath of the received signal in terms of distance between transmitter and receiver. Therefore, for a certain distance between the transmitter and the receiver, it is possible to determine the degree of attenuation of the transmitted signal amplitude compared to the first multipaths received in the receiver. So, by having the received signal amplitude, the amplitude of the sent signal can be obtained.

Another issue is the instant when the transmitter sends the UWB pulse relative to the signal received at the receiver. Using the distance between the transmitter and receiver antennas, the instant time of sending the pulse by the transmitter can be obtained by the first recorded multipaths in the receiver. Thus, the transmitted signal is as far ahead of the first multipath as the signal travels from the transmitter to the receiver. If the distance between the transmitter and the receiver is d, this traveling

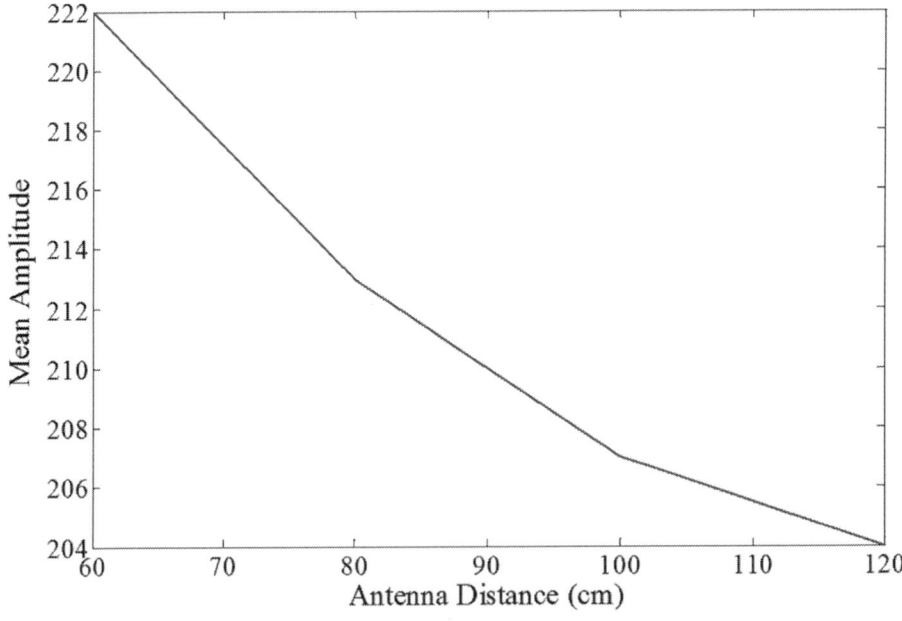

FIGURE 4.16

Attenuation of received signal amplitude based on distance between transmitter and receiver.

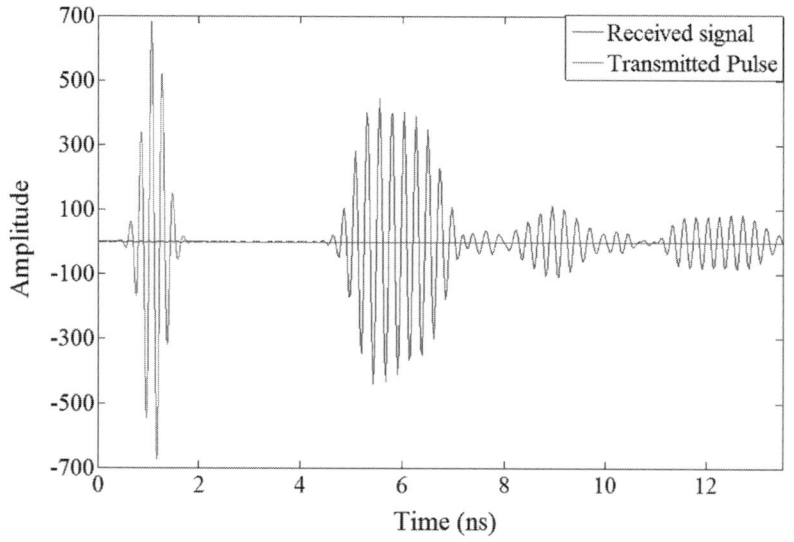

FIGURE 4.17

Pulse sent by transmitter and recorded signal at receiver.

period of the signal will be d/c, where c is the speed of the UWB wave in the air environment. Fig. 4.17 shows the pulse sent by the transmitter (input) and the pulse recorded by the receiver (output) for a specific configuration relative to a time origin. Using input and output, the transfer function between transmitter and receiver can be obtained.

There are various methods for obtaining the transfer function, as mentioned in Sections 4.3 and 4.4. In all the methods, the degree of the model, i.e., the number of zeros and poles of the transfer function, must be specified. The number of conjugate poles of the system is equal to the number of peak points in the PSD of the system output signal. The PSD of the received signal in the healthy state of the transformer winding model is shown in Fig. 4.18, which is estimated by the Burg method (Kay, 1988).

According to Fig. 4.18, it can be seen that there are two maximum points around the frequency of 4 GHz. So, we can guess that there are two-pole pairs in the transfer function, and the denominator order of the transfer function between transmitter and receiver is considered to be four.

The algorithms presented in Section 4.4 have been implemented in the MATLAB environment to estimate the transfer function. The results for each algorithm will be presented in the following sections. In all the following shapes, the blue waveform is the measured signal, and the red one corresponds to the estimated signal.

4.6.2.1 Estimation using least squares method

At first, it is assumed that the number of zeros is equal to one. The estimated transfer function is as follows:

$$H\left(z^{-1}\right) = \frac{-1 + 0.65\, z^{-1}}{0.003 - 0.002\, z^{-1} + 0.006\, z^{-2} - 0.002 z^{-3} + 0.003 z^{-4}} \tag{4.52}$$

Fig. 4.19 shows the signal estimated by the above transfer function and the measured signal. There is a remarkable difference between the measured signal and the estimated one.

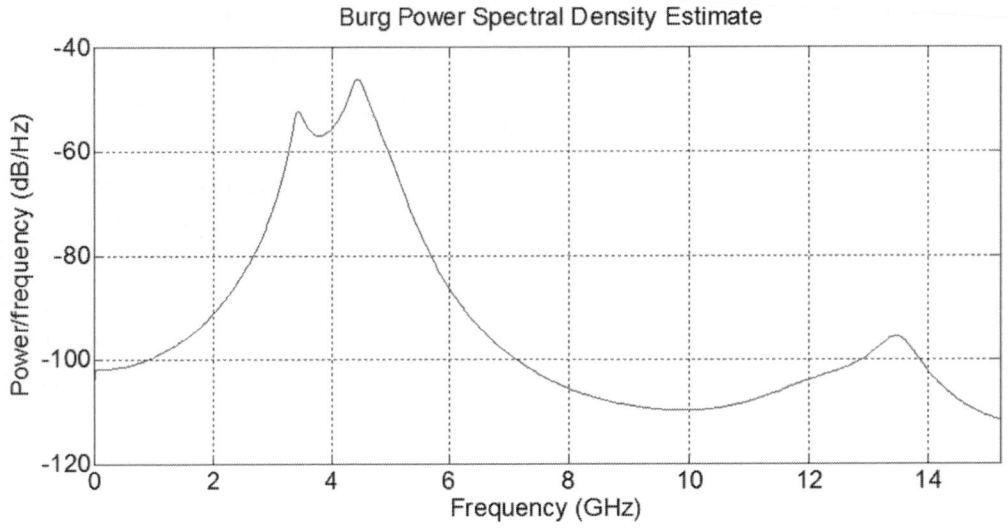

FIGURE 4.18

Power spectrum density of received signal.

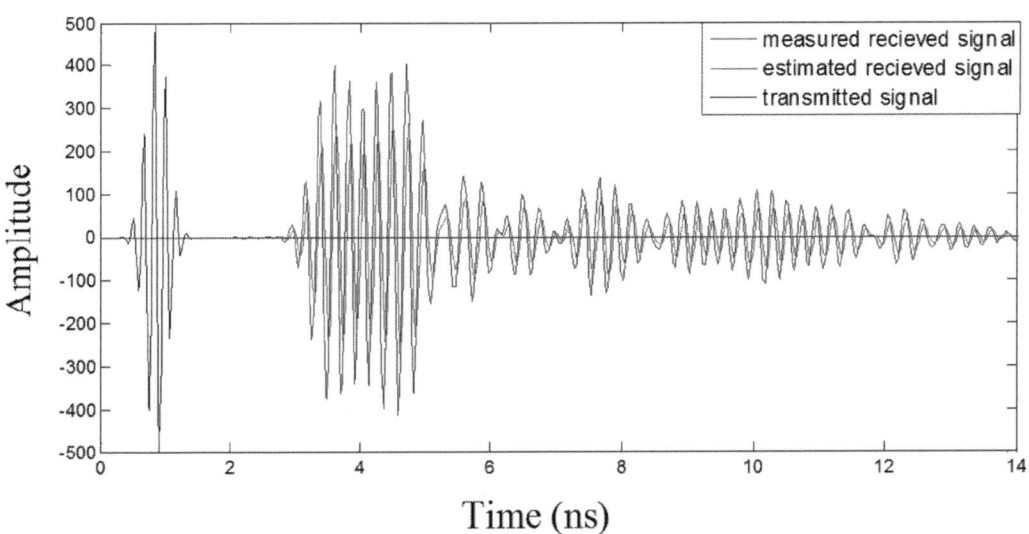

FIGURE 4.19

Estimated and measured received signals along with transmitted signal.

Fig. 4.20 shows the PSD of the estimated and measured received signals. It is also clear that the estimated PSD signal does not correctly follow the maximum points of the measured PSD signal.

In the following equation, the number of zeros of the transfer function is considered equal to 2:

$$H(z^{-1}) = \frac{1 - 1.3z^{-1} + .99z^{-2}}{0.0016 - 0.0027\,z^{-1} + 0.0042\,z^{-2} - 0.0028z^{-3} + 0.0017z^{-4}} \tag{4.53}$$

In this case, the signal and estimated PSD with the above transfer function and the signal and measured PSD are shown in Figs. 4.21 and 4.22. By this transfer function, following measured signal is improved compared with the single zero transfer function.

It can be concluded that in the least squares method, the transfer function of four poles and two zeros should be considered. However, en-bloc estimation methods work well if the error signal of the noise system is white. In the following section, adaptive methods are used to estimate the parameters of the system transfer function.

4.6.2.2 Estimation using adaptive estimation methods
At first, a nonzero model is considered for the transfer function as follows:

$$H(z^{-1}) = \frac{a_0}{1 + b_1\,z^{-1} + b_2\,z^{-2} + b_3\,z^{-3} + b_4\,z^{-4}} \tag{4.54}$$

The parameters of the transfer function are estimated by the methods presented in Section 4.4.2. The following are the results for some of them.

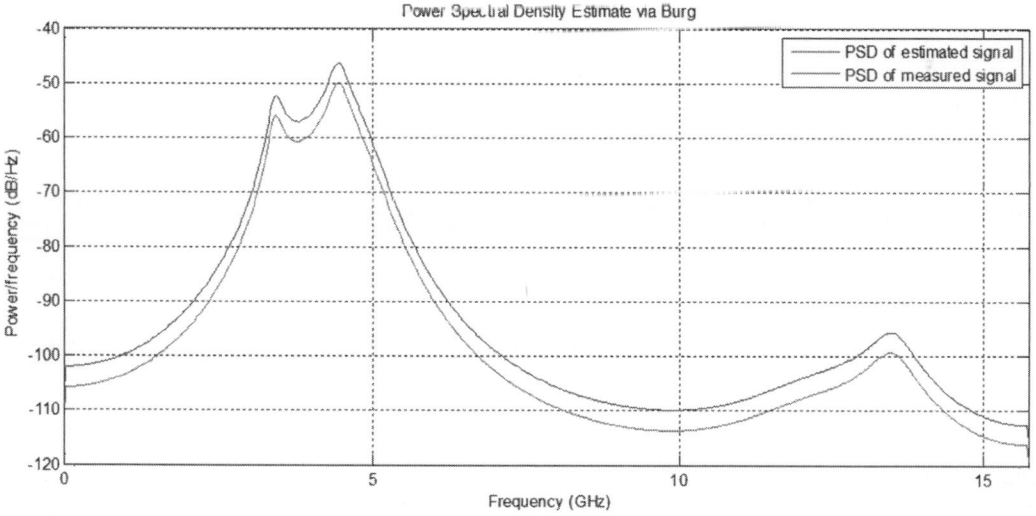

FIGURE 4.20

Power spectrum density of estimated and measured received signals.

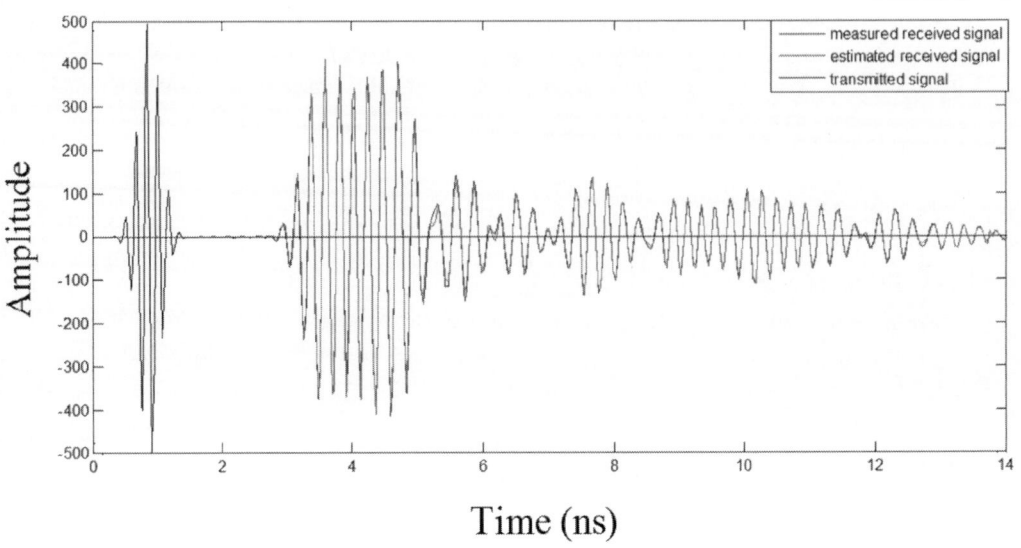

FIGURE 4.21

Estimated and measured received signal along with transmitted signal.

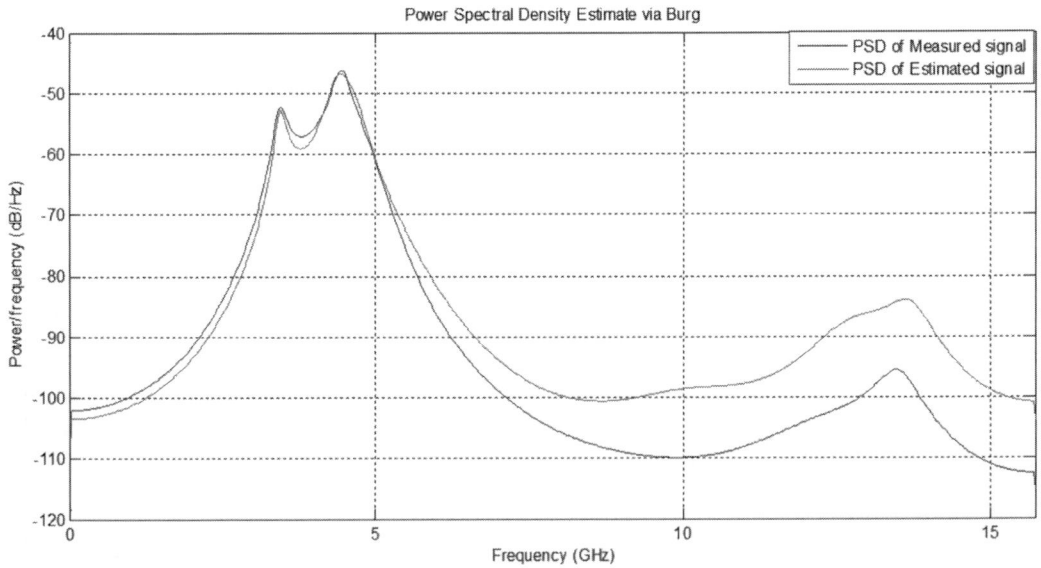

FIGURE 4.22

PSD of measured and estimated received signals by least squares method.

4.6.2.2.1 Results of projection algorithm

For the PA algorithm, the initial estimate is zero. Parameters a and c are set to 0.1 and 1, respectively. Fig. 4.23 shows the estimated parameters b_4, b_3, b_2, b_1, a_0. As can be seen, in the PA algorithm the parameters have converged, but there is a little error after convergence of the parameters. The system parameters do not accurately converge to their original values.

Fig. 4.24 shows the actual and estimated signals. According to this figure, the estimated output follows the measured output of the system, but the amount of error between them is significant. Fig. 4.25 shows the PSD of estimated and measured received signals. The PSD of the estimated signal correctly follows the PSD maximum points of the measured signal.

4.6.2.2.2 Results of RLS algorithm

As shown in Fig. 4.26, in the RLS algorithm, the system parameters converge very precisely. Fig. 4.27 shows that the estimated output follows the actual output of the system and the error between them is very small. In addition, the PSD of the estimated signal correctly follows the PSD maximum points of the measured signal (Fig. 4.28).

4.6.2.2.3 Results of RLSCR algorithm

The initial estimate is assumed to be zero and $K_t = 10$, $k_0 = 5$. The results are shown in Figs. 4.29–4.31. It can be seen that the convergence speed is lower, but the results at PSD peak points are better compared with RLS.

4.6.2.2.4 Results of RLSSDW algorithm

Like other algorithms, the initial estimate is zero ($\hat{\theta}(0)$). Parameters ε, k_2, k_1 are considered equal to 20, 0.01, and 0.01, respectively. The following results are obtained for this algorithm. The convergence speed and following the measured signal seem to be good (Figs. 4.32–4.34).

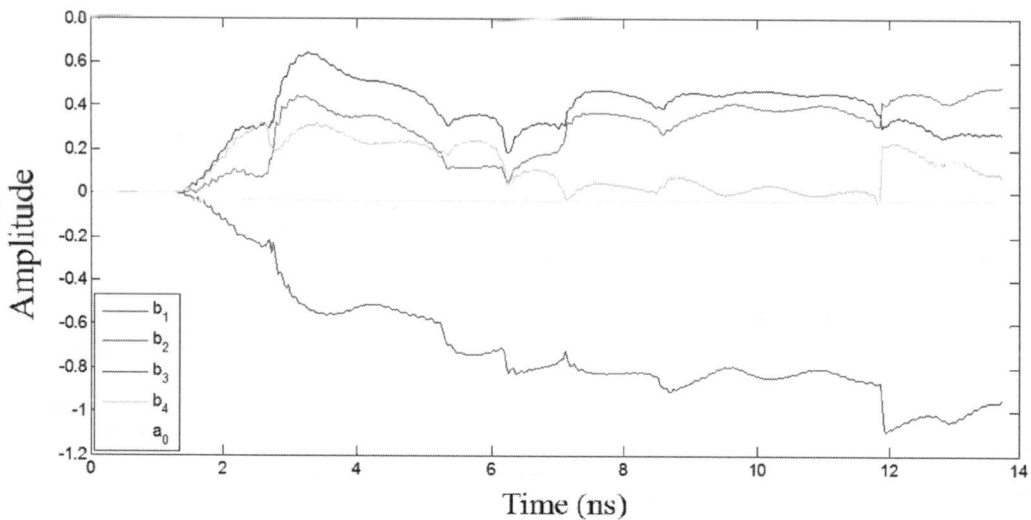

FIGURE 4.23

Convergence curve of system parameters in PA algorithm.

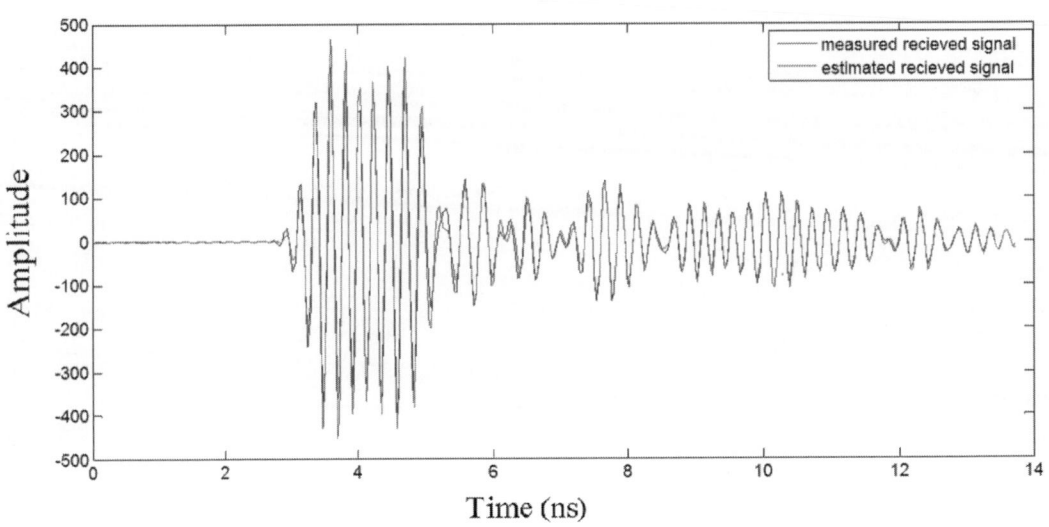

FIGURE 4.24

Received signal measured and estimated by PA algorithm.

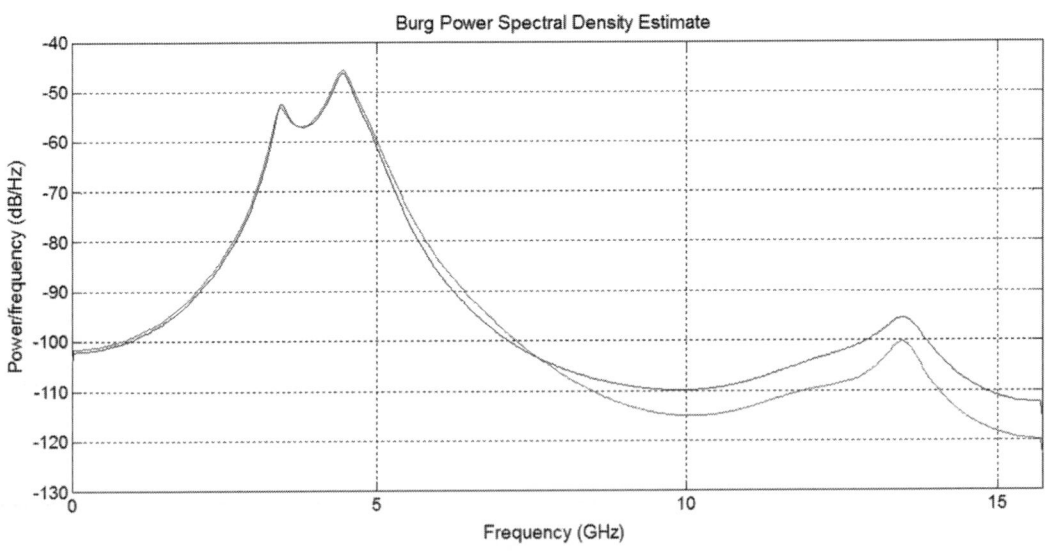

FIGURE 4.25

Power spectrum density of measured and estimated received signals by PA algorithm.

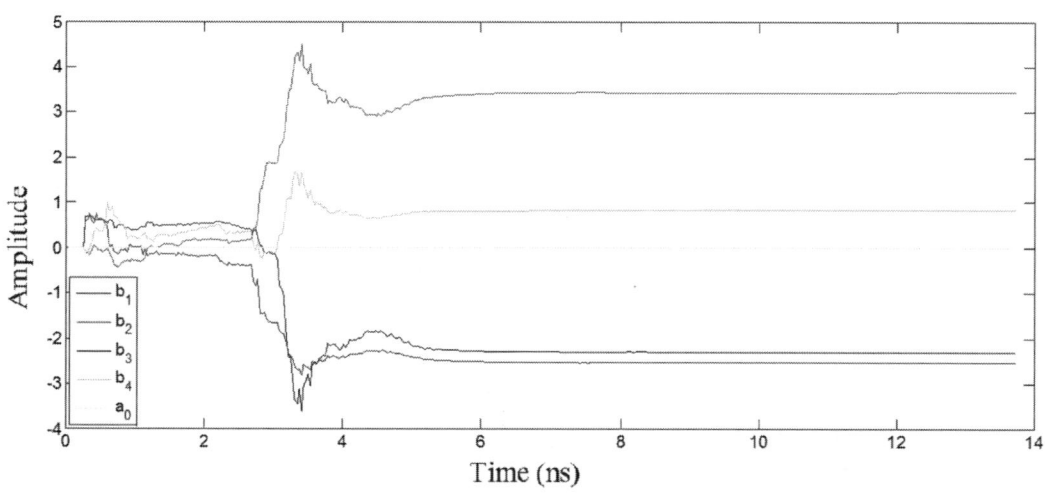

FIGURE 4.26

Convergence curve of system parameters by RLS algorithm.

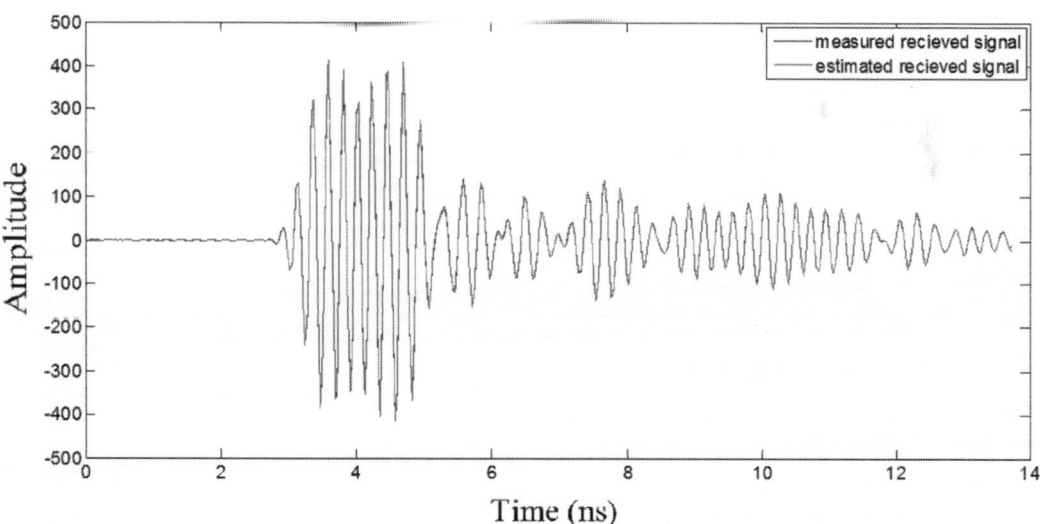

FIGURE 4.27

Measured and estimated signals by RLS algorithm.

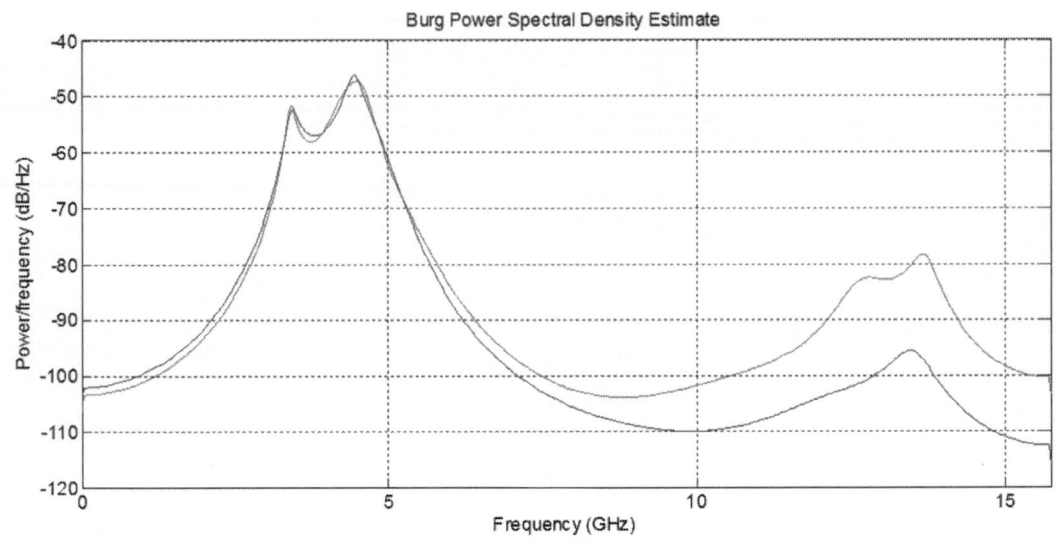

FIGURE 4.28

PSD of measured and estimated received signals by RLS algorithm.

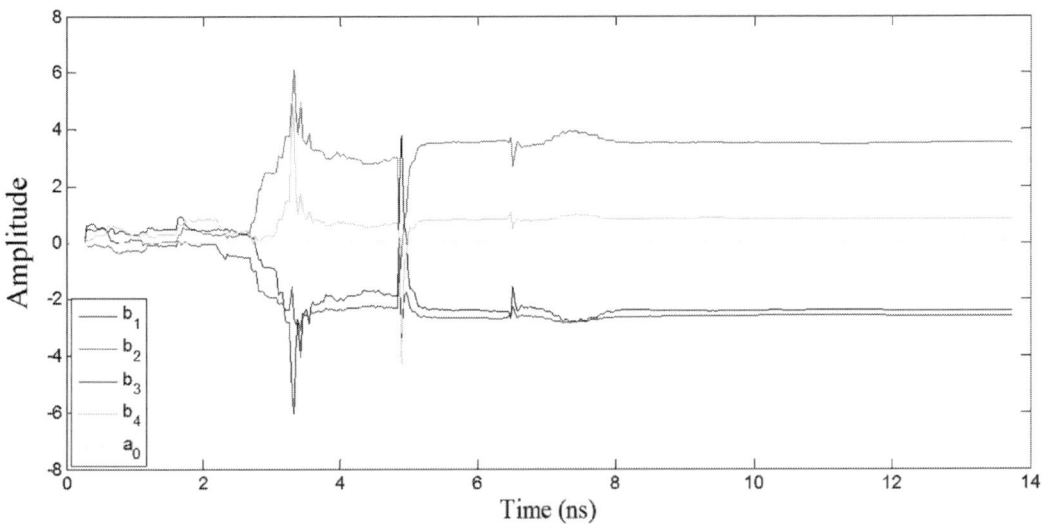

FIGURE 4.29

Convergence curve of system parameters by RLSCR algorithm.

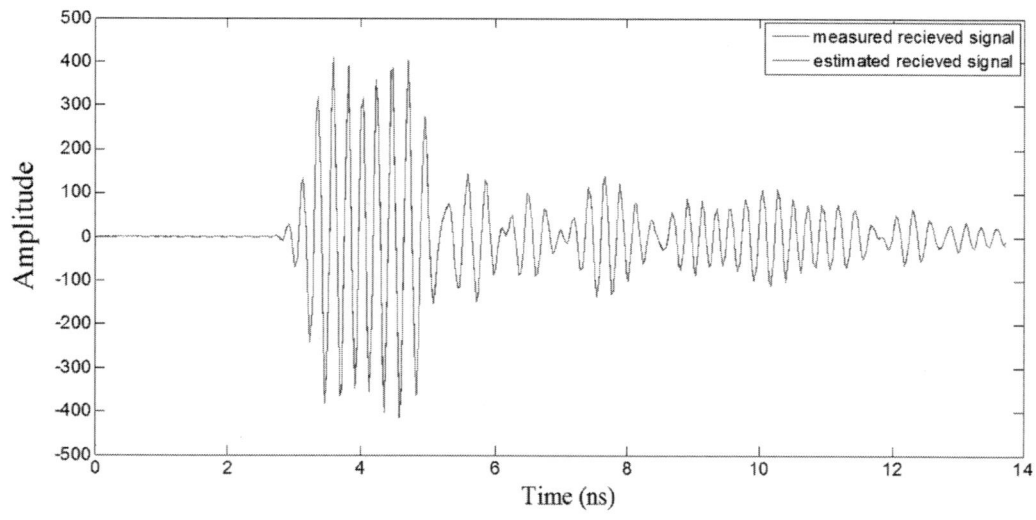

FIGURE 4.30

Received measured and estimated signals by RLSCR algorithm.

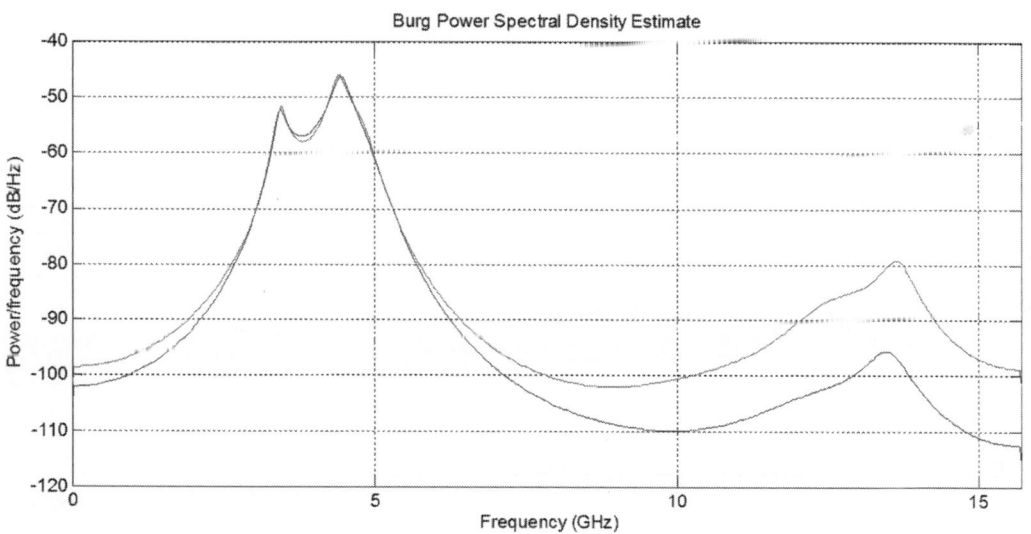

FIGURE 4.31

Power spectrum density of measured and estimated received signals by RLSCR algorithm.

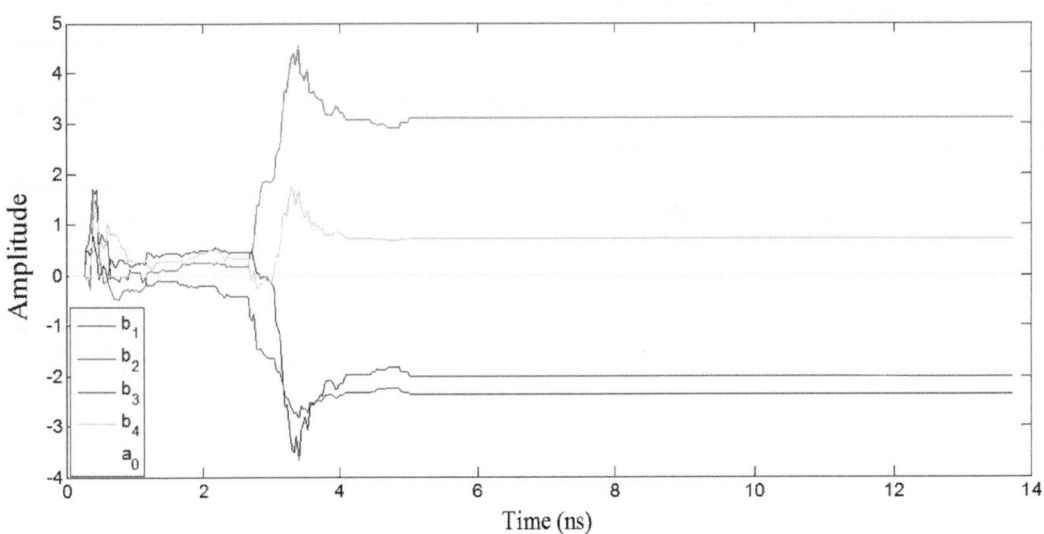

FIGURE 4.32

Convergence curve of system parameters by RLSSDW algorithm.

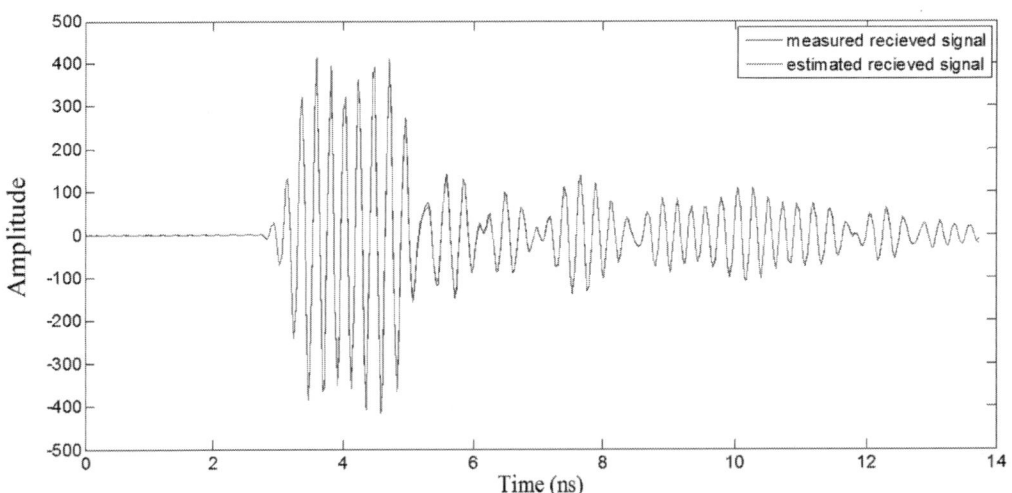

FIGURE 4.33

Measured and estimated received signals by RLSSDW algorithm.

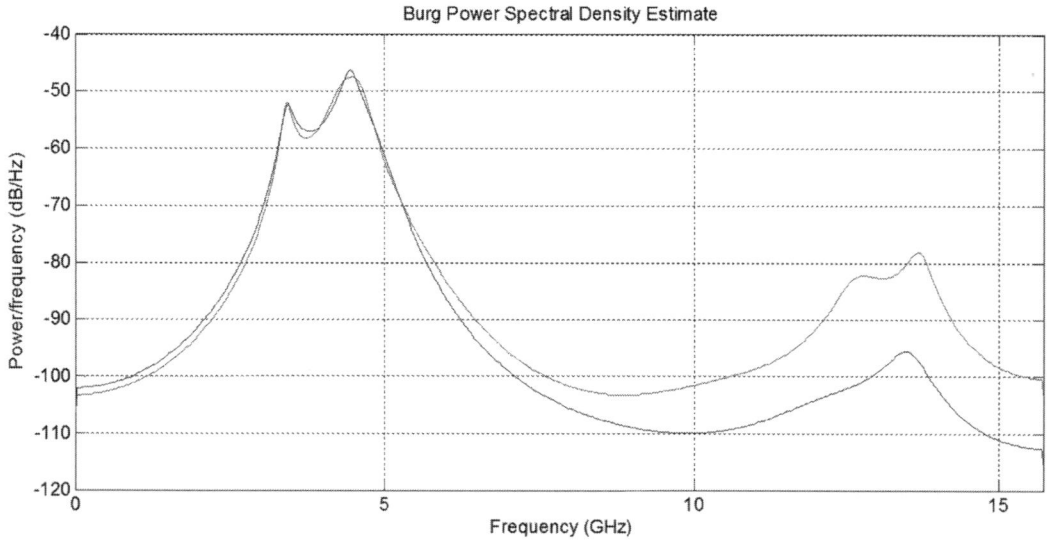

FIGURE 4.34

PSD of measured and estimated received signals by the RLSSDW algorithm.

4.6.2.3 Results comparison of estimation methods

According to the above results, it can be said that in adaptive estimation methods, except for PA, the transfer function parameters converge well, and the PSD of the estimated output signal is in accordance with the PSD of the measured output signal at maximum points around the frequency of 4 GHz. Therefore, the RLS estimation method, which is simpler than other methods, is chosen to estimate the system parameters.

To investigate the zeros of the transfer function in adaptive methods, according to the results of the en-bloc method, two zeros for the transfer function are estimated. Therefore, the desired transfer function should be in the following form.

$$H\left(z^{-1}\right) = \frac{a_0 + a_1\,z^{-1} + a_2\,z^{-2}}{1 + b_1\,z^{-1} + b_2\,z^{-2} + b_3\,z^{-3} + b_4\,z^{-4}} \tag{4.55}$$

To determine the unknown parameters using the RLS method, the results will be as follows.

From Fig. 4.35, it is clear that the coefficients related to the zeros of the system have converged to zero, and this indicates that in the methods of adaptive estimation, the system transfer function has no zero and its model is all-pole. In this case, the output signal is estimated, and its PSD is similar to the all-pole state.

Maybe more poles create better results. To examine this idea, assume that the number of system poles is more than four. Considering eight poles for the system and estimating the parameters of the RLS transfer function, the results are shown in Figs. 4.36 and 4.37.

In the eight-poles condition, the estimated PSD of the received signal at the maximum points near the frequency of 4 GHz is close to the PSD of the measured received signal. As it has not remarkable improvement compared with the four-pole transfer function, it can be concluded that the system transfer function with four poles is preferred.

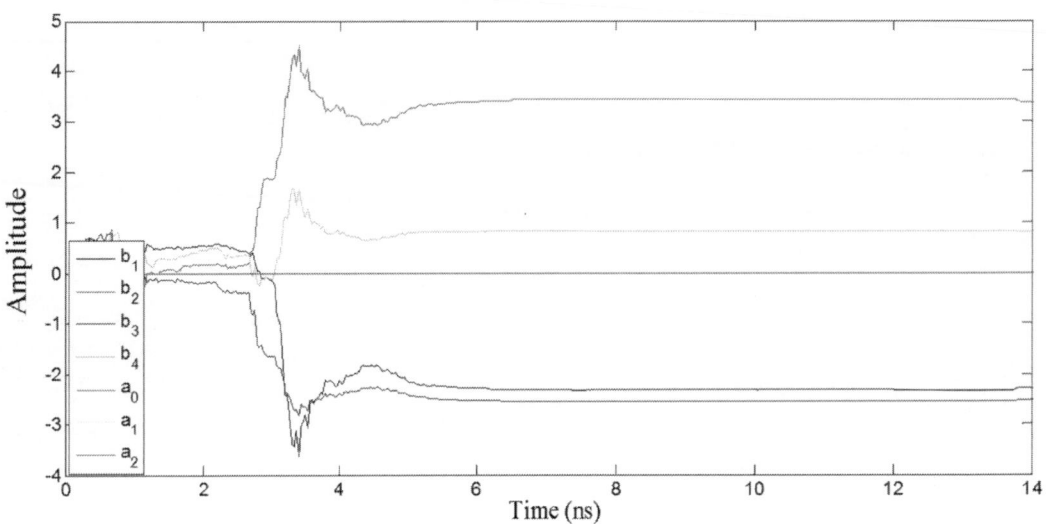

FIGURE 4.35

Convergence curve of system parameters including two zeros by RLS algorithm.

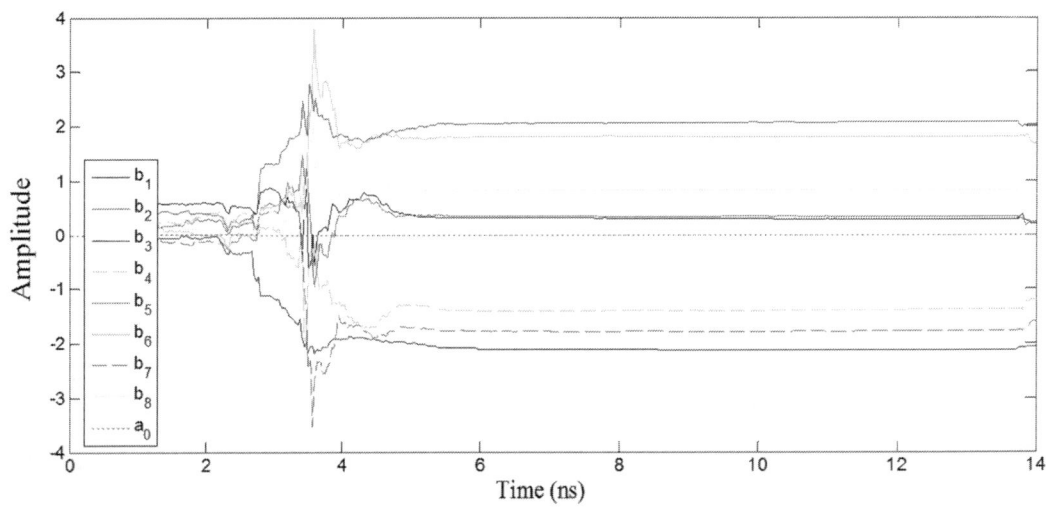

FIGURE 4.36

Convergence curve of system parameters including eight poles by RLS algorithm.

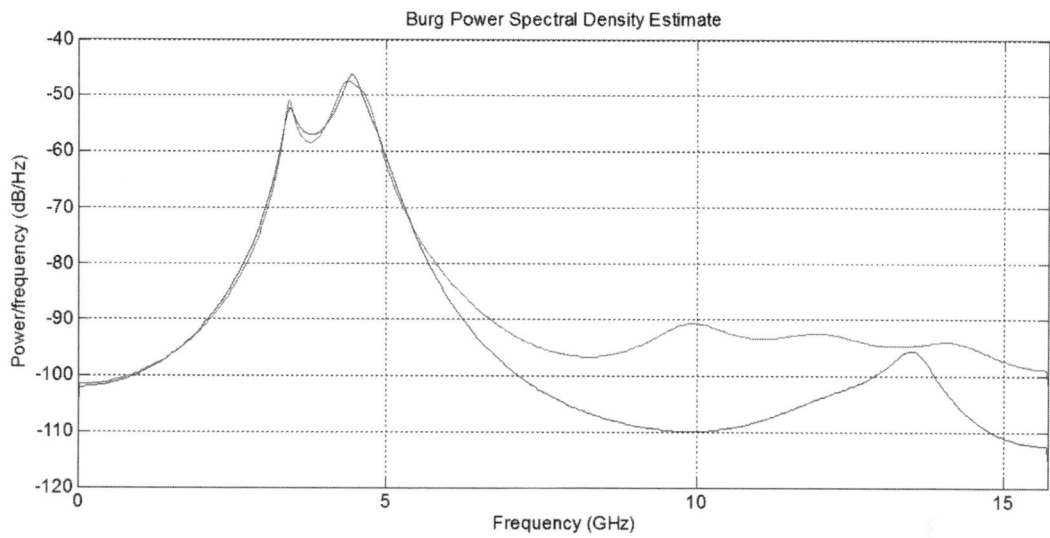

FIGURE 4.37

Measured and estimated PSD of received signal considering eight poles.

4.6.3 ARX model estimation of transfer function

First, a model with four poles and two zeros is considered for the transfer function. The mathematical model is as follows:

$$\left(1 + b_1\,z^{-1} + b_2\,z^{-2} + b_3\,z^{-3} + b_4\,z^{-4}\right)y_t = \left(a_0 + a_1\,z^{-1} + a_2\,z^{-2}\right)x_t + v_t \tag{4.56}$$

As mentioned in Section 4.6.2.3, the RLS method has more advantages than other methods, and so the parameters of other models are only estimated by the RLS method. The convergence of the parameters is shown in Fig. 4.38. The parameters related to the system zeros, i.e., a_0, a_1 and a_2, have converged to zero, indicating that the system has no zero and its ARX model is all-pole. The PSD in this model is shown in Fig. 4.39.

4.6.4 ARMAX model estimation of transfer function

In this model, four poles and two zeros are considered for the system. There are also two zeros for the error. The mathematical model is as follows:

$$\left(1 + b_1\,z^{-1} + b_2\,z^{-2} + b_3\,z^{-3} + b_4\,z^{-4}\right)y_t = \left(a_0 + a_1\,z^{-1} + a_2\,z^{-2}\right)x_t + \left(1 + c_1\,z^{-1} + c_2\,z^{-2}\right)v_t \tag{4.57}$$

The parameters of the transfer function are estimated by the RLS method. The results of the convergence curve of ARMA and exogenous parts can be seen in Figs. 4.40 and 4.41, respectively, and the PSD is shown in Fig. 4.42. Nonzero model is obtained from the estimation.

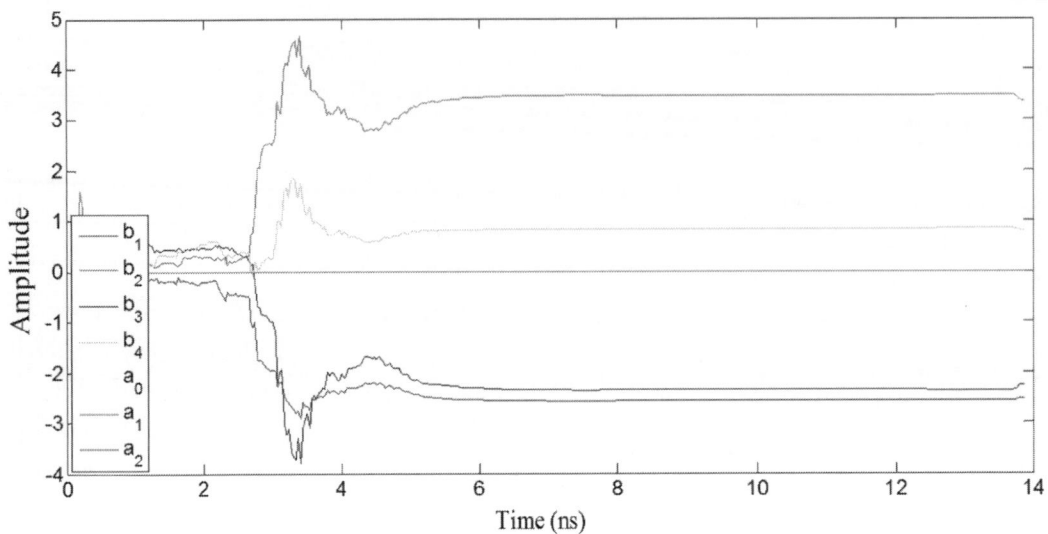

FIGURE 4.38

Convergence curve of ARX model parameters by RLS algorithm.

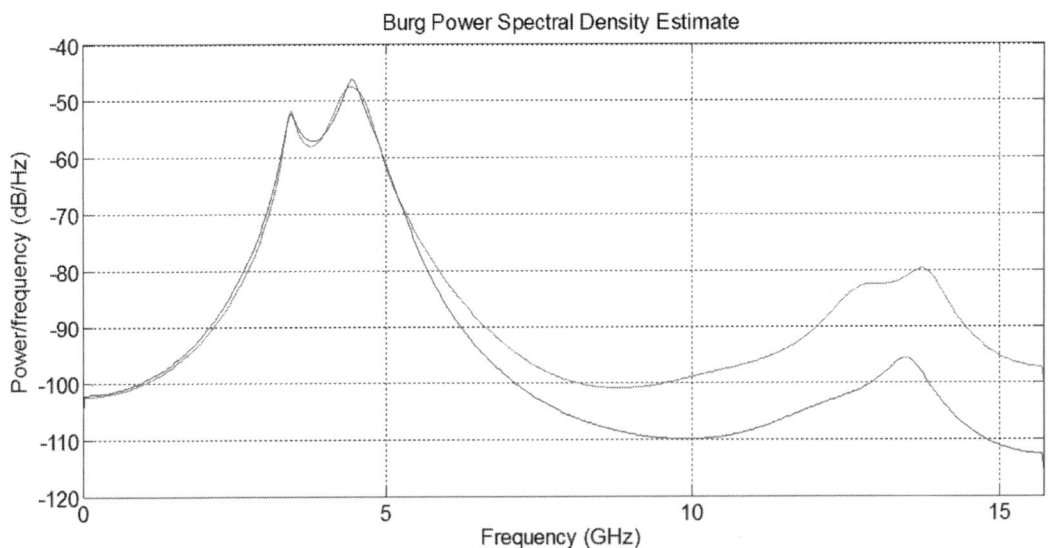

FIGURE 4.39

PSD of measured and estimated received signals in ARX model calculated by RLS algorithm.

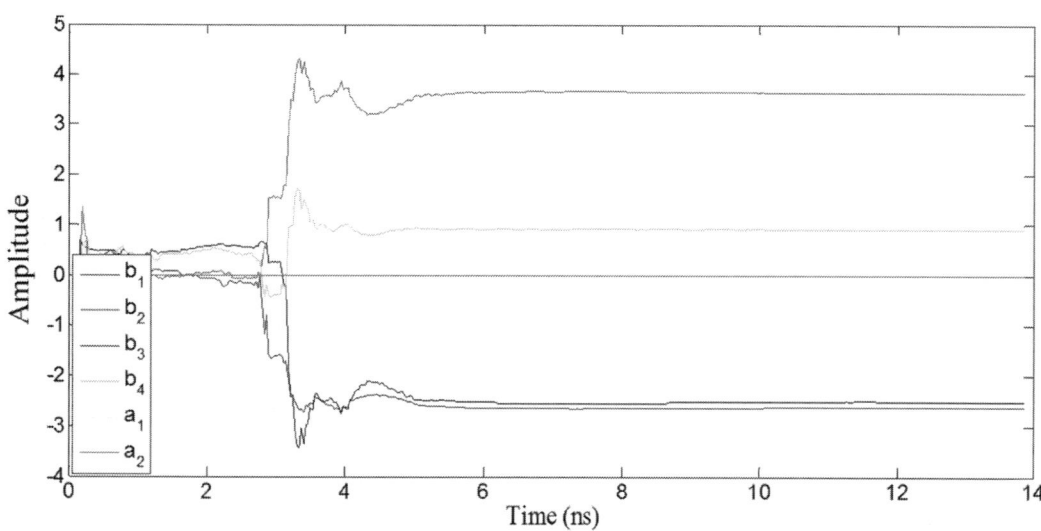

FIGURE 4.40

Convergence curve of ARMAX model parameters by RLS algorithm.

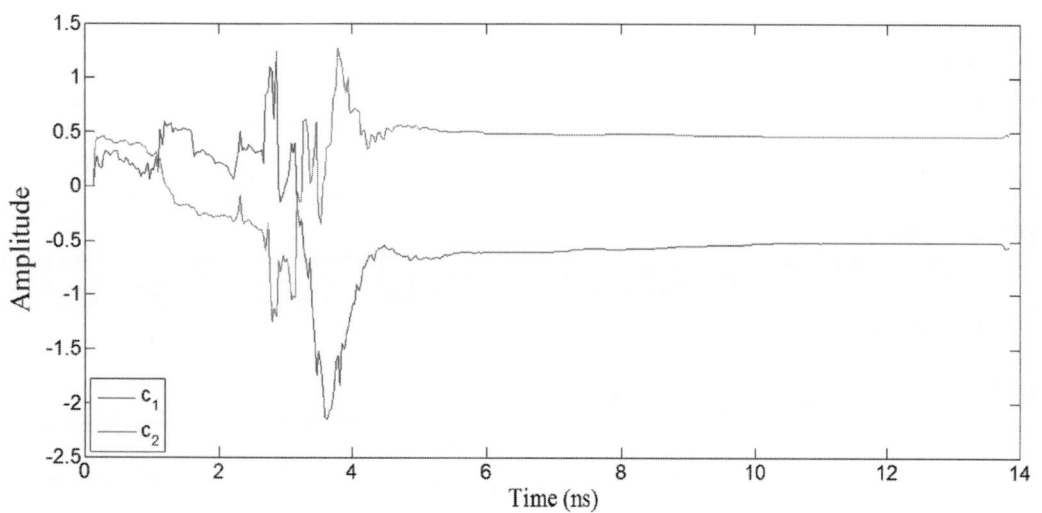

FIGURE 4.41

Convergence curve of error parameters in ARMAX model by RLS algorithm.

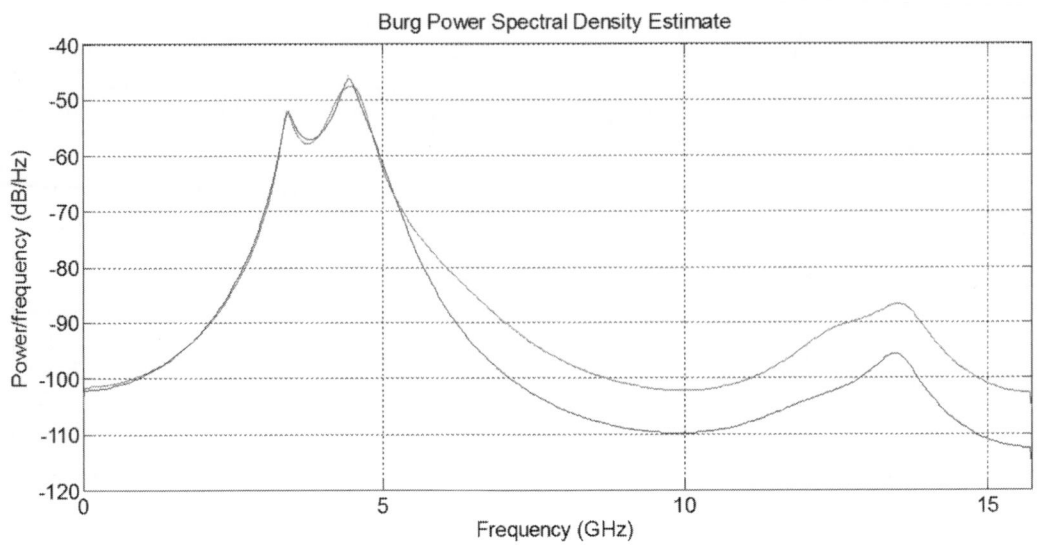

FIGURE 4.42

PSD of measured and estimated ARMAX models by RLS algorithm.

4.6.5 Box-Jenkins model estimation of transfer function

In this model, four poles and two zeros are considered for the system. In addition, four poles and two zeros are considered for error. The mathematical model is as follows:

$$y_t = \frac{\left(a_0 + a_1\, z^{-1} + a_2\, z^{-2}\right)}{\left(1 + b_1\, z^{-1} + b_2\, z^{-2} + b_3\, z^{-3} + b_4\, z^{-4}\right)} x_t + \frac{\left(1 + f_1\, z^{-1} + f_2\, z^{-2}\right)}{\left(1 + d_1\, z^{-1} + d_2\, z^{-2} + d_3\, z^{-3} + d_4\, z^{-4}\right)} v_t \quad (4.58)$$

Using the RLS method, the convergence curves of the coefficients related to x_t are shown in Fig. 4.43. According to this figure, the parameters related to the zeros of the system, i.e., a_0, a_1, and a_2, have converged to zero. Therefore, the system model does not have any zeros. Fig. 4.44 shows the convergence of the parameters related to the error model. The parameters related to the zeros of the error model, i.e., c_1 and c_2, have converged to zero. This indicates that the model intended for the error must have only poles. The final PSD of the estimated model compared to the measured results is shown in Fig. 4.45.

4.6.6 Space state model estimation of transfer function

In this model, the system is expressed by matrices A, B, C, and D. The best model means a model with less state space dimension than others with the same response to the same input. In this condition, the number of system states is equal to the number of its poles. Therefore, matrices A, B, C, and D for the system are obtained by the RLS method.

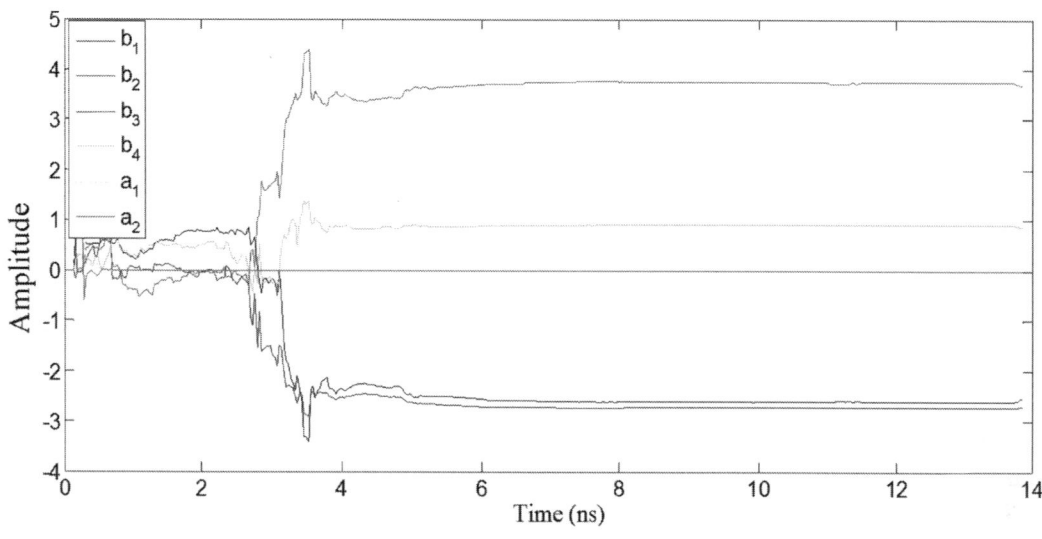

FIGURE 4.43

Convergence curve of coefficients related to x_t in BJ by RLS algorithm.

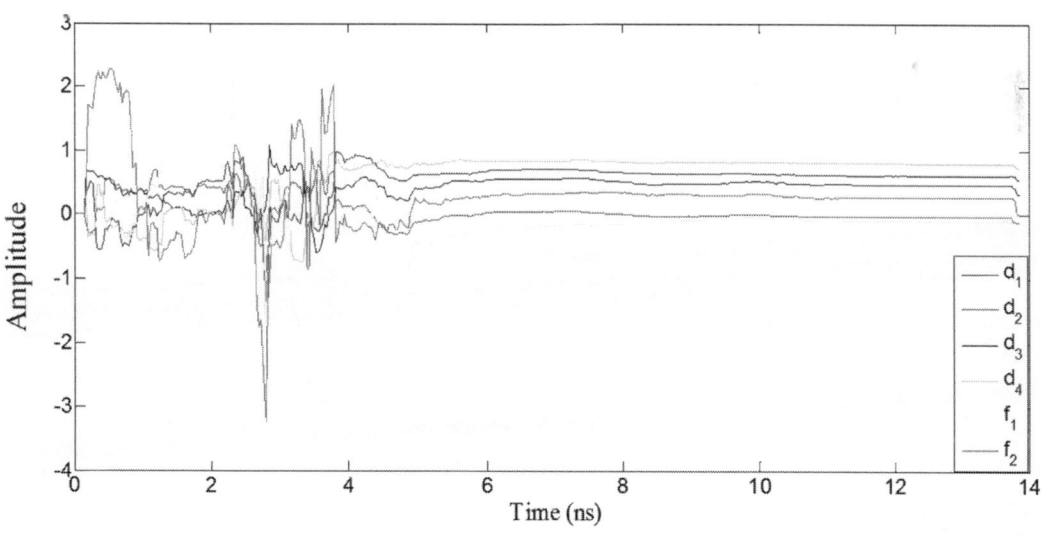

FIGURE 4.44

Convergence curve of error model parameters in BJ by RLS algorithm.

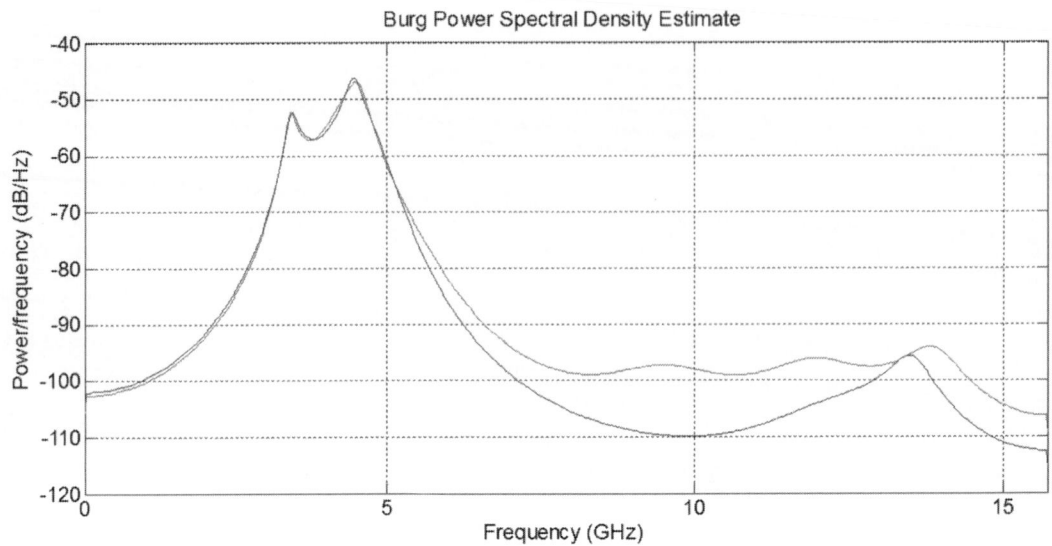

FIGURE 4.45

PSD of measured and estimated output signals in BJ model by RLS algorithm.

$$A = \begin{bmatrix} 1.73 & -2.63 & 1.75 & -1.09 \\ 1 & 0 & 0 & 0 \\ 0 & 1 & 0 & 0 \\ 0 & 0 & 1 & 0 \end{bmatrix} \tag{4.59}$$

$$B = \begin{bmatrix} 1 \\ 0 \\ 0 \\ 0 \end{bmatrix} \tag{4.60}$$

$$C = [0\ 629.1 - 818.1\ 621.3] \tag{4.61}$$

$$D = [0] \tag{4.62}$$

The output estimate and PSD are similar to the figures provided by the RLS method because the mathematical model is the same and the state space model has only changed the way of representation.

4.7 Using transfer function for fault detection
4.7.1 Transfer function evaluation

To evaluate the obtained transfer function, the measured signal related to the healthy state of the winding is compared with the estimated signal related to the healthy state of the winding model obtained from the transfer function. The method used to estimate the transfer function is RLS, and the considered model is BJ. Vivaldi antennas have been used in all experiments in this section. Also, the sampling time is adjusted to 10 bin, or 31.789 ps and the time interval between the pulses sent by the transmitter is 50 ms. The test method and the data matrix formation are according to Section 4.5.

In this section, only the transformer winding model is included, and the effect of the transformer tank is not considered. The distance between the transmitter and receiver antennas is 85 cm and the height of antennas from the surface of the test setup table is 7.32 cm. The distance of the transformer winding model center from the transmitter and receiver antennas is 60 cm. The position of the transformer winding model and the transmitter and receiver antennas is shown in Fig. 4.46:

Fig. 4.47 shows the measurement signals recorded by the receiver (after deleting inappropriate data, eliminating time delays, and averaging as described in Section 4.5.3) in the healthy state of the winding model and the state with 2×4 cm^2 segment displaced by 2.5 cm.

According to the setup configuration specifications and the distance between the transmitter and receiver, multipaths related to the transformer winding model reach the receiver with a time delay of 1.5 nm compared to multipaths of the direct path. The following equation can obtain this value.

$$t_d = (2x - z)/(3 * 10^{10}) \tag{4.63}$$

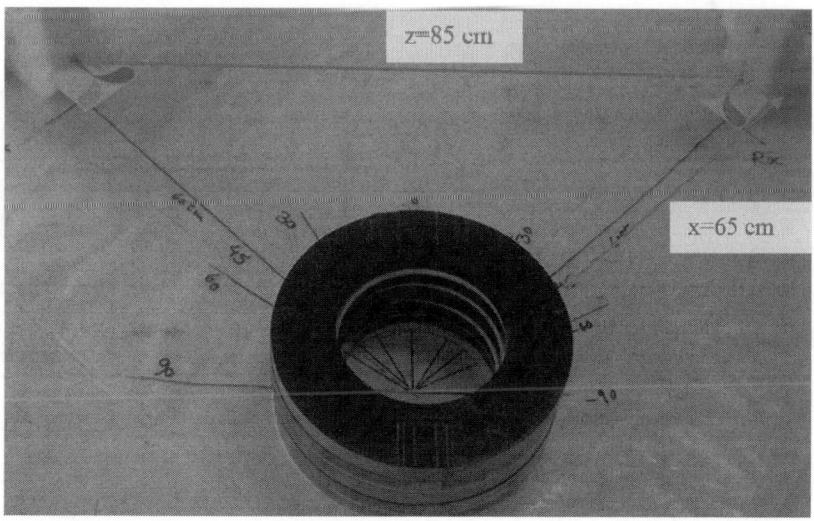

FIGURE 4.46

Configuration of model no. 1.

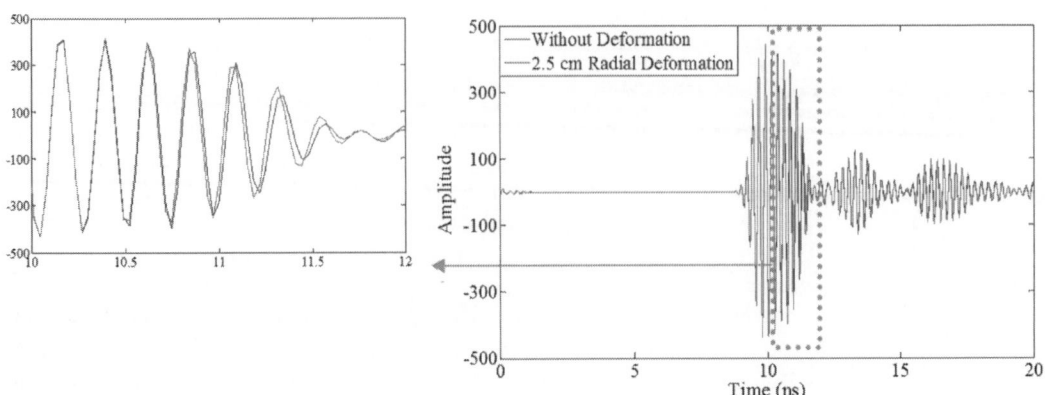

FIGURE 4.47

Measurement signal recorded by receiver in healthy and displaced states.

As the conditions of the experimental environment and positions of the transmitter and receiver antennas are constants relative to each other, it can be expected that the first multipaths recorded at the receiver, which correspond to the direct path between transmitter and receiver, are the same for the healthy and faulty states of the transformer winding model. Due to the time lag of the multipath of the transformer winding compared to the multipath of the direct path, the multipath of the transformer winding model is specified in the figure with a green dotted time window. So, it can be said that the effect of changes in the transformer winding model can be seen in this window.

Having the input signal and signal recorded by the receiver in the experiment, the transfer function between input and output is obtained for this experimental configuration. Fig. 4.48 shows the signal obtained from the measurements and the transfer function. It is clear from this figure that the difference is neglectable.

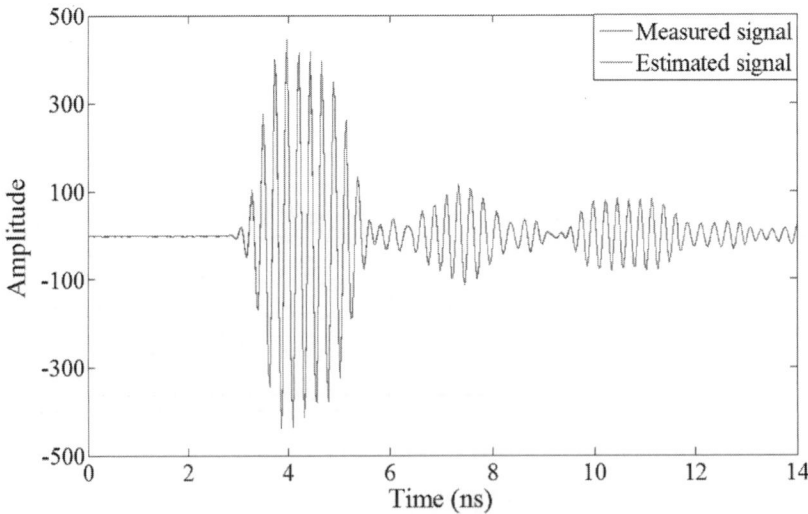

FIGURE 4.48

Comparison of measured and estimated signals of healthy state of transformer winding.

4.7.2 Detection of radial deformation fault

In this section, the aim is to investigate the transfer function ability to detect faults. As described before, the number of conjugate poles of the system equals the number of PSD peak points in the system output signal. Due to two maximum points around the frequency of 4 GHz, there are two-pole pairs in the transfer function. Laboratory configuration specifications are given in Table 4.4.

4.7.2.1 Winding model no. 2 without considering tank

As described in Section 4.5.1, 5 disks with diameters of 30 cm are placed, one of which is sectorized in a radial direction. To create a radial deformation, the sector segments are pulled out, an example of which is shown in Fig. 4.49. The PSD of the recorded signal in the healthy state and the faulty one, i.e., the radial deformation, is shown in Fig. 4.50. As shown in the figure, the PSD has two peak points in all faulty states, so all modes can be estimated with two-pole pairs. Fig. 4.51 shows the location of the poles of the transfer function for the healthy state and the various states of the radial deformation applied on the winding model with a diameter of 30 cm. The transfer function is estimated by the RLSEXP method.

Table 4.4 Specifications of laboratory configuration for radial deformation study.	
Distance between transmitter and receiver antennas	47 cm
Height of antennas from surface of experiment table	24.5 cm
Height of transformer winding model from surface of experiment table	12.5 cm
Distance of transformer winding model from the connecting line between the transmitter and receiver antennas	40 cm
Size of displaced segment	$2 \times 2 \ \mathrm{cm}^2$

A   B

FIGURE 4.49

(A) Configuration of winding model no. 2 without tank, and (B) radial deformation applied on disk.

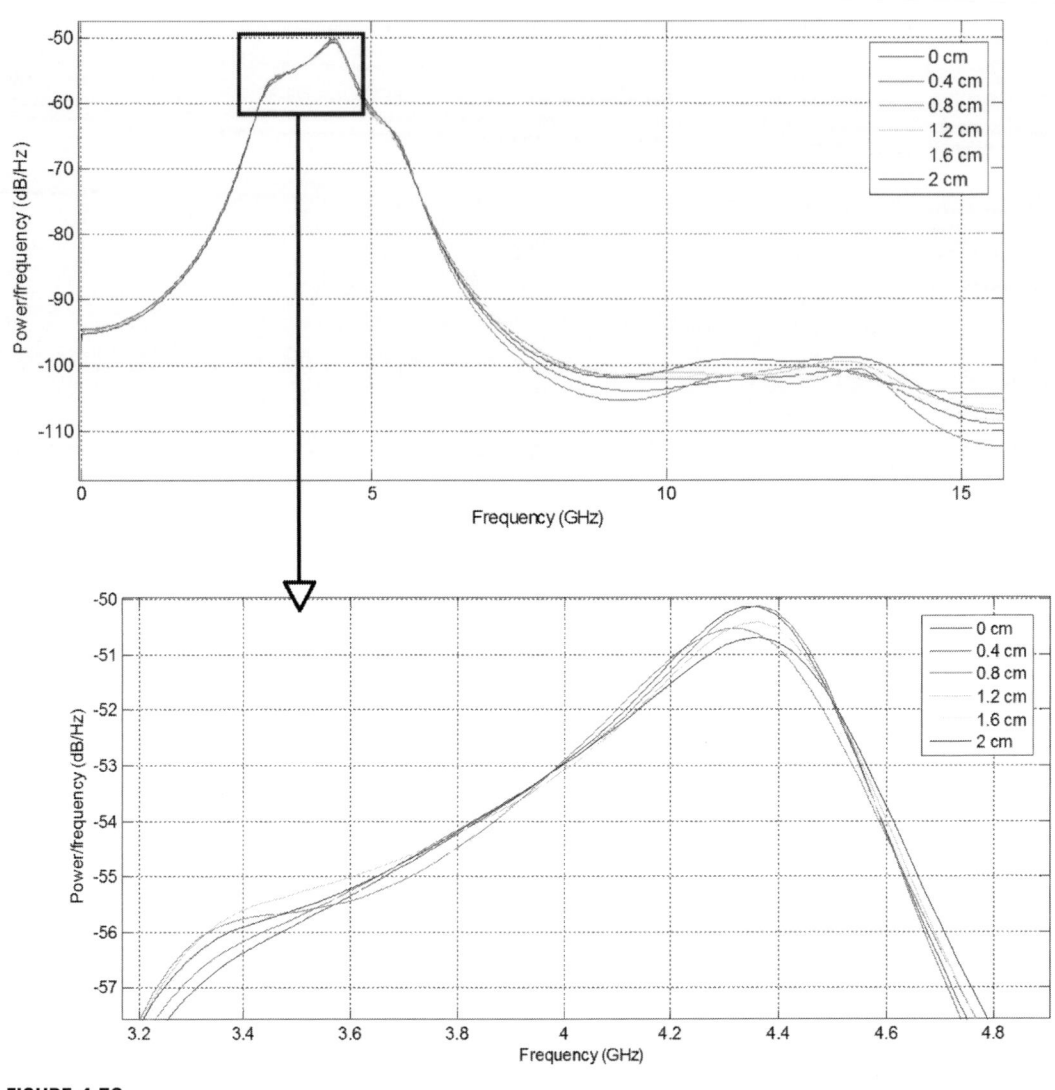

FIGURE 4.50

PSD of measured received signals for various radial deformations on winding model no. 2.

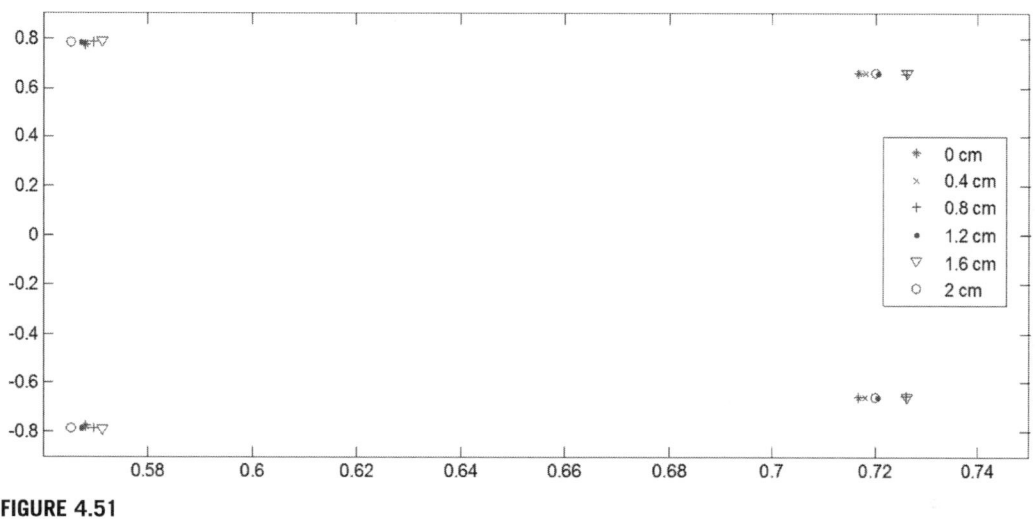

FIGURE 4.51

Location of transfer function poles for various states of radial deformation applied on winding model no. 2.

4.7.2.2 Winding model no. 1 without considering tank

Similar to model no. 2, 5 disks with diameters of 60 cm are placed, one of which is sectorized in the direction of the center of the disk. The radial deformation is applied on the model by pulling out one sector, as shown in Fig. 4.52. The configuration of the setup is similar to that of model no. 2 mentioned in Table 4.4.

The PSD of the received signal in the healthy state and the faulty ones (the radial deformation) of the transformer winding model are shown in Fig. 4.53. As can be seen in this figure, in all faulty states,

FIGURE 4.52

Experimental configuration for winding model no. 1.

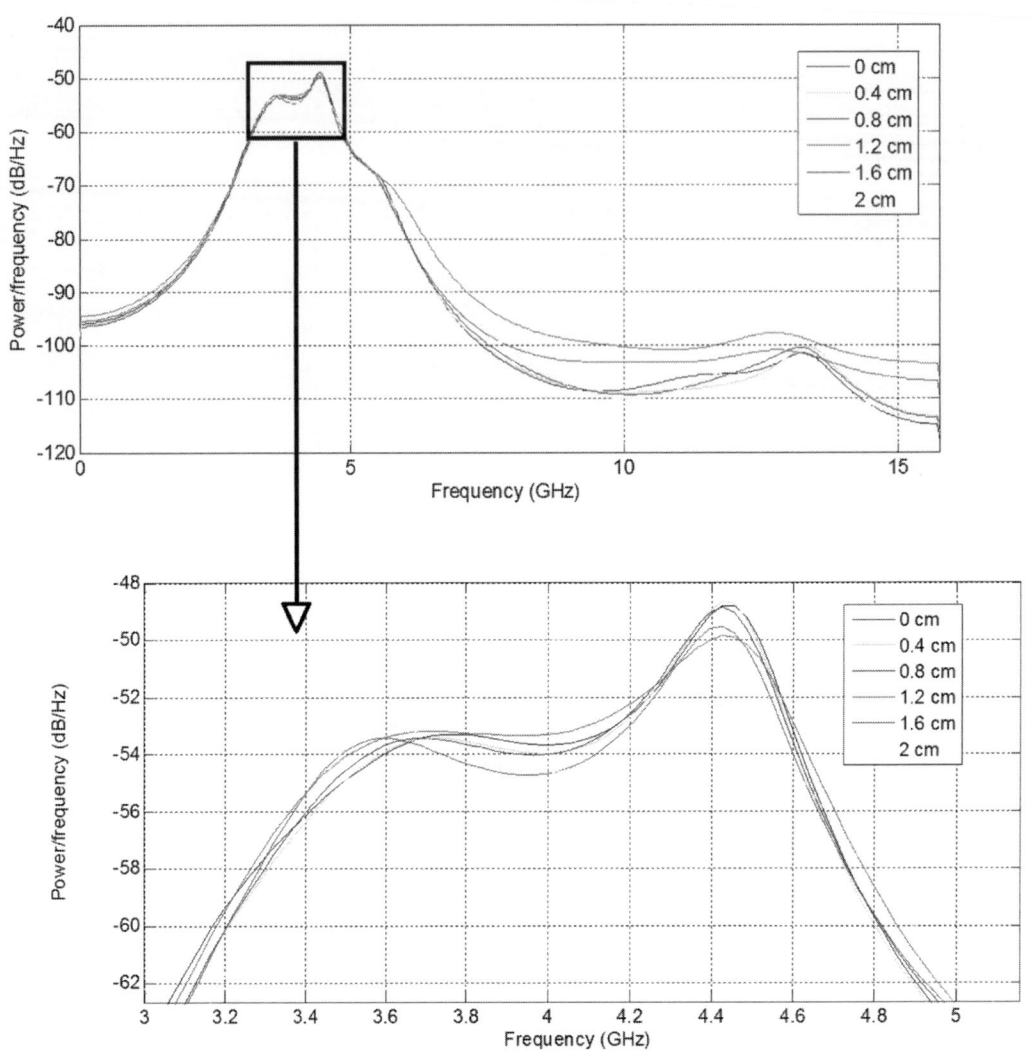

FIGURE 4.53

PSD of measured received signals for various radial deformation faults on the winding model no. 1.

the PSD has two peak points. Fig. 4.54 shows the location of the poles of the transfer function for various states of radial deformation applied on the winding model no. 1. The transfer function is estimated by the RLSEXP method.

4.7.2.3 Considering transformer tank in winding model no. 2

The transformer tank behaves as a reflector for electromagnetic waves. So, the transformer tank can be modeled with an aluminum wall, as shown in Fig. 4.55. The effect of the transformer tank on the received signal will be described further in Chapter 7. In this chapter, only its impact on the transfer function is studied. The model no. 2 is placed inside the transformer tank model with a distance of 25 cm on each side. The specifications for this laboratory configuration are given in Table 4.4.

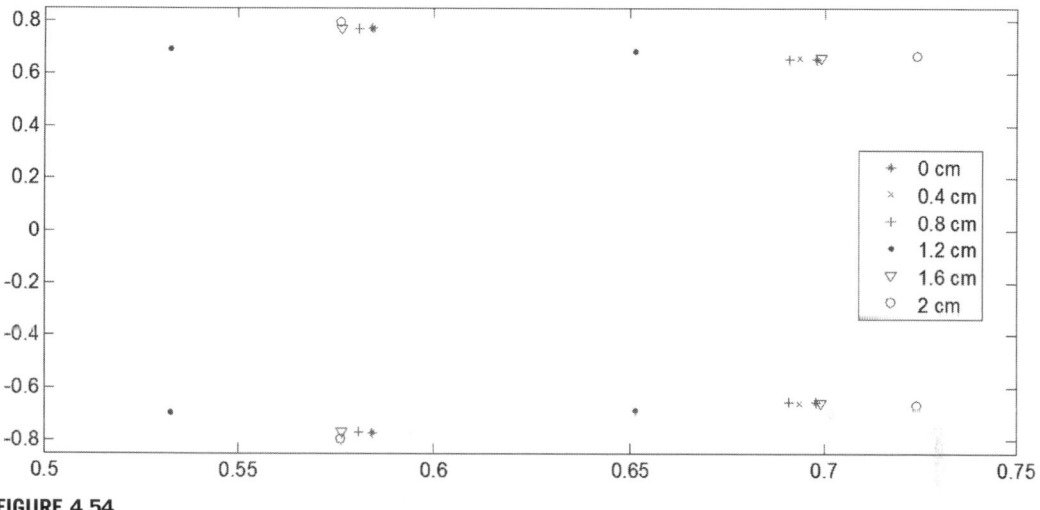

FIGURE 4.54

Location of transfer function poles for various states of radial deformation applied on winding model no. 1.

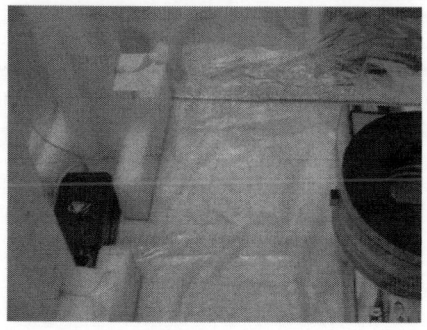 

FIGURE 4.55

Configuration of winding model no. 2 with tank.

The PSD of the received signal in the healthy and faulty states are shown in Fig. 4.56. Fig. 4.57 shows the location of transfer function poles for the healthy state and various radial deformations applied on the winding model no. 2 considering the presence of the tank. The transfer function is estimated by the RLSEXP method.

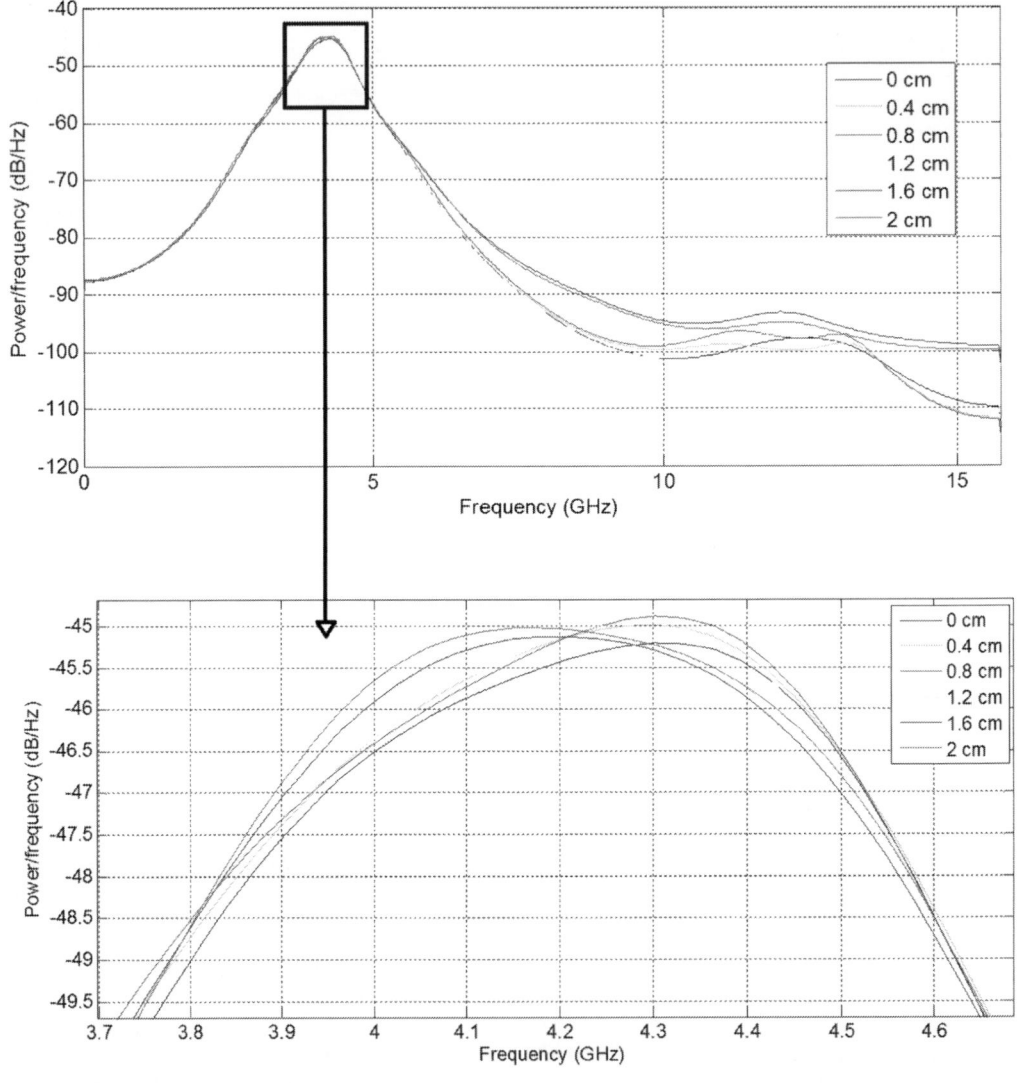

FIGURE 4.56

PSD of measured recorded signals for various radial deformations on winding model no. 2 with tank.

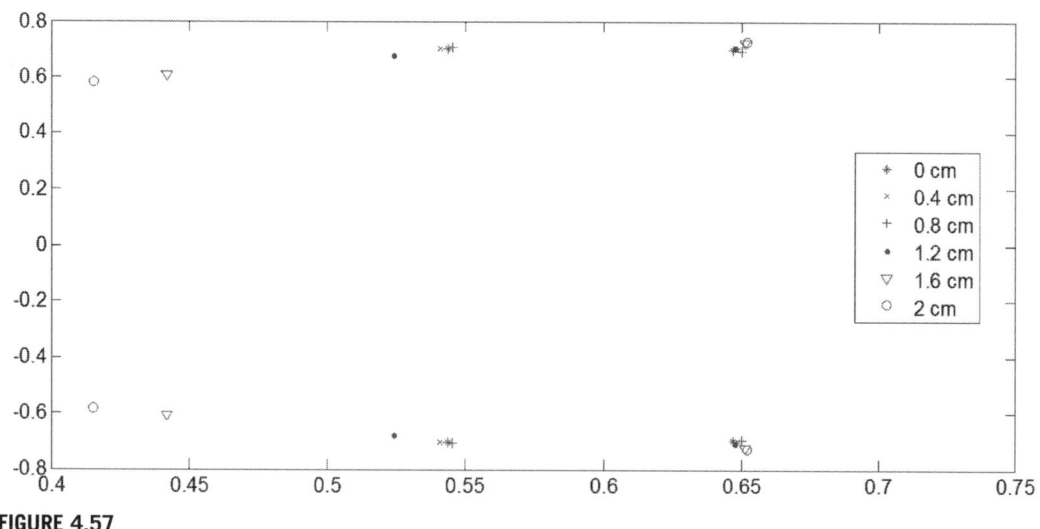

FIGURE 4.57

Location of poles of transfer function for various radial deformations applied on winding model no. 2 with tank.

4.7.2.4 Considering transformer tank for winding model no. 1

The winding model no. 1 is placed inside the model of transformer tank (Fig. 4.58). The specifications of this laboratory configuration are similar to previous tests.

The PSD of the received signals in the healthy and faulty cases are as shown in Fig. 4.59. As can be seen in this figure, in all faulty states, the PSD has two peak points. Fig. 4.60 shows the location of the poles of the transfer function for the healthy state and the various radial deformations applied on the winding model. The transfer function is estimated by the RLSEXP method.

FIGURE 4.58

(A) Configuration of the winding model no. 1 with tank and (B) radial deformation applied on winding model.

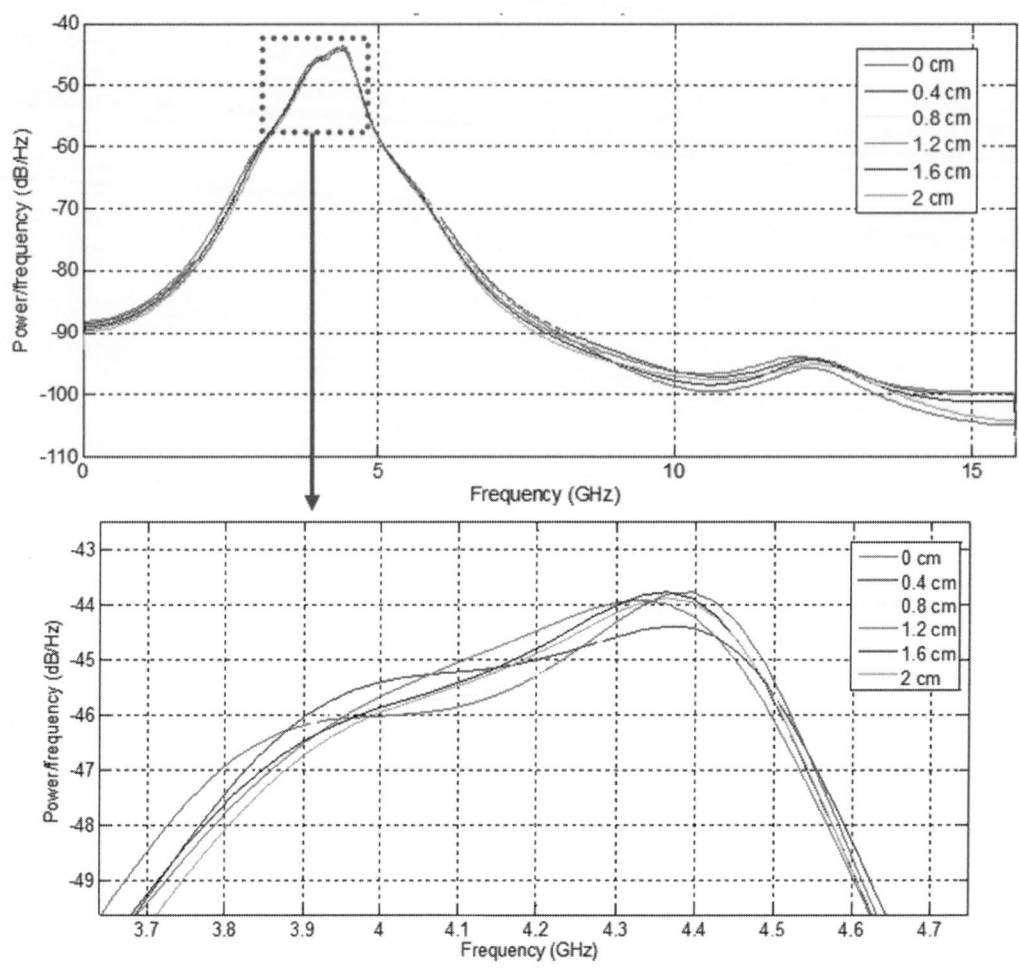

FIGURE 4.59

PSD of measured recorded signals for various radial deformations on winding model no. 1 with tank.

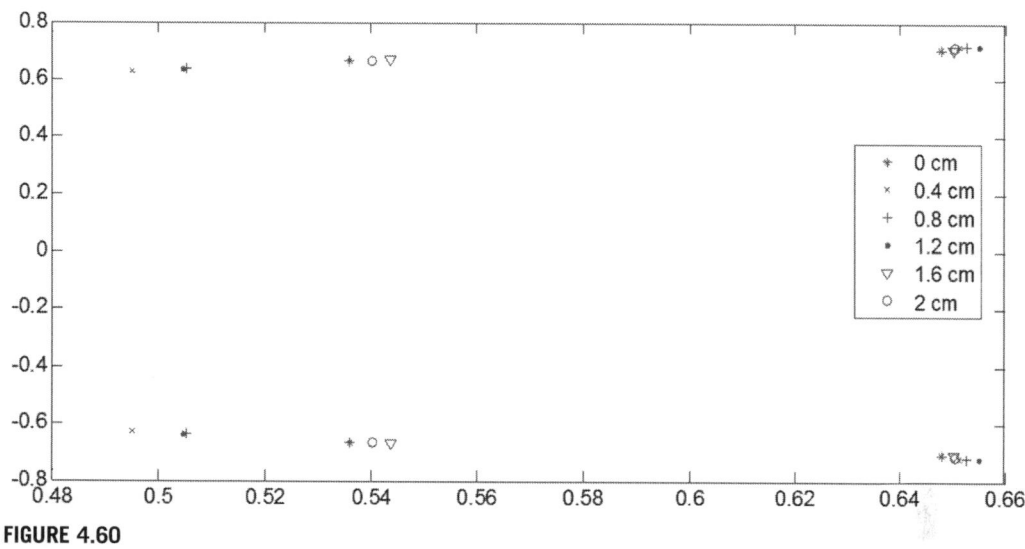

FIGURE 4.60

Location of transfer function poles for various radial deformations applied on winding model no. 1 with tank.

4.7.3 Detection of axial displacement

To perform this experiment, a controllable elevator is used to apply axial displacement, as shown in Fig. 4.61. This elevator has an accuracy of 0.1 mm movement and the range of elevation is from zero to 50 mm.

The specifications of the experimental setup are described in Table 4.5.

Fig. 4.62 shows the location of the poles of the transfer function for different axial displacements. As can be seen in this figure, the locations of the poles of the transfer function are very close to each other for different displacements and in some cases are not separable.

4.7.3.1 Detection of radial deformation extent using zeros and poles as features

Extracting zeros and poles from the transfer function and assigning each state to a limited number of zeros and poles can facilitate storing the data. In this section, using an artificial neural network (ANN), the extent

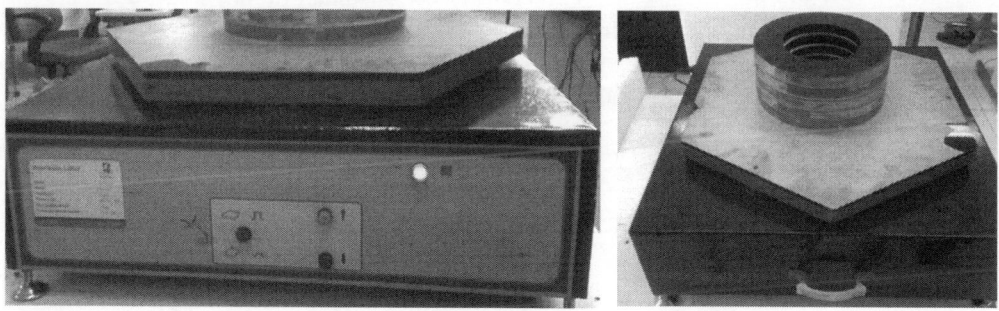

FIGURE 4.61

Elevator for applying axial displacement on winding models.

Table 4.5 Specifications of laboratory configuration for axial deformation study.

Distance between transmitter and receiver antennas	50 cm
Height of antennas from surface of experiment table	65 cm
Height of transformer winding model from surface of experiment table	58 cm
Distance of transformer winding model from the connecting line between the transmitter and receiver antennas	70 cm
Size of axial displacement in each step	0.2 mm

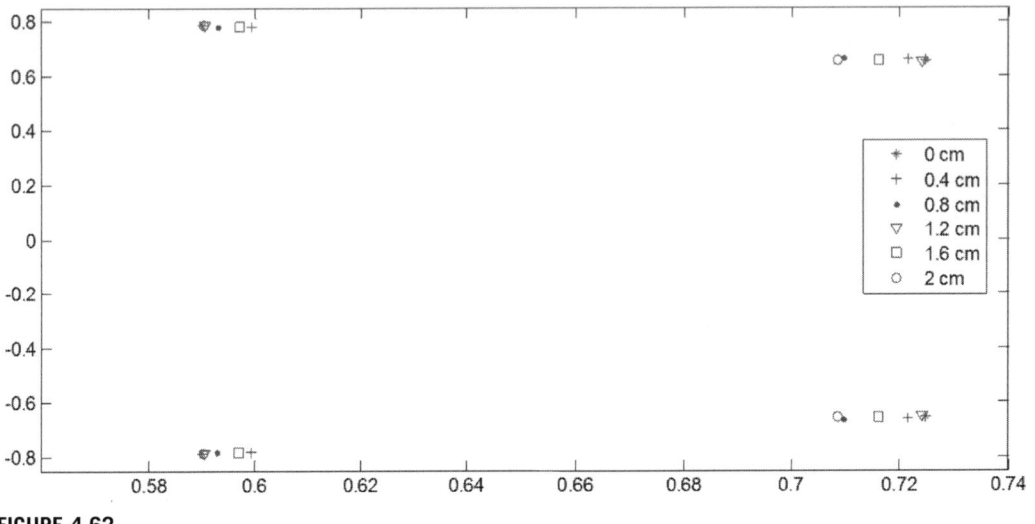

FIGURE 4.62

Location of transfer function poles for various axial displacement applied on winding model.

of radial deformations can be identified. Location of poles is used as the features extracted from the transfer channel function for the input of ANN and the output is the extent of the radial deformation fault.

Experiment studies are carried out for 41 various radial deformations. The $2*2$ cm^2 segment is moved from 0 to 40 mm, and each state is represented by a signal containing 640 samples. The transfer function of the system is estimated as described previously. The poles of the system are used for the input of ANN. Training of ANN is performed by 80% of the total data and 20% are used for ANN performance tests. The used neural network consists of eight layers and ten neurons. The results are shown in Table 4.6.

The average error for the estimated data is 5.7 mm with a standard deviation of 5.41 mm. It is clear that the results are not good and so, the transfer function cannot be used for detecting the radial deformation. Only the existence of a defect can be detected.

4.7.3.2 Detection of axial displacement extent using zeros and poles as features

The total numbers of states for axial displacement are 151. The procedure is similar to the previous section and the results are listed in Table 4.7.

Table 4.6 Results of estimating extent of radial deformation by ANN.

Radial deformation amount (mm)	Estimated extent (mm)
4	16.6845
8	8.0174
12	12.0407
16	14.5552
20	11.9488
24	19.2380
28	43.0480
32	24.1472
36	28.9457
40	40.0122

Table 4.7 Results of estimating extent of axial displacement by ANN.

Axial displacement amount (mm)	Estimated extent (mm)
0.8	3.64
1.8	1.98
2.8	1.44
3.8	4.25
4.8	4.69
5.8	2.80
6.8	6.86
7.8	4.90
8.8	10.97
9.8	11.56
10.8	9.40
11.8	9.56
12.8	10.77
13.8	15.81
14.8	8.43
15.8	15.49
16.8	14.11
17.8	18.52
18.8	8.28
19.8	18.87
20.8	18.36
21.8	21.37
22.8	22.00
23.8	28.52
24.8	26.60
25.8	27.97
26.8	25.22
27.8	26.24
28.8	28.77
29.8	27.02

The average error for the estimated data is 2.08 mm with a standard deviation of 2.12 mm. This error considering the accuracy of displacements, i.e., 0.1 mm, is very large. So, the transfer function can be used only for axial displacement detection.

4.8 Summary

This chapter introduces the transfer function as a tool for fault detection in power transformers. Enbloc methods and adaptive ones estimated the transfer function between the transmitter and receiver antennas, and it was proved that the system has an all-pole model and the number of poles is four. Then the system error dynamics were considered, and finally, an all-pole model with four poles was obtained for error. Then, for a given laboratory setup, different measurements were performed for different fault states. It is concluded that the transfer function can only be used for fault detection, and estimating the extent of faults cannot result in an accurate result.

References

Aström, K. J., & Wittenmark, B. (2013). *Adaptive control.* Courier Corporation.

Foerster, J. (2002). UWB channel modeling contribution from Intel. *IEEE P802,* 15-02/279-SG3a.

Gauss, C. F. (1795). *Method of least squares. unpublished.*

Ghavami, M., Michael, L., & Kohno, R. (2007). *Ultra wideband signals and systems in communication engineering.* John Wiley & Sons.

Giunta, G., Jacovitti, G., & Scarano, G. (1997). Bussgang Gaussianity test for stationary series. In *Proceedings of the IEEE signal processing workshop on higher-order statistics* (pp. 434—437).

Hejazi, M. S. A., Ebrahimi, J., Gharehpetian, G. B., Mohammadi, M., Faraji-Dana, R., & Moradi, G. (2011). Application of ultra-wideband sensors for on-line monitoring of transformer winding radial deformations—A feasibility study. *IEEE Sensors Journal, 12*(6), 1649—1659.

Ikonen, E., & Najim, K. (2001). *Advanced process identification and control.* CRC Press.

Juang, J.-N. (1994). *Applied system identification.* Prentice-Hall, Inc.

Karami, H., Gharehpetian, G. B., Hejazi, M. A. A., & Norouzi, Y. (2020). *Detection of radial deformations of transformers.* Google Patents.

Karami, H., Gharehpetian, G. B., Norouzi, Y., & Akhavan-Hejazi, M. (2020). Simultaneous radial deformation and partial discharge detection of high-voltage winding of power transformer. *IET Electric Power Applications, 14*(3), 383—390.

Kay, S. M. (1988). *Modern spectral estimation: Theory and application.* Pearson Education India.

Mehdipour, A., Mohammadpour-Aghdam, K., & Faraji-Dana, R. (2007). Complete dispersion analysis of Vivaldi antenna for ultra wideband applications. *Progress in Electromagnetics Research, 77,* 85—96.

Parazzoli, C. G., Greegor, R. B., Li, K., Koltenbah, B. E. C., & Tanielian, M. (2003). Experimental verification and simulation of negative index of refraction using Snell's law. *Physical Review Letters, 90*(10), 107401.

PulsON 220 Specifications. (2010). http://www.timedomain.com/datasheets/P220aRD.php.

Zheng, W. X. (1999). A least-squares based algorithm for transfer function identification. In *vol 1. ISSPA'99. Proceedings of the fifth international symposium on signal processing and its applications (IEEE cat. No. 99EX359)* (pp. 183—186).

Analyzing EMWs using wavelet transform

Gevork B. Gharehpetian[1] and Hossein Karami[2]

[1]*Amirkabir University of Technology (AUT), Electrical Engineering Department, Tehran, Iran;* [2]*Niroo Research Institute (NRI), High Voltage Studies Research Department, Tehran, Iran*

5.1 Fourier transform

Generally, applying a mathematical transform to a signal aims to obtain additional information that is not available in the original raw signal (Karami et al., 2019, 2020). In practice, the vast majority of used signals are in the time domain. In many cases, the useful information of the signal lies in its frequency content, which is called the signal spectrum and is of special importance (Tarimoradi et al., 2021). Therefore there must be a tool to obtain the frequency content. The Fourier transform is perhaps the most well-known one, which breaks down a signal into sinusoidal components of different frequencies. On the other hand, it can be said that the Fourier transform provides a mathematical tool for transforming signals from the time domain to the frequency domain. From a mathematical point of view, the Fourier transform is defined as follows:

$$F(\omega) = \int\limits_{-\infty}^{+\infty} f(t)e^{-j\omega t}dt \tag{5.1}$$

The above formulation is the sum of the $f(t)$ signal multiplied by the complex exponential function over time from minus infinity to positive infinity. The result of this transform is the Fourier coefficients of $F(\omega)$, which, if multiplied by sinusoidal waves with the frequency of ω, show the sine components that make up the original waveform.

Fourier analysis has a major drawback in transferring a signal into a frequency domain; its temporal information is lost. The Fourier transform only indicates whether the particular frequency f is present in the signal or not, but does not provide any information about the time interval at which that frequency occurred. In other words, by looking at the Fourier transform of a signal, it is impossible to know the time of occurrence of a particular event. For example, if there is a sudden change in a signal, the effect of this change is spread over the entire frequency spectrum. Therefore the Fourier transform is not suitable at all for analyzing transient signals because we need to determine the time at which each frequency corresponds to such events. So, we are looking for entering some temporal information next to the frequency characteristics of the signal. To solve this problem, instead of taking Fourier transform on the entire time period of the signal, we first select a short period of the signal at a specific time by a window and then take the Fourier transform on that short period.

5.2 Short-time Fourier transform

In this method, the Fourier transform is applied to analyze only a small signal segment in a time window. The short-time Fourier transform maps a signal into a two-dimensional function of time and frequency.

By selecting a large time window, the resolution of frequency increases, while the time resolution of a large window is low. While selecting a small time window, although the time resolution will be good, the frequency resolution will not be desired.

Since the window used to calculate the short-time Fourier transform is constant, we must find a compromise between time and frequency resolution according to the signal being analyzed. This means that high time and frequency accuracy cannot be achieved simultaneously. In most cases, the signal sometimes changes softly and sometimes rapidly. Such signals cannot be processed with a constant time window because, in regions where signal changes are slow, we need a wide window, and in regions where signal changes are rapid, we need a small window.

So, we actually need a smart window. But as we know, in the short-time Fourier transform, the width of the window is constant. In fact, in the short-time Fourier transform, the time-frequency plane is evenly distributed, and this is one of the disadvantages of this transformation.

To solve this problem, the idea of using a transform with a changeable resolution came to mind, which led to the emergence of the wavelet transform. The wavelet transform has successfully been applied to a wide range of signals that are not periodic and may include sinusoidal and impulse components (Naderi et al., 2007; Nafar et al., 2004). In particular, the ability of the wavelet transform to focus on small time intervals and high-frequency components, as well as large time intervals and low frequencies can improve the analysis.

5.3 Wavelet transform

Wavelet theory is a mathematical tool developed in the early 1980s to analyze and synthesize nonstationary signals. Its mathematical foundations date back to Joseph Fourier's research in the 19th century. Fourier established the principles of this theory with his work on frequency analysis (Coifman & Wickerhauser, 1992). A wavelet is a small wave whose energy is concentrated in time and is a means of transient analysis of nonstationary or time-varying signals (Ghanizadeh & Gharehpetian, 2013). Due to its oscillating nature, the wavelet provides the conditions to have a frequency-time analysis of a signal.

5.3.1 Continuous wavelet transform

Continuous wavelet transform was proposed as an alternative method to the short-time Fourier transform, and its purpose is to improve the resolution of problems solved by the short-time Fourier transform. In wavelet analysis, similar to the short-time Fourier transform, the signal is multiplied by a function (wavelet) that acts the same as the window function. Also, similar to the previous one, the wavelet transform is separately applied to different time segments of the signal.

In other words, wavelet transform is the application of a window with variable sizes that allows us:

- To use longer time-based properties where obtaining low-frequency information with high accuracy should be considered.
- To use shorter time-based properties where high-frequency information should be considered.

The continuous wavelet transform of a signal $x(t)$ to a mother wavelet $\psi(t)$ is defined as follows:

$$CWT_x^\psi(a,b) = \Psi_x^\psi(a,b) \frac{1}{\sqrt{a}} \int_{-\infty}^{+\infty} x(t)\psi^* \left(\frac{t-b}{a}\right) dt \qquad (5.2)$$

where, a and b are the scale and translation parameters, respectively. The concept of translation is exactly the same as the concept of time transfer in the short-time Fourier transform, which determines the amount of the window displacement and contains the time information of the transform. But unlike the short-time Fourier transform, we do not directly have any frequency parameter in the wavelet transform. Instead, we have the scale parameter, which is inversely related to frequency. In other words, $s = 1/f$. Thus we will have the details at the high scales, where the signal expands, and at the low scales, where the signal contracts, we will have the generalities. Note that the scale variable appears in the denominator in the definition of the wavelet transform. Therefore the signal is compressed for $s < 1$ values. In Eq. (5.2), ψ is a window function called the mother wavelet, from which all transferred and scaled versions are obtained.

In fact, the mother wavelet is a template function for producing other windows. In other words, the $CWT_x^\psi(a,b)$ wavelet transform coefficients with specific scale and translation coefficients show how similar the main $x(t)$ signal is to the sum of the signals from expanding and translating the mother wavelet.

5.3.2 Discrete wavelet transform

The reason for using discrete wavelet transform (DWT) is that, first the measured samples are usually discrete in the problems we face. Also, if a and b are continuous, a large amount of data is generated in the wavelet transform. Similar to the ratio between the continuous and discrete Fourier transform, continuous wavelet transform has an equivalent called DWT, which is defined as follows (Misiti et al., 1996):

$$DWT(j,k) = \frac{1}{\sqrt{a_0^j}} \sum_n f(k)\psi \left(a_0^{-j}n - kb_0\right) \qquad (5.3)$$

$a_0 = 2$ and $b_0 = 1$ are generally selected. By selecting these values, DWT can be implemented using a multi-step filter.

DWT can be a called wavelet series because a function of a continuous variable is thought of as a sequence of coefficients similar to what is done in the Fourier series. This expansion has two indices: k and j, which are time translation and scale index, respectively. Each function can be expanded using a wavelet, and its scaled and translated versions are given in Eq. (5.4).

$$f(t) = \sum_{j,k} a_{j,k} \cdot \psi_{j,k} \tag{5.4}$$

where, $a_{j,k}$ is the wavelet coefficient and $\psi_{j,k}$ is wavelet function. If ψ is the mother wavelet selected for transform, $\psi_{j,k}$ can be obtained according to the following equation.

$$\psi_{j,k} = 2^{j/2} \psi \left(2^j t - k \right) \tag{5.5}$$

where, 2^j holds the scale of t, $2^{-j}k$ holds the translation in t and the $2^{j/2}$ coefficient keeps the norm of wavelet function constant at different scales. The function generated by Eq. (5.5) is called the mother wavelet, which is a template for generating the functions required in the wavelet expansion.

The sum of the terms in the form of Eq. (5.4) creates the wavelet expansion of the function.

$$f(t) = \sum_{k} a_k \cdot \phi(t - k) + \sum_{j=0}^{M} \sum_{k=0}^{2^j - 1} d_{j,k} \cdot \psi \left(2^j t - k \right) \tag{5.6}$$

In the above equation, the first term produces a function with a low-resolution approximation of $f(t)$. With each increase in j in the second term, a higher-precision function is added to the previous section, which increases the total accuracy. If the bases are orthogonal, the coefficients can be obtained by internal multiplication, as follows.

$$a_k = \langle f(t), \phi_k(t) \rangle = \int f(t) \phi_k(t) \, dt \tag{5.7}$$

$$d_{j,k} = \langle f(t), \psi_{j,k}(t) \rangle = \int f(t) \psi_{j,k}(t) \, dt \tag{5.8}$$

In many signals, the low-frequency component is an essential part of the signal and expresses the original identity of the signal. The low-frequency part contains the general specifications of the signal, and the high-frequency part contains the minor specifications of the signal. Therefore approximation and detail concepts are used in the wavelet transform. In wavelet transform, approximations interpret high scales or low frequencies, and details represent high frequency or low scales components. Signal approximations and details are obtained by passing the signal through low-pass and high-pass filters, respectively. This doubles the number of output data. For example, if the original signal has 1000 samples, the signal will have an estimate and details of each 1000 samples. That is, after one filtering step, 2000 samples are produced. The number of samples is corrected for the next step by reducing one among the obtained data. Therefore we pass the outputs of two filters through a downsample. This action is shown in Fig. 5.1 as ⬇

The above process produces discrete wavelet coefficients.

5.3.3 Consecutive wavelet decompositions

By continuing the decomposition process on the approximation obtained from the first level, the second-level approximation and detail are obtained. The decomposition process is reproducible. It means that the approximation signal can again be divided into two parts with approximately equal

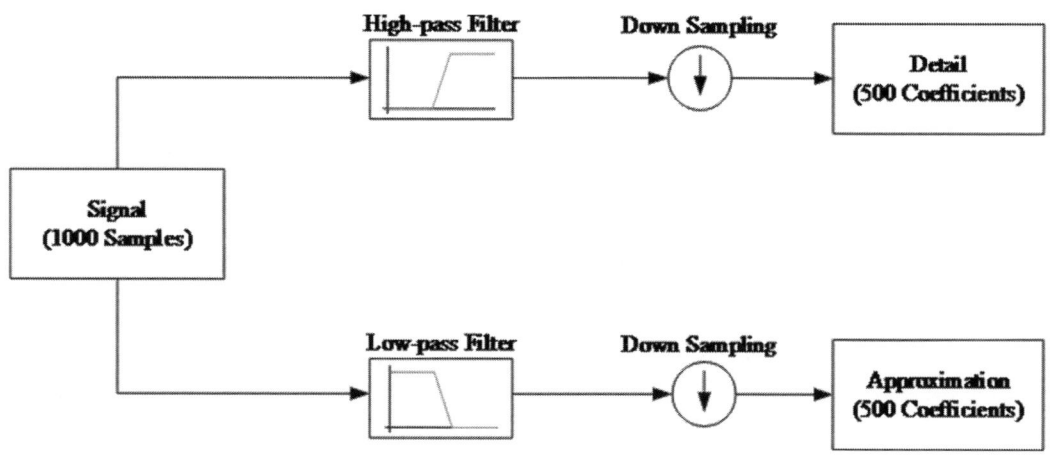

FIGURE 5.1

Correction of number of samples after signal filtering.

lengths, and this can be continued until the lengths of the vectors reach zero. In practice, based on the nature of the main signal waveform, a certain number of levels are selected or appropriate criteria such as entropy are used (Kia et al., 2009). Fig. 5.2 shows the decomposition of a signal up to the third order, which is also called the wavelet decomposition tree.

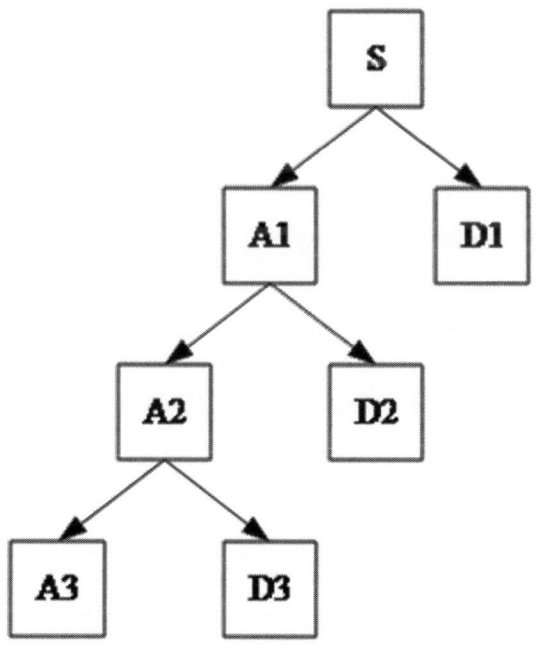

FIGURE 5.2

Tree diagram of decomposition of main signal into details and approximations.

According to Nyquist theory, in the DWT method, if the original signal is sampled at f_s, the highest correct frequency that can be achieved using this sampling frequency is $f_s/2$. Therefore in the output of the first detail, the obtained bandwidth will be from $f_s/2$ to $f_s/4$. In the second detail, the bandwidth is between $f_s/4$ and $f_s/8$, and in the k-th detail, in general, the bandwidth is between $f_s/2^k$ and $f_s/2^{k+1}$. Fig. 5.3 shows the frequency decomposition domain up to level 3.

5.3.4 Wavelet functions used for analysis

There are many types of mother waves that are used in practice. Haar and Morlet wavelets are orthogonal wavelets, while Daubechies and Symlet wavelets are nonorthogonal. Also, the mother wave of the Mexican hat is continuous and lacks a scale function, so the analysis cannot be orthogonal (Nasiri et al., 2004). The most important characteristic of wavelets is their oscillating and rapid damping behavior, along with their location in the time and frequency domains. Smooth wavelets, such as the Symlet, usually provide better frequency accuracy than wavelets, such as the Haar, which have sharp steps. Wavelets such as Haar are applied to signals that require more time accuracy. One of the most commonly used mother wavelets is the Daubechies wavelet. These wavelets do not have a closed formula, and the scale function and the wavelet are approximately calculated. These wavelets are suitable for detecting signals with small amplitudes and short periods of time. In this family of wavelets, naming is based on zero torques. For example, a Daubechies wavelet is an order of m waves with m of the first zero torque.

We must first select an appropriate wavelet function to apply wavelet transform to data. Therefore we tested some well-known wavelets on the signals obtained from the experiments to determine the proper wavelet or wavelets.

5.4 Applications of wavelet transform on measured data and analysis of results

Making measured samples in the dimensions of a real transformer is very expensive and practically difficult. The model introduced in the previous chapter is used as a test object in this section.

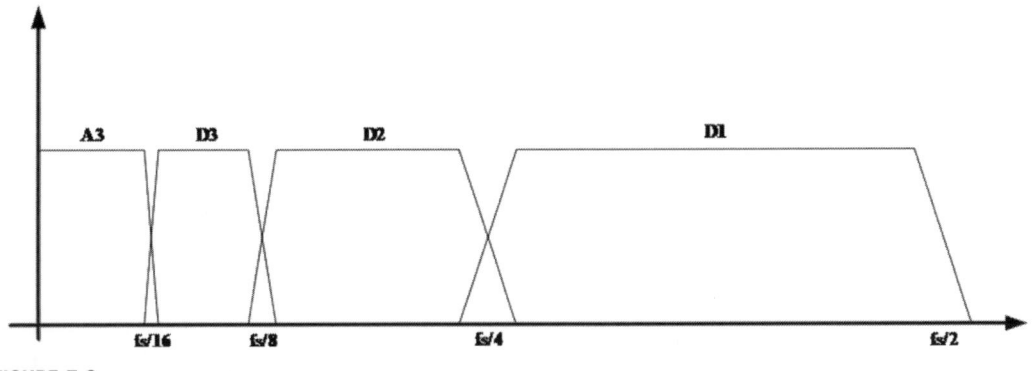

FIGURE 5.3

Frequency range of details and approximations for decomposition up to level 3.

To perform the measurements, the layout of Fig. 5.4 is implemented with the parameters presented in Table 5.1. After finishing the experiments and forming the data matrix accordingly, the wavelet transform is applied to the measured signals recorded by the receiver. Then, using wavelet transform coefficients, the detail and approximation components of the received signal are obtained. Fig. 5.5 shows the a_5 estimate and d_1, d_2, and d_5 details of the signal received by the receiver for the intact states of the transformer winding model. The mother wavelet function used in this case is db16.

After applying the wavelet transform and obtaining the decomposition coefficients, it should be possible to select the appropriate features. Selecting an appropriate feature category is the basis for categorization. These features must be such that, in addition to containing the important characteristics of the signal, they should create a proper distance between the various defects. The more carefully the features are selected, the easier is to categorize them. One of the criteria for fault diagnosis is the difference in energy of the component signals and the approximation of the healthy state and the faulty state of the transformer winding model, which is discussed in the following.

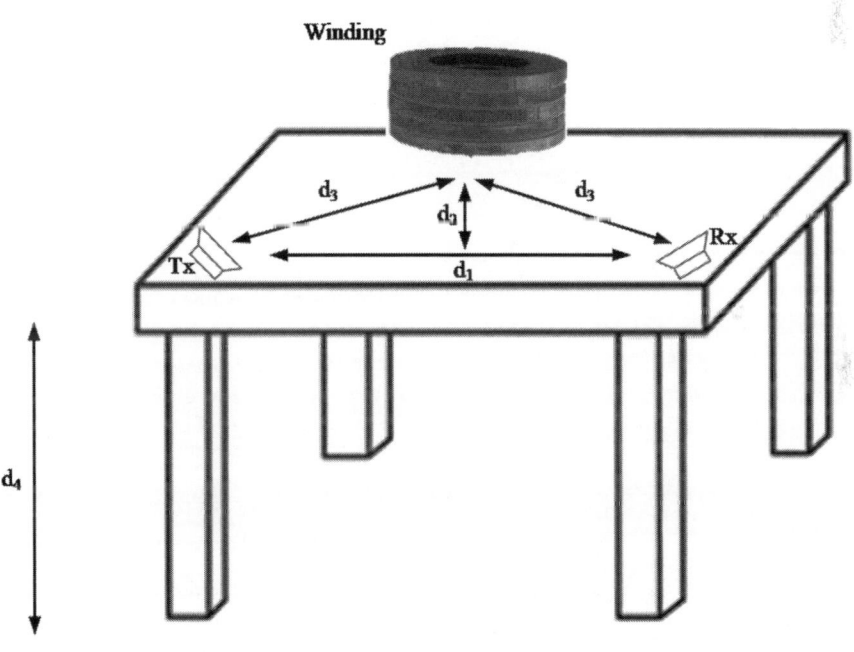

FIGURE 5.4

Measuring equipment and model of transformer winding.

Table 5.1 Configuration parameters of experiment.

Parameter	d_1	d_2	d_3	d_4	T_I	T_s
Value	130 cm	80 cm	103 cm	100 cm	50 ms	31.79 ns

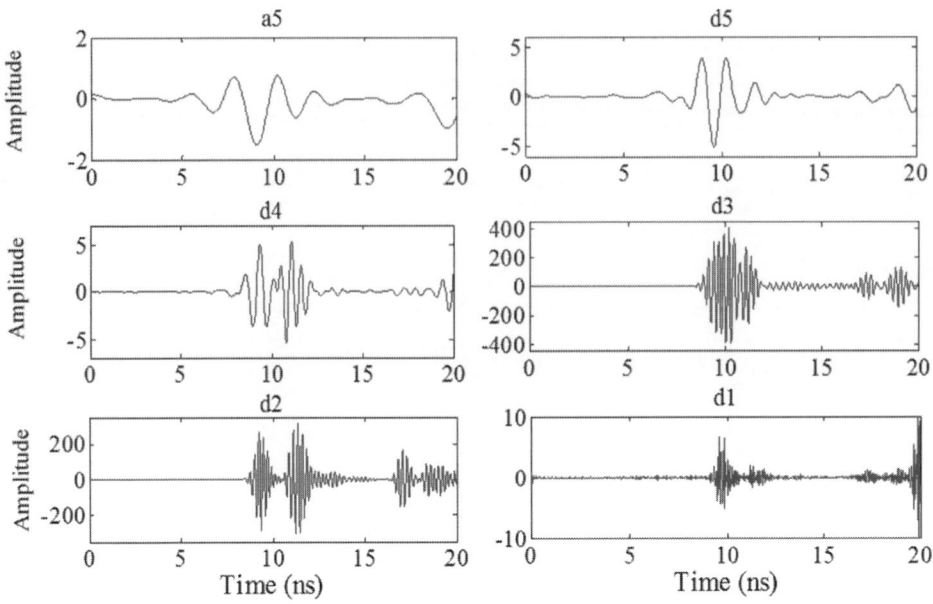

FIGURE 5.5

Detail and approximate components of received signal.

5.4.1 Details energy difference criterion

In Fourier analysis, the power of a signal is expressed by the sum of its power spectral density (PSD), which is equal to the square of absolute of Fourier transform. The power carried by a particular spectral band is defined as the sum of the PSD in that whole band. In this regard, an equivalent definition can be used to convert a wavelet. Power detail density (PDD) in a wavelet transform is expressed as the square of the coefficients of a particular detail (CusidÓCusido et al., 2008). The PDD function obtained from the wavelet transform has been used as a suitable method for classifying defects (Kia et al., 2009). The energies of the detail and approximation of a signal decomposed by a mother wavelet, are expressed by Eqs. (5.9) and (5.10), respectively.

$$E_j^D = \sqrt{\frac{1}{N_l} \sum_{i=1}^{N_l} (D_j)^2 [i]} \tag{5.9}$$

$$E_j^A = \sqrt{\frac{1}{N_l} \sum_{i=1}^{N_l} (A_j)^2 [i]} \tag{5.10}$$

In the above equations, N_l are equal to the length of the data vector in detail l and are obtained from Eq. (5.11)

$$N_l = \frac{N}{2^l} \qquad (5.11)$$

where, N is the length of the original signal data vector.

The difference in PDD in a particular detail can be used as a criterion for separating the healthy state from the faulty one. The next question is what type of mother wavelet function should be chosen for analysis?

5.4.2 Selection of mother wavelet

Mother wavelet selection is one of the most critical issues in wavelet analysis. The method presented in the literature (CusidÓCusido et al., 2008; Misiti et al, 1996) for selecting the mother wavelet is that from the signals obtained for the healthy state and the various fault states, the wavelet transform with different mother functions is obtained. Then the difference between the signal energies in the healthy state and the fault state at different levels for each mother wavelet function is obtained. Thus the appropriate mother wavelet function and the suitable detail for detecting and classifying faults are functions that maximize the energy difference in that detail.

In this project, several wavelet functions such as db4, 8, 16, 28, 32, sym2, 4, 8, and coif3 were applied to the healthy state signals and also several faulty states to determine the appropriate wavelet function. Table 5.2 lists the total energy difference for levels 2, 3, 4, and 5 for a winding model deformation test. According to Table 5.2, it is clear that the db16 function has maximized the criterion for the error in the fourth detail. However, the magnitude of the energy difference for the signals of detail one and approximation five is small compared to the other details. Therefore the mother wavelet function db16 is considered for wavelet analysis.

Fig. 5.6 shows the fourth detail of the received signal for the healthy and faulty states of the transformer winding model for a bulgy radial deformation of $1-4$ cm of 2 cm^2 sector in the case of applying the db16 wavelet function. As shown in the figure, the signals in this detail are well separated from each other.

Table 5.2 Values of the sum of the energy differences in various components.

| Mother wavelet | $\sum Dif(PDD)$ | | | |
	d_5	d_4	d_3	d_2
db4	1.89	3.88	1.01	0.64
db16	2.43	6.45	0.60	1.06
db28	2.67	6.17	0.60	1.06
db32	2.77	5.99	0.68	0.58
sym2	3.44	2.7	0.86	0.33
sym4	4.06	4.18	0.50	0.84
sym8	3.56	5.44	0.56	0.76
coif3	4.73	3.77	0.85	0.66

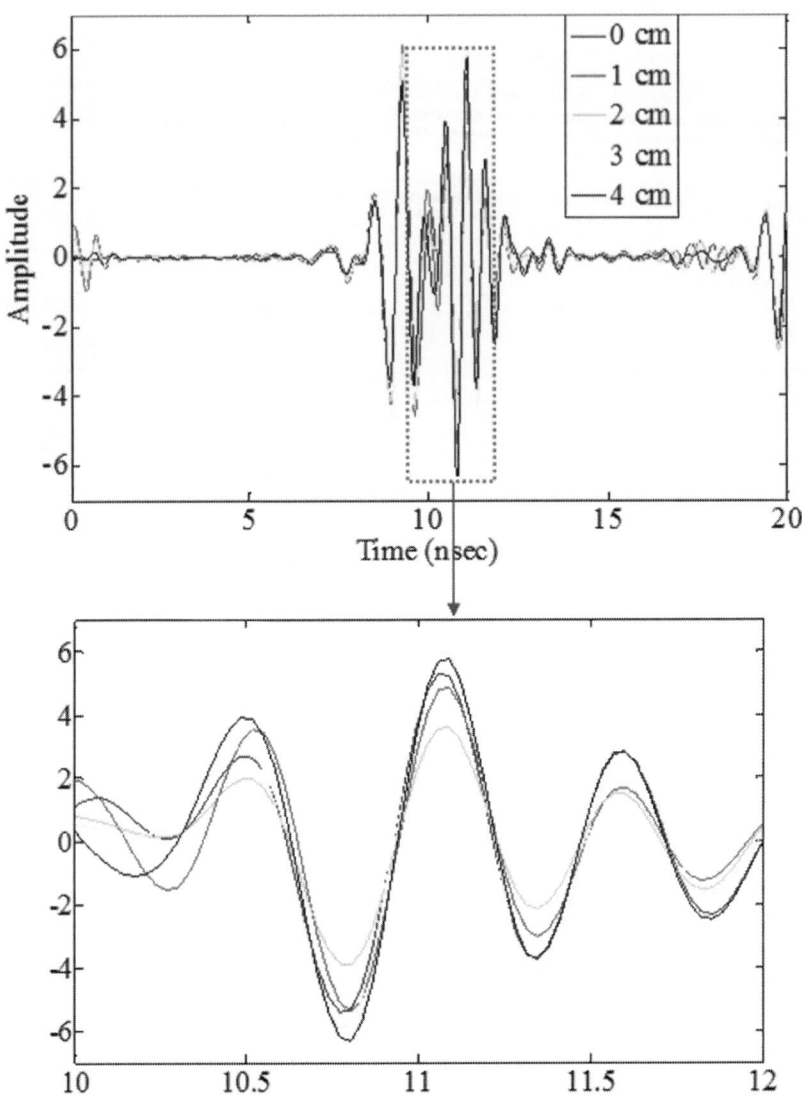

FIGURE 5.6

Signals of fourth detail of healthy and faulty state of transformer winding model.

Table 5.3 presents the results of applying the PDD difference criterion to the fourth detail for various faults. This criterion has different values for different faults, which can be used to distinguish different faults.

Figs. 5.7—5.9 show the energy differences of details 2, 3, and 4, respectively, regarding the radial deformation volume in the experiments of model no. 2 of the transformer winding.

Table 5.3 Values of energy difference in fourth detail for different faults.

PDD difference	Displacement length (cm)	Displaced sector area (cm^2)
−0.1472	1	2
−0.1110	2	2
−0.2604	3	2
−0.2962	4	2
0.5487	1	4
0.4369	2	4
0.2439	3	4
0.3029	4	4
0.0166	1	6
0.2188	2	6
0.3483	3	6
0.0100	4	6
0.8065	1	8
0.7506	2	8
0.8443	3	8
0.8876	4	8

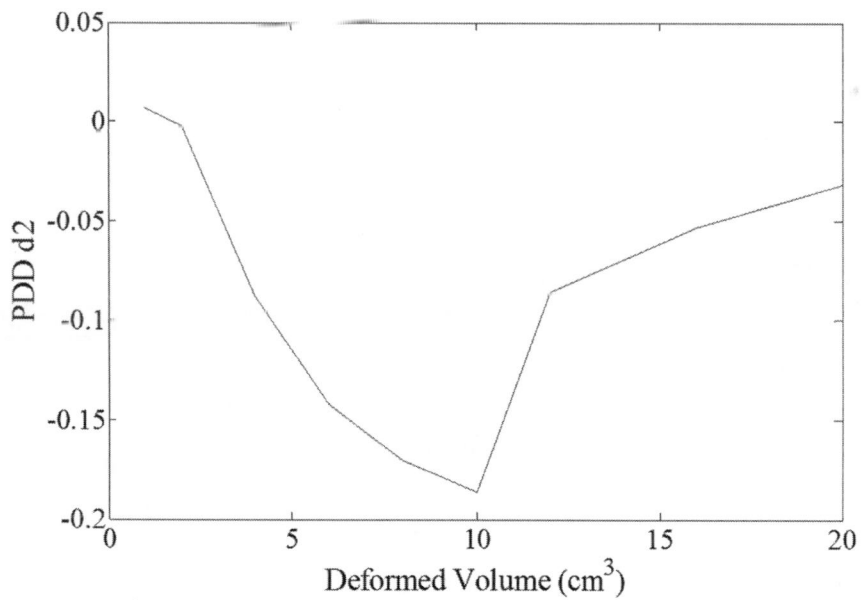

FIGURE 5.7

PDD difference of second detail of received signals in terms of radial deformation volume.

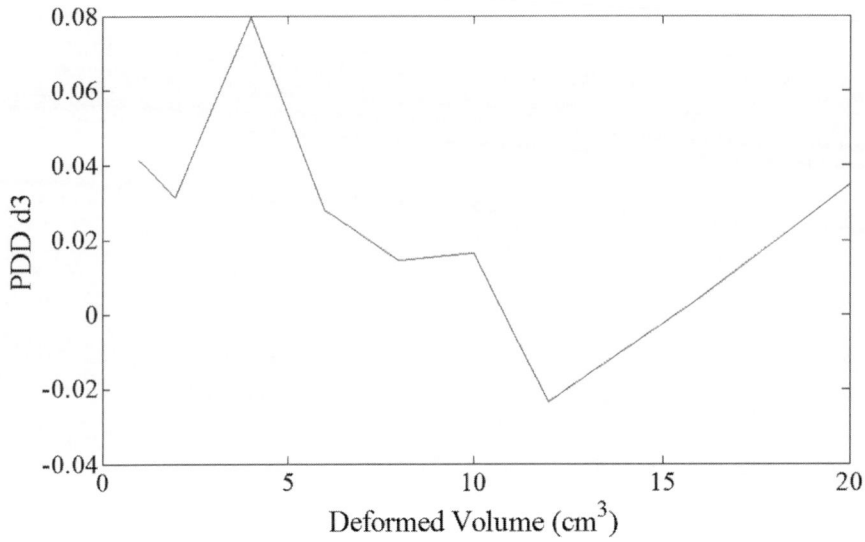

FIGURE 5.8

PDD difference of third detail of received signals in terms of radial deformation volume.

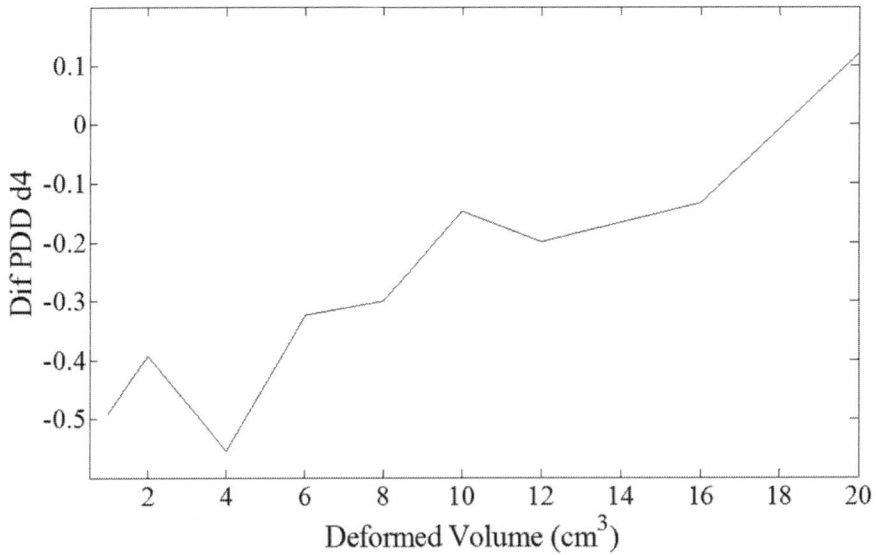

FIGURE 5.9

PDD difference of fourth detail of received signals in terms of radial deformation volume.

From the above figures, it can be concluded that applying the PDD difference criterion of the fourth detail of the received signals shows better results than details two and three. It is clear from Fig. 5.8 that the energy difference of the fourth detail of the received signal increases with the increase in fault extent. Therefore it can be a good criterion for detecting the fault and its degree. But the level 2 and 3 details of the relationship between fault rate and energy difference are not logical. Therefore only four components of the received signals can be used to process the received signals. Considering the PDD difference criterion, this detail contains information that increases with increasing the fault degree of the mentioned criterion.

5.4.3 Using Fourier transform in wavelet analysis

As mentioned in the previous sections, the wavelet transform of a signal in the time domain can be considered as a filter that displays a signal at different frequency intervals in the time domain. This section examines the possibility of detecting faults by applying the Fourier transform to the details and approximation of the signals obtained from the experiments and their analysis. The function of the selected mother wavelet is db1. The results presented in this section are related to the measurements of winding model no. 2.

Figs. 5.10 and 5.11 show the magnitude of the Fourier transform of approximation a_5 and the details d_1, d_2, and d_5 of the receiver signal to displace the 2×2 cm^2 and 2×4 cm^2 sectors of the

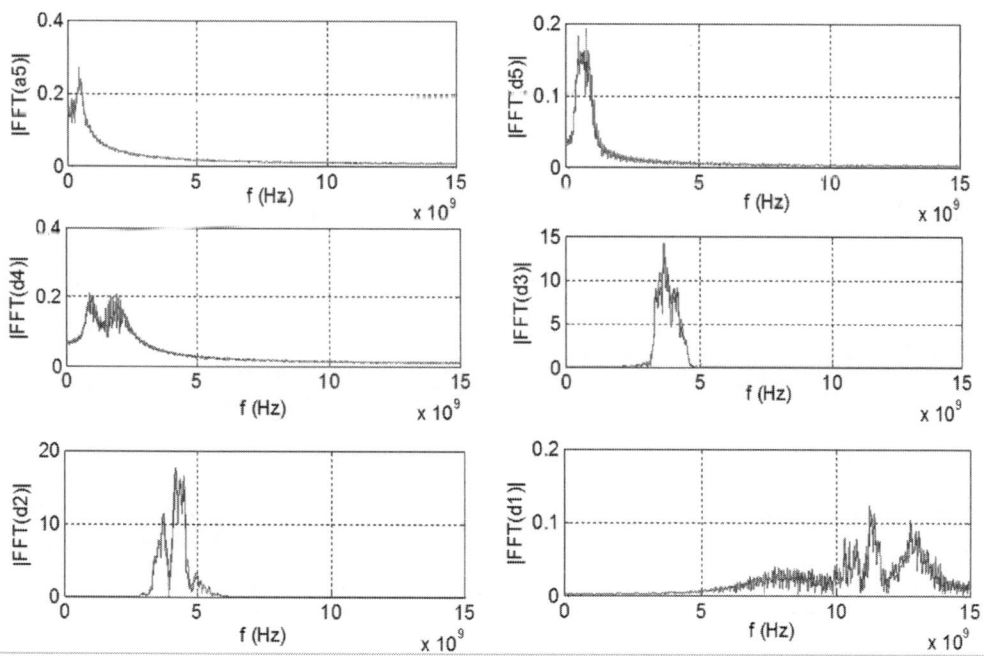

FIGURE 5.10

Magnitude of Fourier transform of fourth detail of received signals. Healthy and (2×2 cm^2 sector) displaced modes, blue and red curves, respectively. For interpretation of the references to color in this figure legend, please refer online version of this title.

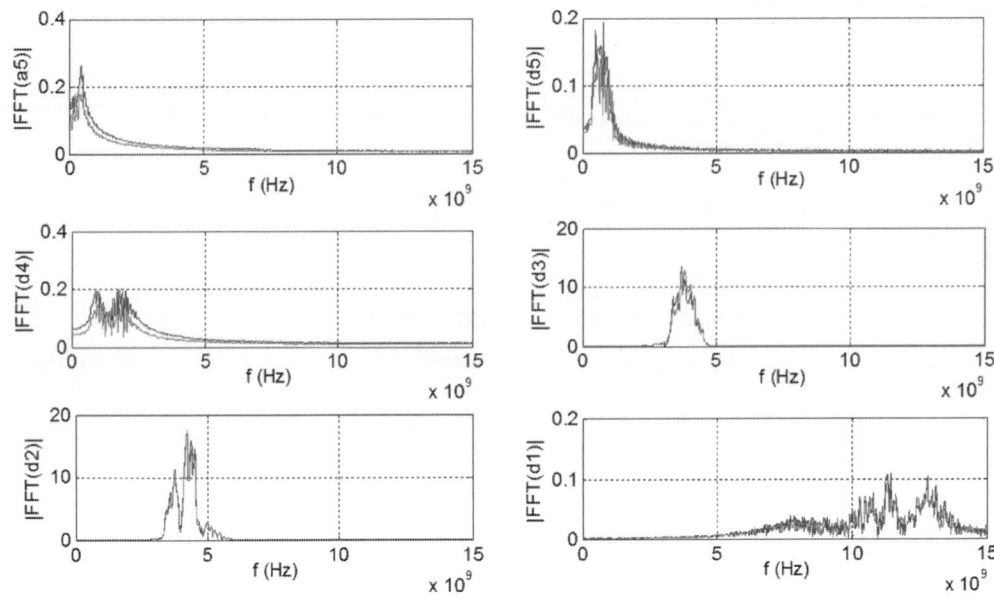

FIGURE 5.11

Magnitude of Fourier transform of fourth detail of received signals. Healthy and (2×4 cm^2 sector) displaced modes, blue and red curves, respectively. For interpretation of the references to color in this figure legend, please refer online version of this title.

transformer winding model by 1 cm. The blue waveform is related to the healthy state, and the red waveform is related to the faulty state. From the following figures, it is clear that the signal of the detail four detects the faults better. As the extent of the fault increases, the difference in the magnitude of Fourier transform of the detail d_4 in the healthy and faulty states increases. Also, the difference between the Fourier transform magnitude of the detail d_4 of the healthy state signal and the faulty state of the transformer winding model for larger faults is larger than the other details. Therefore the magnitude of the Fourier transform of detail d4 can be used as a suitable feature for differentiating faults.

Figs. 5.12 and 5.13 show the Fourier transform phase of approximation a_5 and the details d_1, d_2, and d_5 of the received signal, respectively, for 1 cm displacement of the 2×2 cm^2 and 2×4 cm^2 sectors of the transformer winding model. The blue waveform is related to the healthy state and the red waveform is related to the faulty state. It is clear in these figures that firstly, the signal of detail two detects the faults from each other better. With increasing the amount of the fault, the difference between the phases of Fourier transform of detail d_2 in the healthy and faulty states increases. Therefore the Fourier transform phase of detail d_2 can be considered as a suitable feature for separating different states.

According to the above results, it can be said that the wavelet transform can be used as a tool to separate various faulty states from the healthy state. However, with the increase in the number of faulty cases, it is impossible to find out which faulty state is related to the measured value of the PDD difference of the signal and the healthy state signal of the transformer winding model. Therefore an

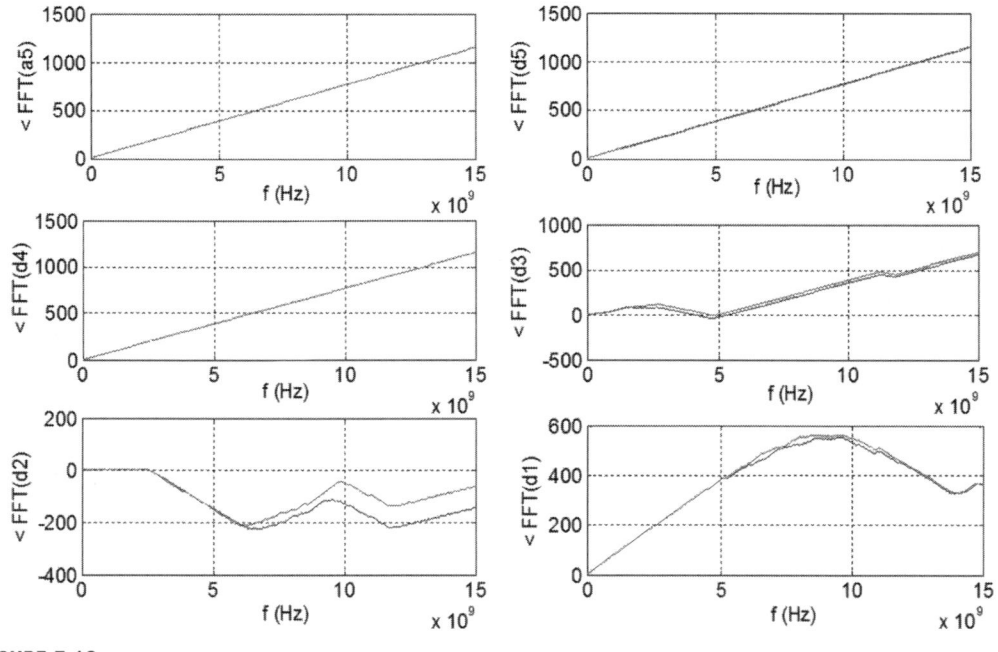

FIGURE 5.12

Phase of Fourier transform of fourth detail of received signals. Healthy and (2×2 cm^2 sector) displaced modes, blue and red curves, respectively. For interpretation of the references to color in this figure legend, please refer online version of this title.

artificial intelligence learning system is needed to be applied and trained for various faults of the transformer winding to detect the fault and express its extent for a measured signal. According to the analyses presented in this chapter, the important point is that the input of this system is part of the signal wavelet transform that distinguishes different faults. For example, suppose a neural network is used. In that case, the input of this neural network for training could be the difference in PDD signal or the differences in Fourier transform magnitude of detail four or phase of third detail of wavelet decomposition in the healthy state and the faulty state of the transformer winding model. After training the neural network, various fault states are applied to it, and the neural network accuracy is checked. The features applied to the intelligent learning system using the wavelet transform, are the magnitude, phase and power of Fourier transform of the approximation and detail of the received signal. Table 5.4 lists the features.

In the designing of the applied classifier:

- Almost all approximations and details of the wavelet transform give acceptable results for separating the axial fault from the radius one.
- The fourth detail level of the Fourier transform gives the best answer to determine the extent of the axial fault.
- To determine the extent of the radial fault, the magnitude of the Fourier transform of the second detail gives the best answers.

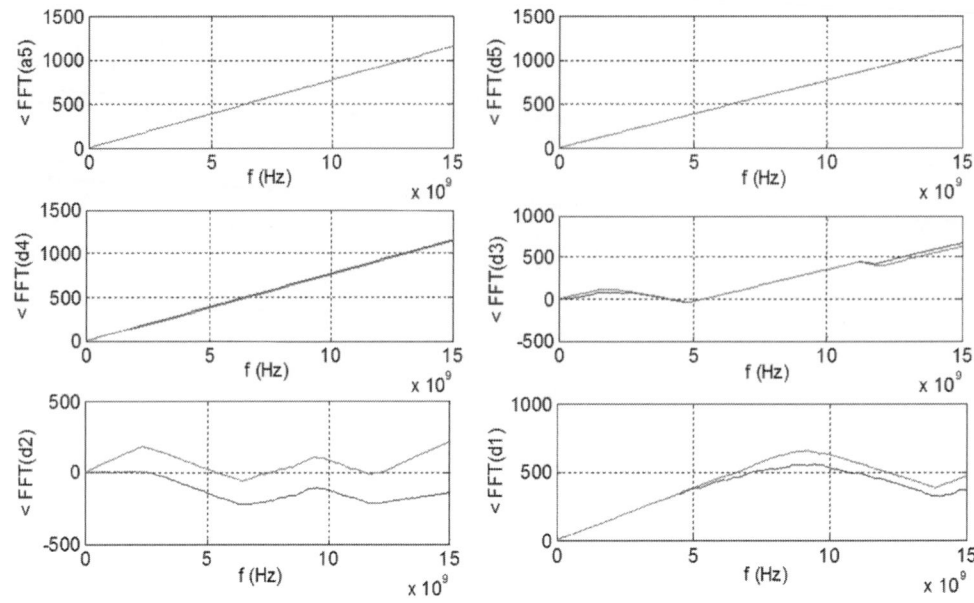

FIGURE 5.13

Phase of Fourier transform of fourth detail of received signals. Healthy and (2×4 cm^2 sector) displaced modes, blue and red curves, respectively. For interpretation of the references to color in this figure legend, please refer online version of this title.

Table 5.4 Extracted features using wavelet transform.

Magnitude		Phase		Power	
Feature	**Symbol**	**Feature**	**Symbol**	**Feature**	**Symbol**
a_7 (f_0:f_n)	X_1: X_{45}	a_7 (f_0:f_n)	X_{361}: X_{405}	a_7	X_{721}
d_1(f_0:f_n)	X_{46}:X_{90}	d_1(f_0:f_n)	X_{406}:X_{450}	d_1	X_{722}
d_2(f_0:f_n)	X_{91}:X_{135}	d_2(f_0:f_n)	X_{451}:X_{495}	d_2	X_{723}
d_3(f_0:f_n)	X_{136}:X_{180}	d_3(f_0:f_n)	X_{496}:X_{540}	d_3	X_{724}
d_4(f_0:f_n)	X_{181}:X_{225}	d_4(f_0:f_n)	X_{541}:X_{585}	d_4	X_{725}
d_5(f_0:f_n)	X_{226}:X_{270}	d_5(f_0:f_n)	X_{586}:X_{630}	d_5	X_{726}
d_6(f_0:f_n)	X_{271}:X_{315}	d_6(f_0:f_n)	X_{631}:X_{675}	d_6	X_{727}
d_7(f_0:f_n)	X_{316}:X_{360}	d_7(f_0:f_n)	X_{676}:X_{720}	d_7	X_{728}

5.4.4 Impact of radial defect on Fourier transform changes

Winding model no. 2, presented in the previous chapter, has been used to model the radial displacement. As shown in Fig. 5.14, the segments are placed at -30, 0 and $+30$ angles by rotating the disc. Then, the radial deformation is modeled by extracting the sections in different sizes. The test results are shown in Figs. 5.15—5.20.

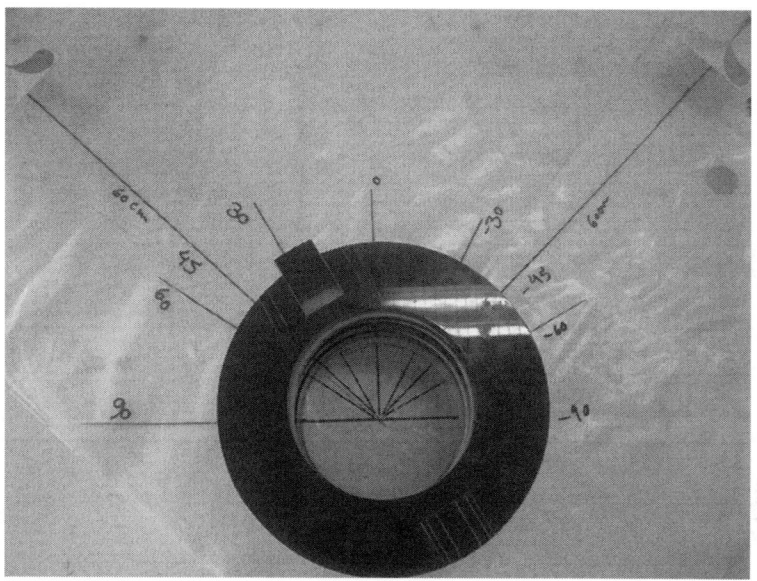

FIGURE 5.14

Top view of the experiment layout.

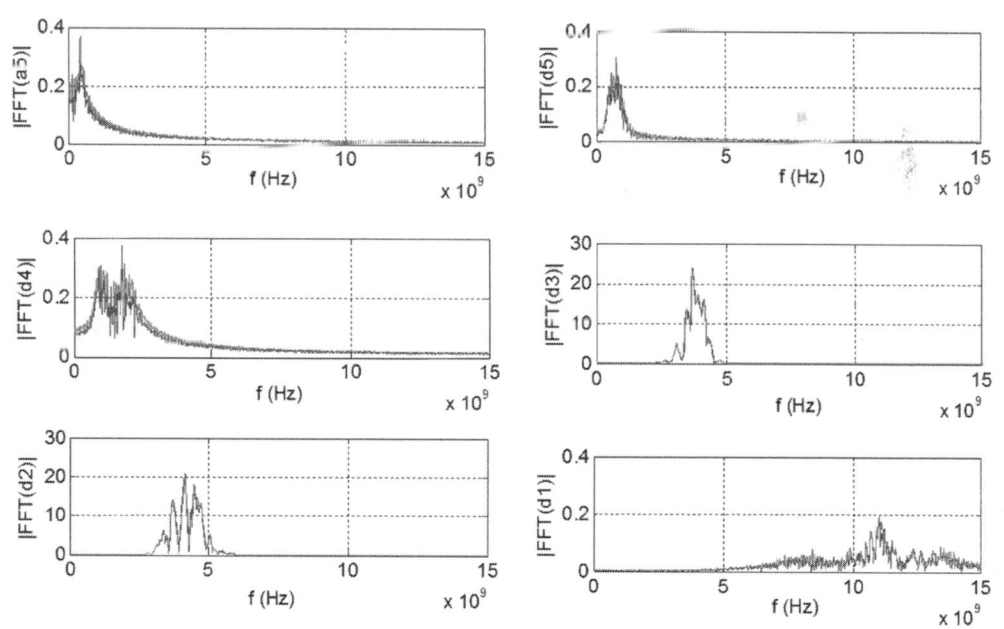

FIGURE 5.15

Magnitude of Fourier transform of fifth approximation and first to fifth details. Healthy and (2×4 cm^2 sector) displaced modes at the angle of $+30$ degree, blue and red curves, respectively. For interpretation of the references to color in this figure legend, please refer online version of this title.

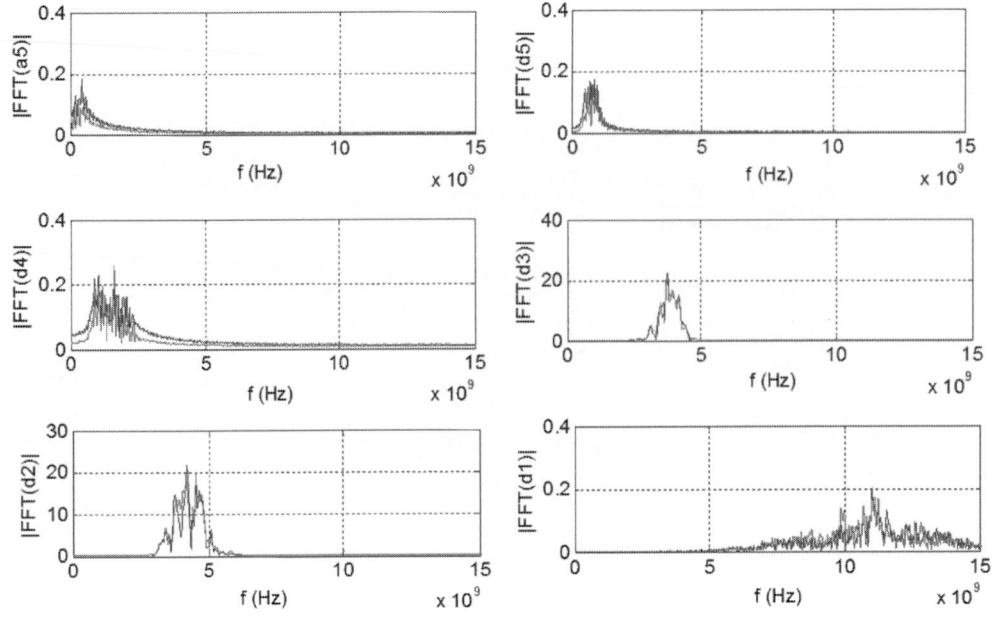

FIGURE 5.16

Magnitude of Fourier transform of fifth approximation and first to fifth details. Healthy and (2×4 cm^2 sector modes) displaced modes at the angle of 0 degree, blue and red curves, respectively. For interpretation of the references to color in this figure legend, please refer online version of this title.

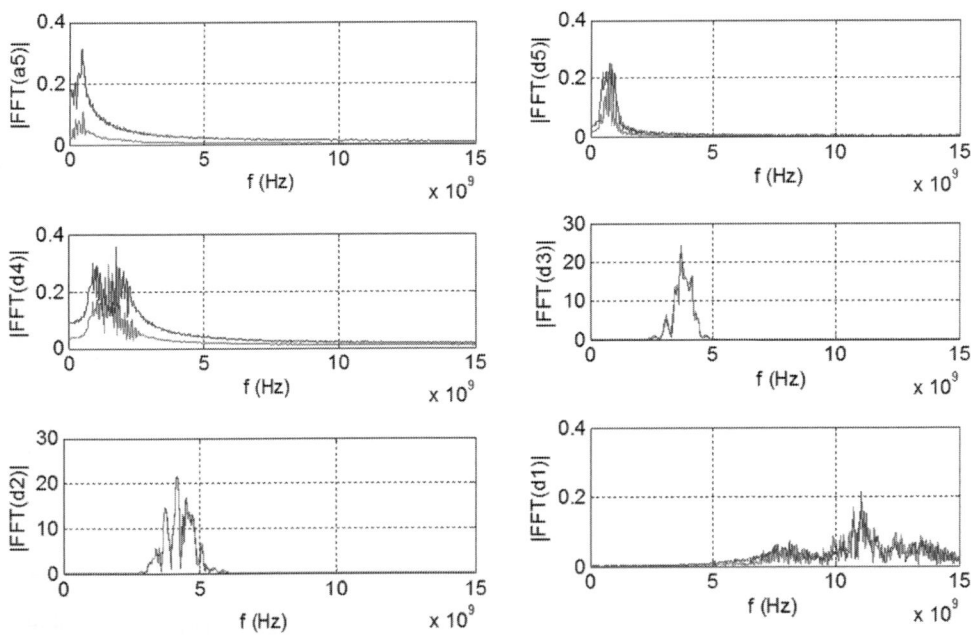

FIGURE 5.17

Magnitude of Fourier transform of fifth approximation and first to fifth details. Healthy and (2×4 cm^2 sector) displaced modes at the angle of -30 degree, blue and red curves, respectively. For interpretation of the references to color in this figure legend, please refer online version of this title.

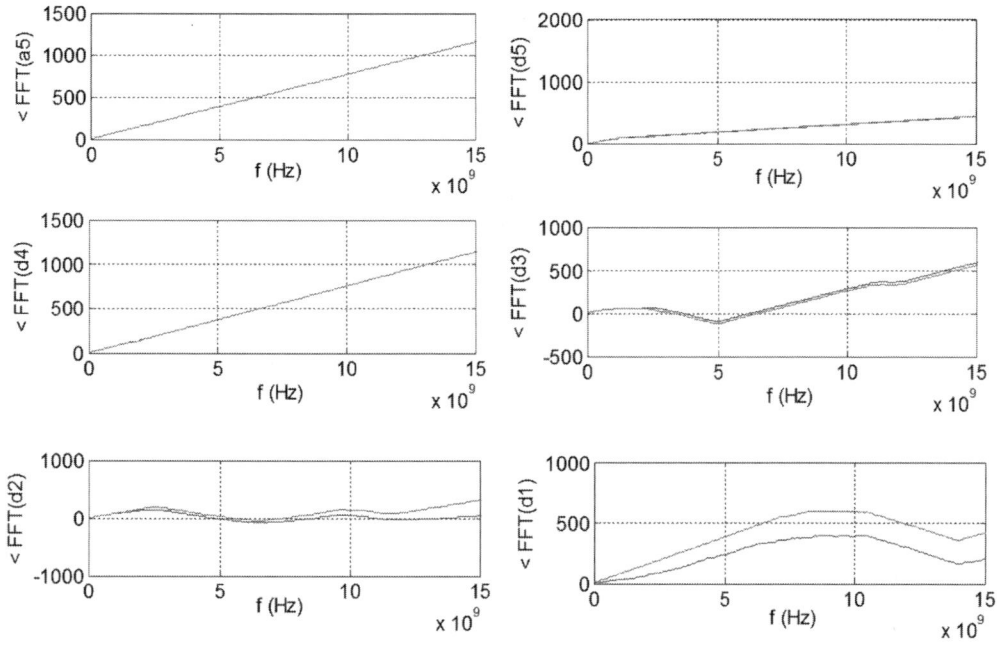

FIGURE 5.18

Phase of Fourier transform of fifth approximation and first to fifth details. Healthy and $(2 \times 4 \, cm^2$ sector) displaced modes at the angle of $+30$ degree, blue and red curves, respectively. For interpretation of the references to color in this figure legend, please refer online version of this title.

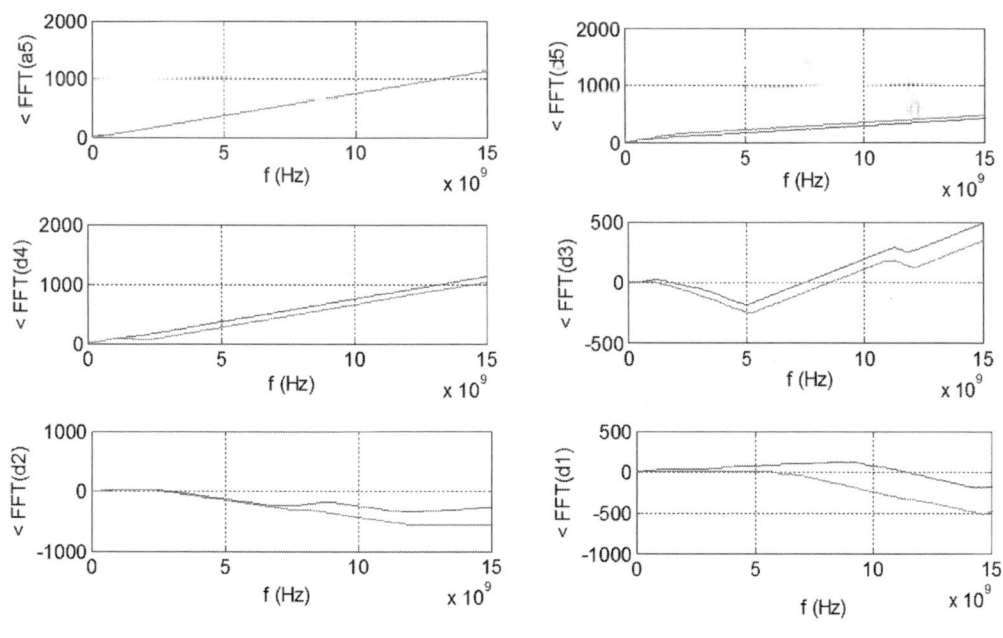

FIGURE 5.19

Phase of Fourier transform of fifth approximation and first to fifth details. Healthy and $(2 \times 4 \, cm^2$ sector) displaced modes at the angle of 0 degree, blue and red curves, respectively. For interpretation of the references to color in this figure legend, please refer online version of this title.

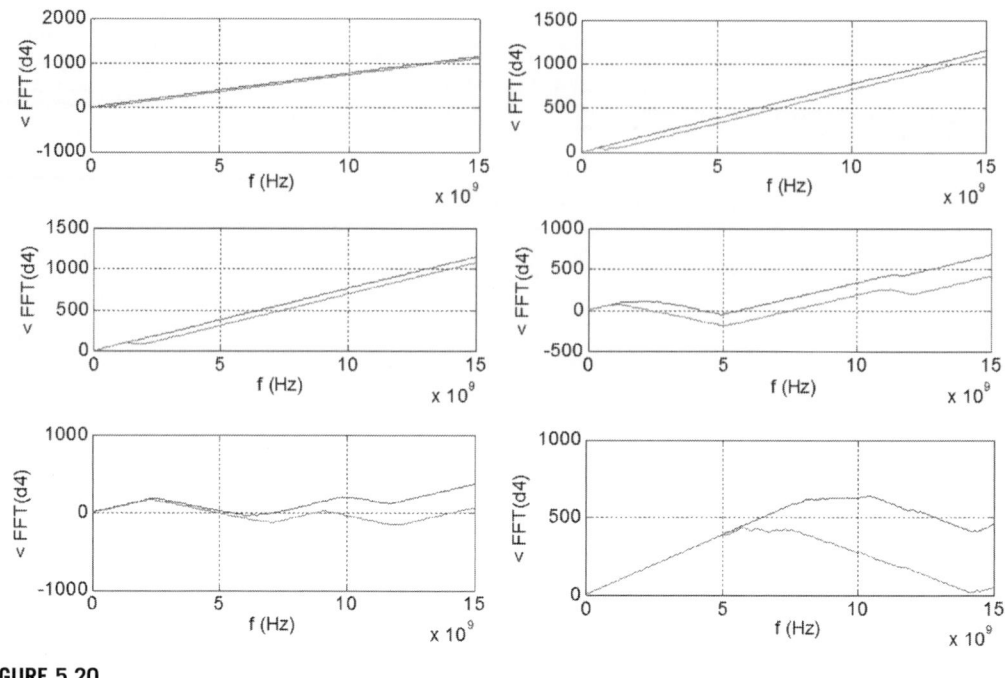

FIGURE 5.20

Phase of Fourier transform of fifth approximation and first to fifth details. Healthy and (2×4 cm^2 sector) displaced modes at the angle of -30 degree, blue and red curves, respectively. For interpretation of the references to color in this figure legend, please refer online version of this title.

According to Figs. 5.15–5.17, it is clear that by increasing the angle of the radial deformation location from -30 to $+30$, the magnitude of the red waveform has decreased compared to the blue waveform in detail four. This can be used as a feature to differentiate the location of various faults.

According to Figs. 5.18–5.20, it is clear that by increasing the angle of the radial deformation location from -30 to $+30$, the phase of the red waveform has decreased compared to the blue waveform in detail three. This can be used as a feature to differentiate the location of various faults.

5.4.5 Radial and axial fault detection using Fourier transform

As stated in Section 5.4.3, an artificial intelligence learning system is needed to be applied and trained for various faults of the transformer winding to detect the fault and express its extent for a measured signal. In this section, we will diagnose the defect with different artificial intelligence methods.

The experiment studies for axial displacement were applied on 151 different states, with each state containing 640 samples. Considering the waveform of the received signals, it can be seen that the main reflections related to the transformer winding model are in the range where the amplitude of the received signal is large compared to the rest of the signal. Based on these explanations, the part of the signal related to the winding model includes 200 samples from sample numbers 300–500. This part of the signal is used for processing. Therefore the total available data for training and testing procedures in classification methods is a matrix of 151×200. We randomly separate 70% of the total data for training and 30% for testing.

The total number of experiments for radial deformation is 161 different states, each state containing 640 samples. According to the waveform of the received signal, it can be seen that the main reflections related to the transformer winding model are in the range where the amplitude of the received signal is large compared to the rest of the signal. Therefore the part of the signal related to the winding model includes 200 samples in the range of 300–500. This part of the signal is used for processing. Therefore the total core data for training and testing in the classification methods is a matrix of 161 × 200. Like the previous case, we randomly separate 70% of the total data (108 cases, a matrix dimension of 108 × 200) for training and 30% (53 cases, which is a matrix of 53 × 200) for testing.

First, it is assumed that the expert system should only estimate the extent of axial or radial faults, and the defect type is known. Tables 5.5 and 5.6 present the results of estimating the extent of the radial deformation and axial displacement using the decision tree method.

Table 5.5 Result of estimating radial deformation using decision tree method.

Real radial deformation (mm)	Estimated radial deformation (mm)	Absolute error (mm)
2	2	0
4.5	4.5	0
7	7.5	0.5
9.5	10	0.5
12	12	0
14.5	14	0.5
17	17.5	0.5
19.5	20	0.5

Table 5.6 Result of estimating axial displacement using decision tree method.

Real axial displacement (mm)	Estimated axial displacement (mm)	Absolute error (mm)	Error (%)
4	2	2	0.016
16	14	2	0.016
22	20	2	0.016
30	32	−2	−0.016
42	40	2	0.016
52	54	−2	−0.016
68	66	2	0.016
72	74	−2	−0.016
86	86	0	0
98	96	2	0.016
102	104	−2	−0.016
114	112	2	0.016
128	126	2	0.016
132	130	2	0.016
146	148	−2	−0.016

In Table 5.6, the error (%) is defined as the ratio of the absolute error to the length of the winding model. As can be seen, the absolute error is equal to the measurement step. Therefore in the event of an error, the winding position in the adjacent class is estimated.

In the next step, it is assumed that detecting the fault type is also the responsibility of the expert system. Therefore the expert system designed must have the ability to solve three problems:

- Separation of the axial and radial faults from each other,
- Detect the amount of axial fault, and
- Detect the amount of radial defect.

Concerning these three issues, the classes used in pattern recognition methods will be defined as described in the following.

For the first problem, two classes are defined according to the data (Table 5.7). In the second problem, 15 classes (Table 5.8) and for the third problem, 16 classes are defined (Table 5.9).

Table 5.10 lists the percentage of the classification accuracy in different methods for the measured data for all three problems.

Table 5.7 Classes of first problem.

Description of the class	Class no.
Axial	Class 1
Radial	Class 2

Table 5.8 Classes of second problem.

Class	Number of positions	Start of interval (mm)	End of range (mm)
1	9	−15 (lowest)	−13.4
2	9	−13.2	−11.6
3	10	−11.4	−9.6
4	10	−9.4	−7.6
5	10	−7.4	−5.6
6	10	−5.4	−3.6
7	10	−3.4	−1.6
8	15	−1.4	+1.4
9	10	+1.6	+3.4
10	10	+3.6	+5.4
11	10	+5.6	+7.4
12	10	+7.6	+9.4
13	10	+9.6	+11.4
14	9	+11.6	+13.2
15	9	+13.4	+15 (highest)

Table 5.9 Classes of third problem.

Description of the class	Class no.
The protrusion of the first sector is 0—5 mm	Class 1
The protrusion of the first sector is 6—10 mm	Class 2
The protrusion of the first sector is 11—15 mm	Class 3
The protrusion of the first sector is 16—20 mm	Class 4
The full protrusion of the first sector and protrusion of the second one of 1—5 mm	Class 5
The full protrusion of the first sector and the protrusion of the second one of 6—10 mm	Class 6
The full protrusion of the first section and the protrusion of the second one of 11—15 mm	Class 7
The full protrusion of the first sector and the protrusion of the second one of 16—20 mm	Class 8
The full protrusion of the first and second sectors and the protrusion of the third one by 1—5 mm	Class 9
The full protrusion of the first and second sectors and the protrusion of the third one of 6—10 mm	Class 10
The full protrusion of the first and second sectors and the protrusion of the third one of 11—15 mm	Class 11
The full protrusion of the first and second sectors and the protrusion of the third one of 16—20 mm	Class 12
	Class 13

Continued

Table 5.9 Classes of third problem.—cont'd

Description of the class	Class no.
The full protrusion of the first, second, and third sectors and the protrusion of the fourth one by 1–5 mm	
The full protrusion of the first, second, and third sectors and the protrusion of the fourth one of 6–10 mm	Class 14
The full protrusion of the first, second, and third sectors and the protrusion of the fourth one of 11–15 mm	Class 15
The full protrusion of the first, second, and third sectors and the protrusion of the fourth one of 16–20 mm	Class 16

Table 5.10 Classification accuracy in different methods for measured data.

	Neural network	SVM	K-NN method	Parzen window method	Bayesian method
1st problem	100	100	100	100	100
2nd problem	92.15	96.07	76.70	80.39	94.11
3rd problem	98.11	100	86.79	84.90	96.22

5.5 Summary

In this chapter, the wavelet transform was introduced to analyze measured signals. The important point in using the wavelet transform is the type of mother wavelet function used. It was found that the wavelet function suitable for investigating various faults is the 16-order Daubechies function. Then, the signal details were evaluated using the PDD difference criterion to detect and separate various faults. Next, the adequacy of the magnitude and phase of the Fourier transform of details and approximations of the measured signals was investigated in the signal processing. Based on the results, in general, it can be concluded that using wavelet transform alone cannot separate the various faults and determine the extent of the faults according to the received signal. Therefore it was suggested to use other methods to detect the fault and the amount of the fault, which will be examined in the next chapter.

References

Coifman, R. R., & Wickerhauser, M. V. (1992). Entropy-based algorithms for best basis selection. *IEEE Transactions on Information Theory, 38*(2), 713–718.

CusidÓCusido, J., Romeral, L., Ortega, J. A., Rosero, J. A., & Espinosa, A. G. (2008). Fault detection in induction machines using power spectral density in wavelet decomposition. *IEEE Transactions on Industrial Electronics, 55*(2), 633–643.

Ghanizadeh, A. J., & Gharehpetian, G. B. (2013). Application of characteristic impedance and wavelet coherence technique to discriminate mechanical defects of transformer winding. *Electric Power Components and Systems, 41*(9), 868–878.

Karami, H., Gharehpetian, G. B., Norouzi, Y., & Akhavan-Hejazi, M. (2020). Simultaneous radial deformation and partial discharge detection of high-voltage winding of power transformer. *IET Electric Power Applications, 14*(3), 383–390.

Karami, H., Tabarsa, H., Gharehpetian, G. B., Norouzi, Y., & Hejazi, M. A. (2019). Feasibility study on simultaneous detection of partial discharge and axial displacement of HV transformer winding using electromagnetic waves. *IEEE Transactions on Industrial Informatics, 16*(1), 67–76.

Kia, S. H., Henao, H., & Capolino, G.-A. (2009). Diagnosis of broken-bar fault in induction machines using discrete wavelet transform without slip estimation. *IEEE Transactions on Industry Applications, 45*(4), 1395–1404.

Misiti, M., Misiti, Y., Oppenheim, G., & Poggi, J. M. (1996). *MATLAB help, wavelet toolbox*. The Mathworks Inc.

Misiti, M., Misiti, Y., Oppenheim, G., & Poggi, J.-M. (1996). *Wavelet toolbox* (Vol 15, p. 21). Natick, MA: The MathWorks Inc.

Naderi, M. S., Gharehpetian, G. B., Blackburn, T. R., & Abedi, M. (2007). Pattern classification of internal incipient faults during impulse tests using continuous wavelet analysis. *Electrical Engineering, 90*(2), 79–85.

Nafar, M., Abedi, M., Gharehpetian, G. B., Taghipour, S., & Yousefpour, B. (2004). Locating partial discharge in transformer by wavelet. *WSEAS Transactions on Circuits and Systems, 3*(6), 1499, 1503.

Nasiri, A., Poshtan, J., Kuhaei, M. H., & Taringoo, F. (2004). Wavelet packet decomposition as a proper method for fault detection in three phase induction motor. In *Proceedings of the IEEE international conference on mechatronics, 2004. ICM'04* (pp. 13–18).

Tarimoradi, H., Karami, H., Gharehpetian, G. B., & Tenbohlen, S. (2021). *Sensitivity analysis of different components of transfer function for detection and classification of type, location and extent of transformer faults. Measurement* (p. 110292).

Frequency domain analysis of scattering parameters in transformers

Gevork B. Gharehpetian[1], Hossein Karami[2] and Seyed-Alireza Ahmadi[3]

[1]*Electrical Engineering Department, Amirkabir University of Technology (AUT), Tehran, Iran;* [2]*High Voltage Studies Research Department, Niroo Research Institute (NRI), Tehran, Iran;* [3]*School of Electrical and Computer Engineering, College of Engineering, University of Tehran, Tehran, Iran*

6.1 Concept of scattering parameter

Nondestructive electromagnetic testing has been used in many error detection applications, including measuring the dielectric properties of materials, determining the thickness of dielectric plates (Jawad & Akbar, 2021), and detecting surface cracks in metals (Mirala & Sarraf Shirazi, 2017). The idea of detecting deformation and displacement of transformer windings based on the usage of electromagnetic antennas can be another example of a nondestructive test (Karami et al., 2019; Karami, Gharehpetian, et al., 2020; Rahbarimagham et al., 2015).

There are various methods for modeling electromagnetic linear time-invariant networks. Some of them are classical methods with many disadvantages at high frequencies (Pozar, 2011). Practically, the scatter parameter, shown by S, describes how energy is distributed between different parts of a network (Akhavanhejazi et al., 2011; Hejazi, Gharehpetian, Moradi, Alehosseini, & Mohammadi, 2011). Fig. 6.1 shows the transformer model as an n-port network. The ports are installed antennas on the tank wall.

If a wave with an amplitude of V_1^+ is transmitted to port one, the reflected wave with the amplitude of V_1^- is achieved at that port, where $V_1^- = S_{11}V_1^+$, and S_{11} is the reflection coefficient. The waves are scattered across other ports whose amplitude corresponds to the amplitude of V_1^+, i.e., $V_n^- = S_{n1}V_1^+$, where d' is the transmission coefficient from port one to port n. Generally, by transmitting and receiving from all the ports, we have:

$$
\begin{bmatrix} V_1^- \\ V_2^- \\ \cdot \\ \cdot \\ \cdot \\ V_n^- \end{bmatrix} = \begin{bmatrix} S_{11} & S_{12} & \cdots & S_{1N} \\ S_{21} & S_{22} & \cdots & S_{2N} \\ \cdot & \cdot & \cdots & \cdot \\ \cdot & \cdot & \cdots & \cdot \\ \cdot & \cdot & \cdots & \cdot \\ S_{N1} & S_{N2} & \cdots & S_{NN} \end{bmatrix} \begin{bmatrix} V_1^+ \\ V_2^+ \\ \cdot \\ \cdot \\ \cdot \\ V_n^+ \end{bmatrix}
\tag{6.1}
$$

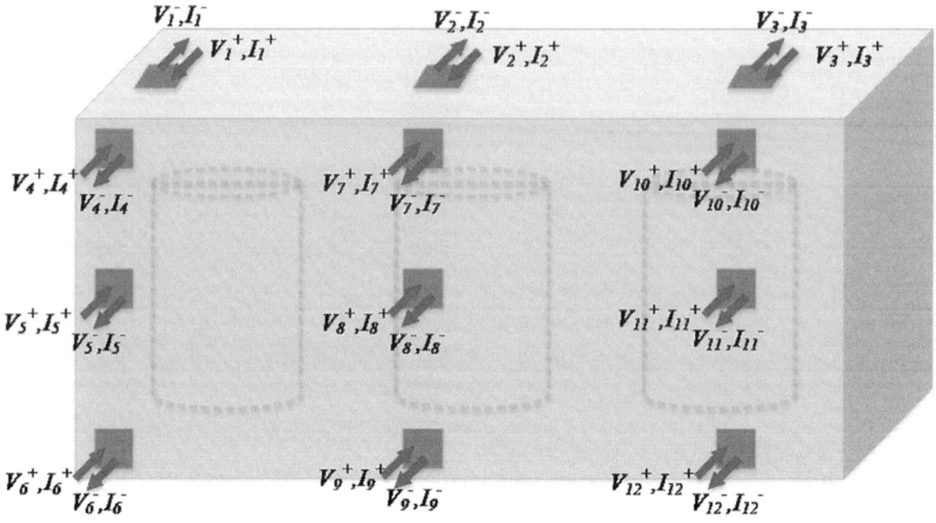

FIGURE 6.1

Transformer model as n-port network.

$$\left\lfloor V^- \right\rfloor = \left\lfloor S \right\rfloor \times \left\lfloor V^+ \right\rfloor \tag{6.2}$$

In this study, only one antenna is used as a port, so its scattering matrix contains one element; in this case, we have:

$$S_{11} = \frac{V_1^-}{V_1^+} \tag{6.3}$$

where, $|S_{11}|$, the absolute value of S_{11}, can be calculated in a simpler way:

$$|S_{11}| = \sqrt{\frac{P_{ref}}{P_{in}}} \tag{6.4}$$

where, P_{in} is the transmitted signal power and P_{ref} is the received power.

6.2 Effect of mechanical defect on *S*-parameter

Using the theory of partial perturbations in a resonant shield, which is the transformer tank in our study, the sensitivity of scattering parameter to displacement and deformation of the transformer winding can be explained. A resonant shield is made of any closed space surrounded by a high-conductivity metal surface. The Maxwell equations must be solved according to the shield boundary conditions to analytically calculate the shield fields and the resonance frequency. The fields must satisfy the boundary conditions that are generally assumed to be an ideal conductor wall, i.e., tangential component of electric field, and the normal component of the magnetic field must be zero. Maxwell equations are established for a shield without free charge or current density (Sadiku, 2000).

The electromagnetic problem of a resonant shield is a three-dimensional eigenvalue problem. This problem in simple shapes such as a rectangular cuboid and a cylinder has an analytical solution.

Usually, we are involved in two problems: the problem without perturbation, which we know how to solve it, and the problem with perturbation, whose solution is unknown, and is slightly different from the problem without perturbation (Sadiku, 2000). Our goal is to find variations in the resonance frequency due to changes in the shield wall. If E_0, H_0, and ω_0 represent the electric field, magnetic fields, and resonant frequency in the main shield, and E, H, and ω correspond to the parameters of the perturbed resonant shield, the satisfying field equations are formulated as Eqs. (6.5)–(6.8):

$$-\nabla \times E_0 = j\omega\mu H_0 \tag{6.5}$$

$$\nabla \times H_0 = j\omega\varepsilon E_0 \tag{6.6}$$

$$-\nabla \times E = j\omega\mu H \tag{6.7}$$

$$\nabla \times H = j\omega\varepsilon E \tag{6.8}$$

where μ and ε are permeability and permittivity, respectively. Now, the Eqs. (6.9) and (6.10) can be written by performing a series of mathematical calculations.

$$\frac{\omega - \omega_0}{\omega_0} \approx \frac{\iiint\limits_{\Delta\tau} \left(\mu|H_0|^2 - \varepsilon|E_0|^2\right) d\tau}{\iiint\limits_{\Delta\tau} \left(\mu|H_0|^2 + \varepsilon|E_0|^2\right) d\tau} \tag{6.9}$$

$$\frac{\omega - \omega_0}{\omega_0} \approx \frac{\Delta\overline{w}_m - \Delta\overline{w}_e}{W} \tag{6.10}$$

where $\Delta\overline{w}_m$ and $\Delta\overline{w}_e$ are the average magnetic and electrical energies stored during $\Delta\tau$, respectively, and W is the total energy stored in the shield. If $\Delta\tau$ is small enough, the above relation can be approximated as Eq. (6.11):

$$\frac{\omega - \omega_0}{\omega_0} \approx \frac{(\overline{w}_m - \overline{w}_e)\Delta\tau}{\widehat{w}\tau} = C\frac{\Delta\tau}{\tau} \tag{6.11}$$

where $\overline{w}_m$ and $\overline{w}_e$ are density of magnetic and electrical energy, and $\widehat{w}$ is the average spatial energy density, respectively. In addition, C depends on the geometry of the shield and the position and shape of the perturbation (Maier & Slater, 1952).

An inward perturbation increases the resonant frequency if it is remarkable in the H position and decreases the resonant frequency if it is remarkable in the E position. If the perturbation is outward, the results will be reversed. Further changes in resonant frequency occur when perturbations occur at the position with a maximum of E and zero of H or vice versa. The frequency shift value is proportional to the value of the field at the perturbation point (Maier & Slater, 1952). Eq. (6.11) shows that the frequency shift of the resonance is a function of $\Delta\tau/\tau$. This indicates that large mechanical faults have greater effects on the resonance frequency. In general, the frequency shift depends on the shape of the deformation and the enclosure. Its value depends on the fields at the perturbation point. Therefore, the deformation position can affect the scattering parameter, so this method can predict the dimensions and position of the fault. The transformer tank can be considered as a large resonant shield and a mechanical deformation acts as a partial disturbance that can shift the resonant frequency.

The perturbation of the material inside the shield can also shift the resonant frequency. Similarly, the change in the resonant frequency of a shield due to the perturbation of the inside material can be obtained as formulated by Eq. (6.12):

$$\frac{\omega - \omega_0}{\omega_0} \approx -\left(C_1 \frac{\Delta\varepsilon}{\varepsilon} + C_2 \frac{\Delta\mu}{\mu} \right) \tag{6.12}$$

The transformer coolant is oil. According to the mentioned description, variation in the dielectric coefficient of oil can change the resonance frequency. It means that the change in resonant frequency may be due to mechanical displacement or oil deterioration.

The best approach for comparing the scattering parameter is time-based comparison, similar to the frequency response analysis (FRA) method for mechanical detection of power transformers (Behkam et al., 2021; Tarimoradi et al., 2021). In this approach, the measured scattering parameter is compared with the latest measurement results of a healthy transformer. Continuous monitoring of transformer winding by scattering parameter provides the necessary data for time-based comparison. Accurate information about the measurement set-up, test steps, and antennas position is required to obtain reproducible and comparable results. This method can be applied to all types of transformers. If data from previous measurements are not available, type-based comparisons can be used for transformers of the same type. The results of the new measurements are compared with those of another transformer. This state is applicable if the construction specifications of both transformers are similar. If both transformers are designed and manufactured in the same factory, the similarity between the core structure and their winding will be maximized. However, in both transformers, the test parameters and the location of the antenna must be the same.

Structure-based comparisons can only be done to a three-phase transformer if the phases have the same geometry. According to research studies, antenna location has an important effect on the measured scattering parameter, so the structure-based comparison method cannot be used for integrated three-phase transformers.

Model-based comparison is based on an electromagnetic model of a transformer that can determine the scattering parameters. The parameters obtained from electromagnetic modeling are compared with those obtained from measurements. Scattering parameters can be obtained by solving Maxwell equations under certain boundary conditions. This problem does not have a closed solution for a complex structure such as a transformer but can be solved numerically.

In the FRA method, altering the transformer tap changer, which physically changes the arrangement of the transformer windings, affects the response. Also, altering the tap changer affects the scattering parameters due to mechanical structure variation in the electromagnetic wave method. Therefore, only the measurements with the same tap changer position are considered for comparison. The winding axial displacement detection accuracy depends on the transmitted signal frequency, i.e., decreasing the wavelength increases the displacement sensitivity.

6.3 Defect detection using classification method

Finding appropriate and optimal features of the received signals to train and test the classification method will improve its performance. The input data are the scattering parameter measurement values in the frequency domain for axial and radial faults with different extents. Gathered information for different states of axial displacement is considered in one class and, similarly, measured data from different states of radial deformation (RD) in another. At first, features should be extracted from signals and the patterns for each class should be identified to distinguish them from each other. The MATLAB software is used to analyze the data.

6.3.1 Feature extraction

Data preprocessing includes all the transformations applied to data to make it easier and more efficient for later processing, such as classification. There are various methods for preprocessing, such as normalization, whitening, or dimension reduction (Idri et al., 2018; Jia & Zhuang, 2021). For a set of measured data, dimension reduction can be examined from two different perspectives. In the first view, the variables that do not help the separation of classes should be determined and ignored. Therefore, we should find desired features from available ones, called feature selection (Wan, 2019).

The second view is the use of a method to transfer data from m-dimensional feature space to a smaller space, called feature extraction (Ahmadi et al., 2020; Zebari et al., 2020). Principal component analysis (Moradzadeh et al., 2020), Fourier transform (Ahmadi et al., 2019), and wavelet transform (Ghanizadeh & Gharehpetian, 2013) are some of the methods used to extract the features. The goal is to replace the original data with fewer data, which have the following benefits:

- Reducing the input data bandwidth, which leads to increased classification speed.
- Providing a more suitable set of features for the classifier, which leads to improved classifier performance.
- Reducing duplicated information.
- Finding new meaningful features for better describing the system.
- Storing data in smaller dimensions (ideally in two dimensions) with the lowest information loss, where the data are seen better and the structures and relations among the data can be determined better).

6.3.2 Data sets

Pattern recognition aims to classify objects into some classes or categories (McClean, 2003). Depending on the application, these objects can be images, signals, or any type of measured data that needs to be classified. In designing pattern recognition systems, we need data from objects (which should be classified) and the data must describe them well. For this purpose, we consider the characteristics of those objects with which we can distinguish between objects related to different classes. The characteristics that are chosen to differentiate classes are features. One feature may not be sufficient for classification with acceptable accuracy and several features may be required for classification.

In pattern recognition systems, three sets are generally used. These sets contain the desired input and output system data. Each of them corresponds to a part of the classification learning process, as follows:

- Training set: This is the largest set used to obtain model parameters.
- Test set: This set is used to adjust the classification parameters and also to test and prevent overlearning of the model. This set is much smaller than the training one.
- Validation set: This set is not used during design (classification learning process). Therefore, the data of this set are a suitable criterion for the final validation of the model (checking the generalizability of the model).

6.3.3 Learning process

Learning methods have many applications in medical diagnosis, image processing, marketing and economic issues, power system security assessment, load forecasting, and monitoring of power system equipment such as transformer and power plant (Ahmadi & Sanaye-Pasand, 2021; Malekizadeh et al., 2020; Sanjari et al., 2016, 2017). These methods usually have the following general steps:

- Representation: Includes selecting appropriate inputs to represent the problem of selecting features, defining output information, and selecting suitable models such as decision tree (DT).
- Selecting appropriate features to reduce input space size: Inputs that do not contain valuable information for estimating the output are removed. In some methods, such as DT, feature and model selection are done simultaneously.
- Model extraction
- Interpretation and evaluation to understand the physical meaning of the extracted model to determine its validity.
- Test the model with a set of unseen inputs (with known output).
- Use the model to get the output of new situations

Fig. 6.2 shows the general performance framework of these methods for evaluating the mechanical condition of the transformer. As shown in this figure, each of these methods consists of three different steps:

The first step prepares a database by simulating a power transformer or off-line measurement. The database is divided into two independent sets, training and test sets. The training set is used to train the selected learning method, such as artificial neural network (ANN), DT, k-nearest neighbor (k-NN), etc., and the test set is also used to check the qualifications of the learning process.

In the second step, the system is trained according to the database prepared in step one. At this stage of the work, choosing the proper machine learning method with appropriate dimensions is important, e.g., the number of layers and neurons in the ANN. In addition, appropriate inputs without any unnecessary data improve the performance of the training. In the third step, the trained system can be used online to monitor the condition of the power transformer.

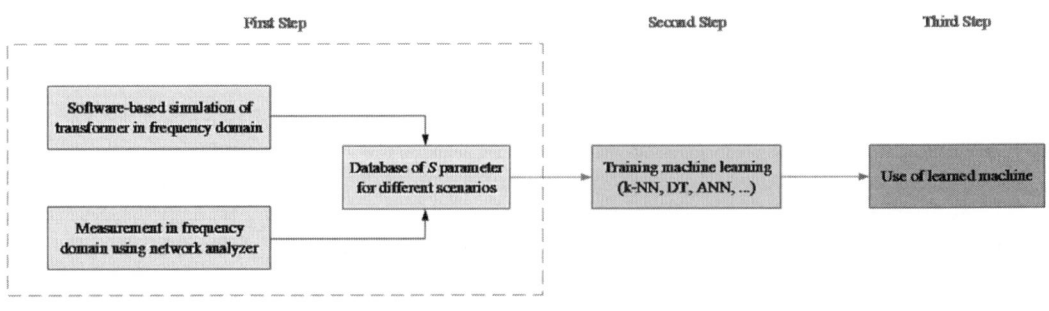

FIGURE 6.2

Framework of machine learning procedure for assessment of mechanical condition of transformer.

6.4 Experimental results

In this section, to evaluate the performance of the proposed method for detecting defects in power transformers, a transformer model is prepared and using a network analyzer, data are gathered. The classification of data to detect the type or extent of faults is investigated and the advantages/disadvantages of the proposed method are discussed.

6.4.1 Transformer model

The power transformer includes a core, inner low voltage, and outer high voltage windings for each phase. Due to symmetry, only one phase can be modeled instead of three. The power transformer winding height is about 2 meters or more. It is not reasonable to build a real-size power transformer model to test the method while the procedure with some modifications can be done in a scaled one. Hence, the transformer model dimension is scaled to 5% of a real transformer.

In the literature, the installation of ultra—high-frequency (UHF) antennas in the power transformer was first proposed to detect partial discharge (PD) (Chai et al., 2019). The location of PD can be estimated using four sensors (Karami et al., 2012) although some new researchers are working using only one sensor to localize PD (Azadifar et al., 2020; Karami, Azadifar, et al., 2020). In addition to detecting the PD, the mechanical fault of the winding can be detected through these antennas (Karami et al., 2016). The UHF sensors must have a wide-band response because the frequency content of PD signals varies with the signal path (Karami et al., 2012, 2013, pp. 1—5).

The frequency bandwidth of experimental studies on the transformer model is chosen between 7 and 12 GHz, to imagine an actual transformer wave performance at the PD frequency between 350 and 600 MHz. In other words, as the model dimension is about 5% of a real transformer, the understudy wavelength (λ) in the model should be considered 5% of PD wavelength in the actual transformer to simulate the behavior of the real environment. Therefore, the frequency should be multiplied by 20 according to $\lambda = c/f$. A standard waveguide is used as an antenna in Ansoft-HFSS software, the simulation environment. Fig. 6.3 shows the simulated model, which is dimensionally equal to the constructed model.

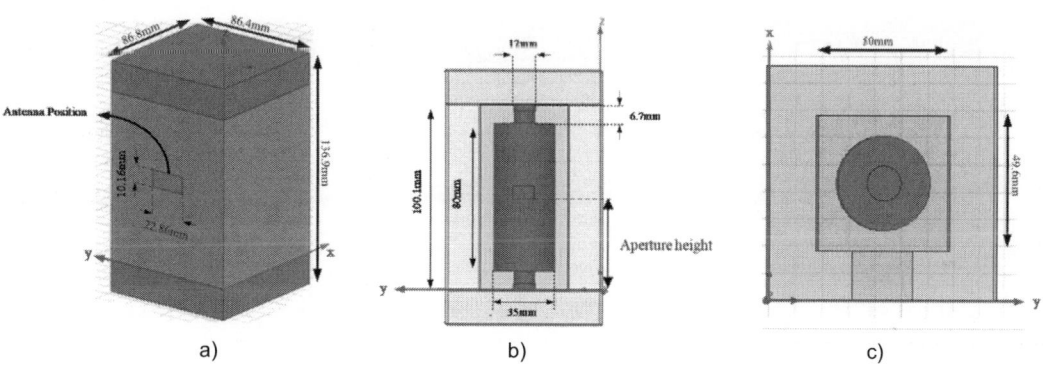

FIGURE 6.3

Transformer model in simulation environment, (A) exterior view, (B) whole model side view, and (C) top view.

Considering the scale ratio of the model and given what we know from the ratio of the winding disk spaces to the HV winding height in a real transformer, the wave will not cross the distance between the model disks. So, the HV and LV windings are modeled as a solid metal cylinder. The simulation wavelength at $f = 9.5$ GHz is approximately equal to $\lambda = 31.6$ mm. As discussed in the previous chapters, since the LV winding is more exposed to axial electromagnetic forces than the HV winding, the probability of axial displacement due to a short circuit is higher than the HV winding.

The tank model is a hollow cuboid metal with an inner space of 100.1 mm height, 50 mm length and 49.6 mm width, and a wall thickness of 18.4 mm (Fig. 6.3D). It means that the dimensions with wall thickness are 136.9 mm for the height, 86.8 mm for the length, and 86.4 mm for the width, as shown in Fig. 6.3A.

In Fig. 6.3C, the larger cylinder is modeled as LV winding and another one as HV winding. The heights of LV and HV winding models are 93.4 and 80 mm, respectively. Their diameters are 6 and 17.5 mm, respectively. The central cylinder or LV winding model can be moved up and down with an accuracy of 0.1 mm. The dimension of the waveguide port is a rectangle with 10.16 and 22.86 mm sides. The interior and exterior views of real transformer model are shown in Fig. 6.4.

6.4.1.1 Axial displacement setup

To test the axial displacement of the winding model, the windings should be moved up in the vertical direction. It can be done by a regulator screw installed at the bottom of the tank model. The winding models are moved from the lowest to the highest position, with steps of 0.1 mm. Therefore, the number of axial displacement cases is 94. For each position, the scattering parameter is recorded by the network analyzer through the waveguide port from the frequency of 7−12 GHz in steps of 25 MHz. Therefore, the number of samples received for each position is 201. The same experiment is repeated in HFSS software for the simulated model. Fig. 6.4 shows the test setup.

a)

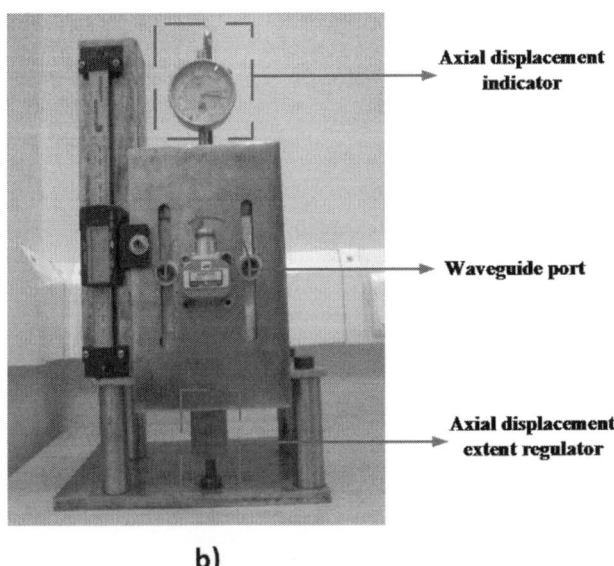

b)

FIGURE 6.4

Real transformer model, (A) interior view and (B) exterior view.

6.4.1.2 Radial deformation setup

The main structure is the same as the axial model in this setup. Only the HV winding model cylinder is replaced with new HV models, including RD. The specifications of four different RD models with different sizes are described in the following.

The RD model No. 1 has a bulgy deformation in one side of winding by cutting a rectangular cuboid from a copy of the original model, as shown in Fig. 6.5. The height of the cutted part is equal to the height of the HV model. The sizes of the other sides are 6.8 and 18.4 mm.

Two RDs in two opposite sides of the winding model are considered in RD model No. 2, as shown in Fig. 6.6. The dimensions of the cutted parts are similar but different from RD model No. 1. The heights are similar to the HV winding model height and the sizes of the other sides are 5.6 and 18 mm.

The RD model No. 3 describes a situation with three bugly deformations in the winding. As shown in Fig. 6.7, three cuts are applied with the same height as previous RD models and the sizes of other sides are 5.8 and 18 mm.

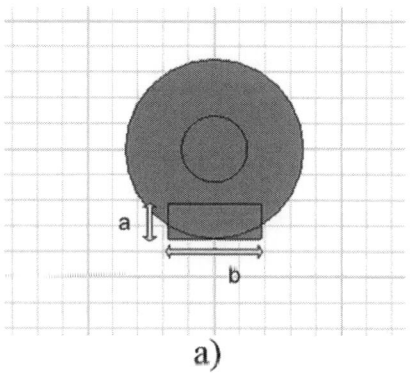
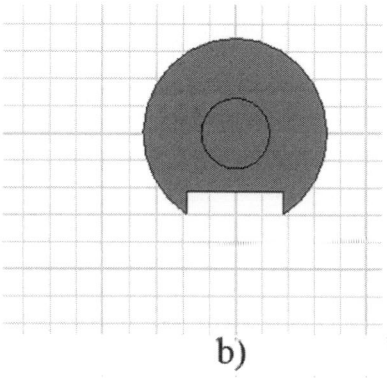

a)   b)

FIGURE 6.5

RD model No. 1, (A) cuboid cut and (B) cutted model.

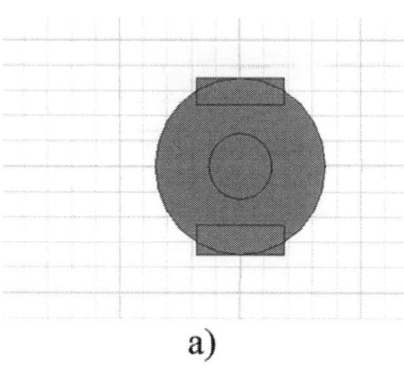
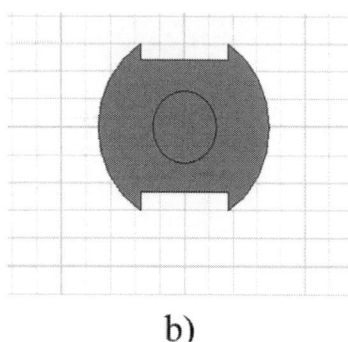

a)   b)

FIGURE 6.6

RD model No. 2, (A) cuboid cut and (B) cutted model.

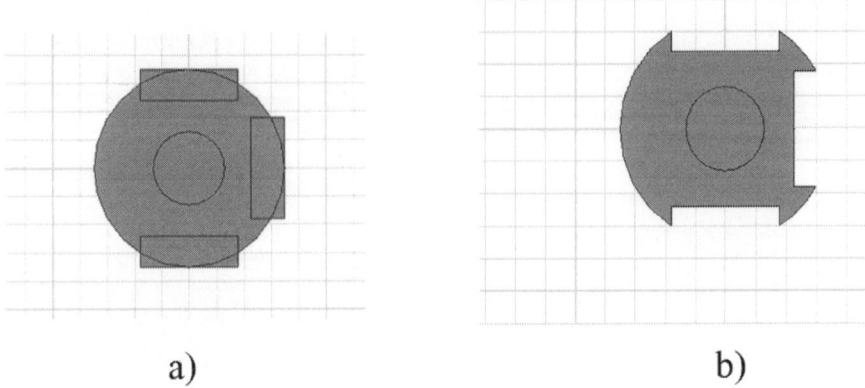

a)   b)

FIGURE 6.7

RD model No. 3, (A) cuboid cut and (B) cutted model.

The RD model No. 4 is the same as RD model No. 3 with a bugly deformation in the intact side (Fig. 6.8). The size of deformations is similar to RD model No. 3, except that 5.8 mm is increased to 6 mm.

Fig. 6.9 shows images of different actual winding models for RD study. The frequency range and steps are similar to the axial displacement study. The scattering parameters are stored as complex numbers.

At first, the winding model in the healthy state should be measured by the network analyzer. Then the RD models should be replaced with the healthy model and the measurement should be repeated. All of them are positioned in the center without any axial displacement. We rotate the winding model from 0 to 360 degrees with 5-degree steps for each model and record the S-parameter. It means that we have 73 S-parameter signals for each model. The same procedure is repeated in HFSS software for the simulated models.

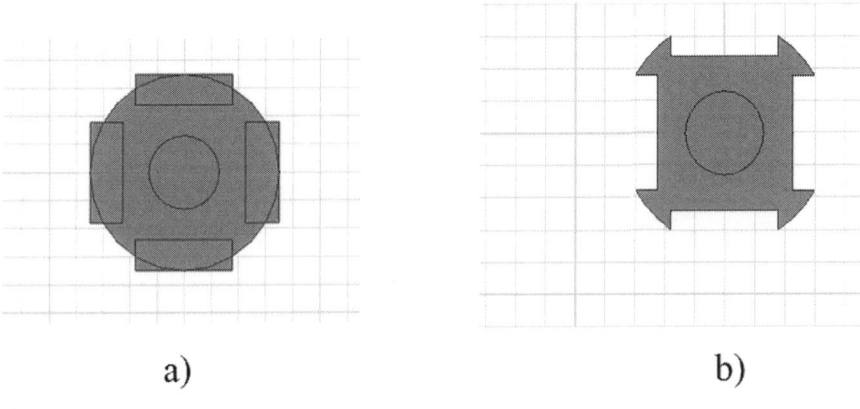

a)   b)

FIGURE 6.8

RD model No. 4, (A) cuboid cut (B) cutted model.

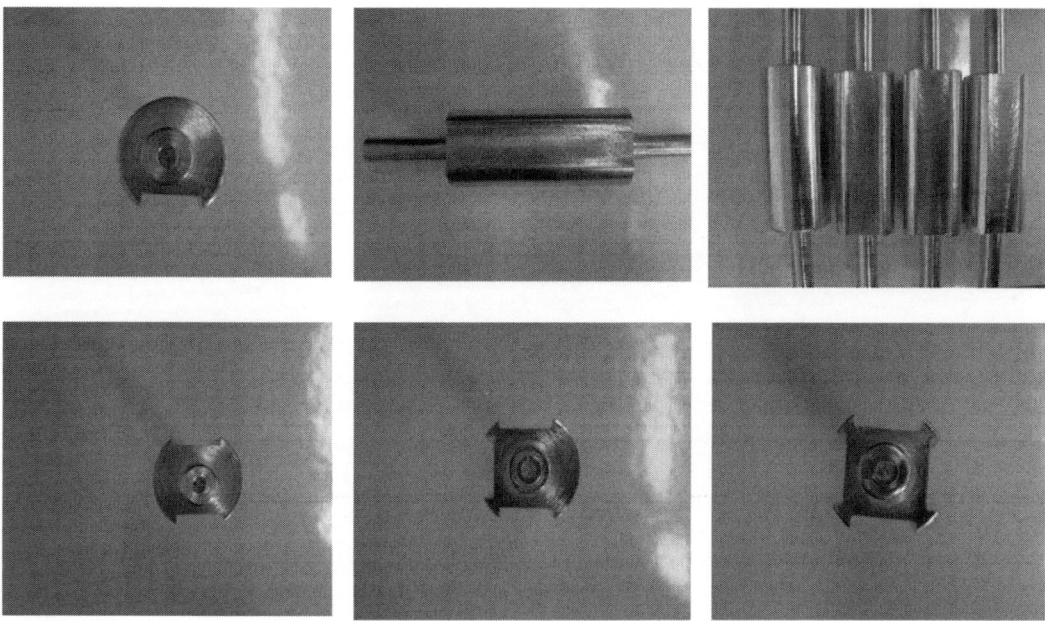

FIGURE 6.9

Different models made for radial deformation.

6.4.2 Axial displacement study

The purpose of a monitoring system is first to detect the presence of a fault and then determine its type, location, and extent. In this regard, we will introduce various indices to detect defects, as explained in the next. The base of the proposed indices is to compare the amplitude and phase of the scattering parameter of the unknown state with the healthy state. For this purpose, the scatter parameter related to the healthy state of the winding must have already been measured.

6.4.2.1 Index definition
6.4.2.1.1 Mean absolute of amplitude/phase difference
This index compares the amplitude and phase of the scattering parameter measured at N different frequencies in faulty and healthy conditions. The mean absolute amplitude difference ($MAAD$) and the mean absolute phase difference ($MAPD$) at the faulty state x are defined as follows:

$$MAAD(x) = \frac{\sum\limits_{i=1}^{N} ||S_x(i)| - |S_0(i)||}{N} \tag{6.13}$$

$$MAPD(x) = \frac{\sum\limits_{i=1}^{N} |\angle S_x(i) - \angle S_0(i)|}{N} \tag{6.14}$$

where $|S_x(i)|$ and $|S_0(i)|$ are the amplitude, and $\angle S_x(i)$ and $\angle S_0(i)$ are the phase of the scattering parameter at the i-th frequency, respectively, at the faulty state x and the reference (healthy) state, zero.

6.4.2.1.2 Cross-correlation index

This index determines the similarity of two waveforms. The similarity in amplitude and phase waveforms is measured separately. The correlation of the S-parameters amplitude (CSA) for faulty state x is defined by Eq. (6.15):

$$CSA(x,0) = \frac{\sum_{i=1}^{N} (|S_x(i)| - |\bar{S}_x|)(|S_0(i)| - |\bar{S}_0|)}{\sigma_1 \sigma_2 N} \tag{6.15}$$

where $|\bar{S}_x|$ and $|\bar{S}_0|$ are the mean scattering parameter at the whole frequency content of faulty state x and the reference (healthy) state 0, respectively. σ_1 and σ_2 are defined as follows:

$$\sigma_1 = \left(\frac{\sum_{i=1}^{N} (|S_x(i)| - |\bar{S}_x|)^2}{N} \right)^{\frac{1}{2}} \tag{6.16}$$

$$\sigma_2 = \left(\frac{\sum_{i=1}^{N} (|S_0(i)| - |\bar{S}_0|)^2}{N} \right)^{\frac{1}{2}} \tag{6.17}$$

Similarly, the correlation of the S-parameters phase (CSP) is defined by Eq. (6.18):

$$CSP(x,0) = corr(\angle S_x, \angle S_0) = \frac{\sum_{i=1}^{N} (|\angle S_x(i)| - \angle \bar{S}_x)(|\angle S_0(i)| - \angle \bar{S}_0)}{\sigma_1 \sigma_2 N} \tag{6.18}$$

where σ_1 and σ_2 are as follows:

$$\sigma_1 = \left(\frac{\sum_{i=1}^{N} (|\angle S_x(i)| - |\angle \bar{S}_x|)^2}{N} \right)^{\frac{1}{2}} \tag{6.19}$$

$$\sigma_2 = \left(\frac{\sum_{i=1}^{N} (|\angle S_0(i)| - |\angle \bar{S}_0|)^2}{N} \right)^{\frac{1}{2}} \tag{6.20}$$

6.4.2.2 Axial displacement detection

Figs. 6.10 and 6.11 show the variations of *MAAD* and *MAPD* for different upward/downward axial displacement extents in measured and simulated studies, respectively. The results show that the increase in indices is proportional to the increase in axial displacement. Therefore, they can be used as an index to detect axial displacement and estimate the amount of displacement approximately. However, these indices cannot distinguish between upward and downward displacements.

FIGURE 6.10

Variations of indices for measured axial displacement, (A) *MAAD* and (B) *MAPD*.

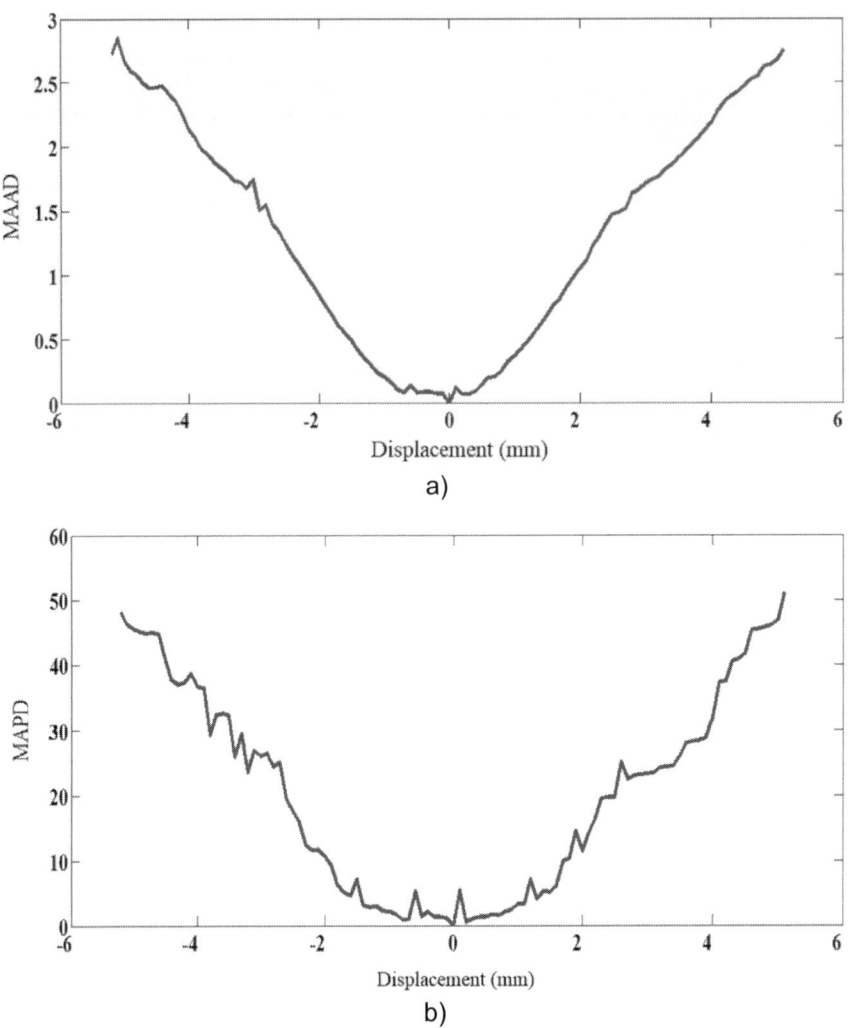

FIGURE 6.11

Variations of indices for simulated axial displacement, (A) *MAAD* and (B) *MAPD*.

Fig. 6.12 shows *CSA* and *CSP* vs. axial displacement extent measured and simulated at 201 frequency points. It can be seen that the decrease in the indices leads to an increase in axial displacement. Similar to *MAAD* and *MAPD*, this index can detect axial displacement but cannot differentiate between upward and downward displacements.

The cross-correlation index has also been used to interpret the FRA test results (Hejazi, Gharehpetian, Moradi, Alehosseini, & Mohammadi, 2011; Nirgude et al., 2008). To compare the scattering parameter results with the FRA method, the sensitivity of the cross-correlation index to axial displacement is formulated by Eq. (6.21):

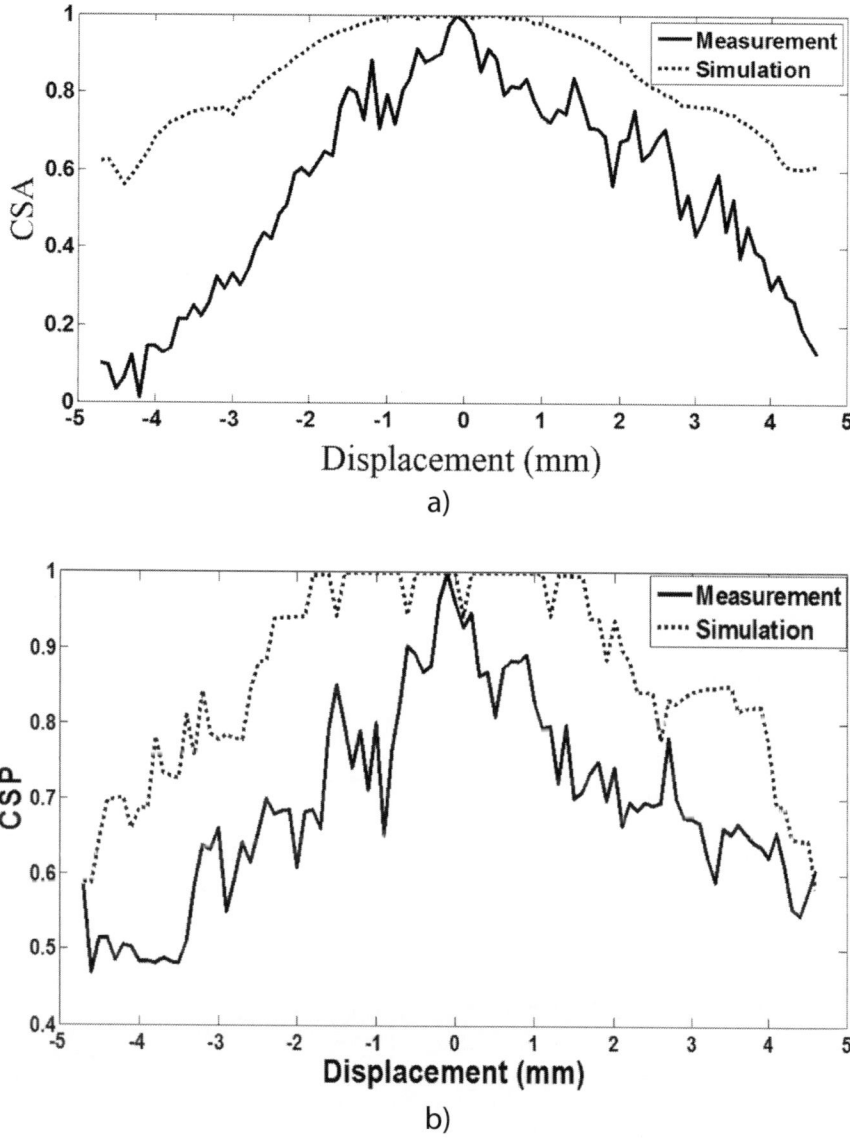

FIGURE 6.12

Measured and simulated results of axial displacement, (A) *CSA* and (B) *CSP*.

$$Sensitivity = \frac{\Delta Cross - correlation}{\Delta x} \qquad (6.21)$$

Table 6.1 compares the scattering parameter method with the FRA method. The proposed method has more sensitivity for axial displacement than the FRA method.

Fig. 6.13 shows the algorithm for detecting axial displacement using the *S*-parameter.

Table 6.1 Comparison of scattering parameter method with FRA method.

	Scattering parameters method		FRA method
Minimum detectable displacement (%)	0.125		0.5
Cross-correlation index	CSA	CSP	CC
Sensitivity	0.2	0.18	0.001

6.4.2.3 Determining axial displacement extent

After detecting the fault in the winding, it is time to obtain its extent. For this purpose, in addition to the S-parameter of the healthy state, data related to various fault states are also saved in the stored database. These data may be provided through software simulations or measurements in a sample transformer. In this subsection, it is assumed that the type of fault is axial and we only determine its extent. For this purpose, several pattern recognition methods are examined to determine the winding axial displacement extent.

6.4.2.3.1 k-NN method

The k-NN algorithm for axial displacement extent determination is as follows:

Step 1: Measurements of different axial displacements are divided into m classes, c_j ($j = 1, 2, \ldots, m$). A class may have only one measure of displacement or more. For example, if there are 10 measurements with 1 mm axial displacement, it can be classified into five steps of 2 mm or two steps of 5 mm.

Step 2: Label a class to each S-parameter amplitude pattern, S_x ($1 \leq x \leq n$), stored in the training database, where n is the number of patterns.

Step 3: Assume that the S-parameter amplitude of unknown axial displacement is S_q. Calculate the amplitude Euclidean distance (AED) between S_q and each S_x at N measurement frequency points as follows:

$$AED_{S_x,S_q} = \sqrt{\frac{\sum_{i=1}^{N} \left(|S_x(i)| - |S_q(i)|\right)^2}{N}} \tag{6.22}$$

Step 4: Select the k nearest patterns based on the AED and gather them in a new set, $l(S_q)$. The members of $l(S_q)$ are named in the next as $S_{q,t}$ where $1 \leq t \leq k$.

Step 5: The class with the highest members in $l(S_q)$ can be assigned as the class of S_q.

The above algorithm can be a discrete estimate of the axial displacement. The minimum possible error (E) of the estimated axial displacement relative to the actual axial displacement is the maximum distance of axial displacements among members of two consecutive classes.

If the total axial displacement of the understudy case is 9.4 mm and divided into $m = 19$ steps, each of which has $E = 0.5$ mm length or can be divided into $m = 10$ steps, each of which is $E = 1$ mm in length. The measured data at different axial displacements are divided to train and test data. The test data are used to achieve the performance of the trained k-NN. The incorrect classification (IC) of axial displacement estimation is defined by Eq. (6.23):

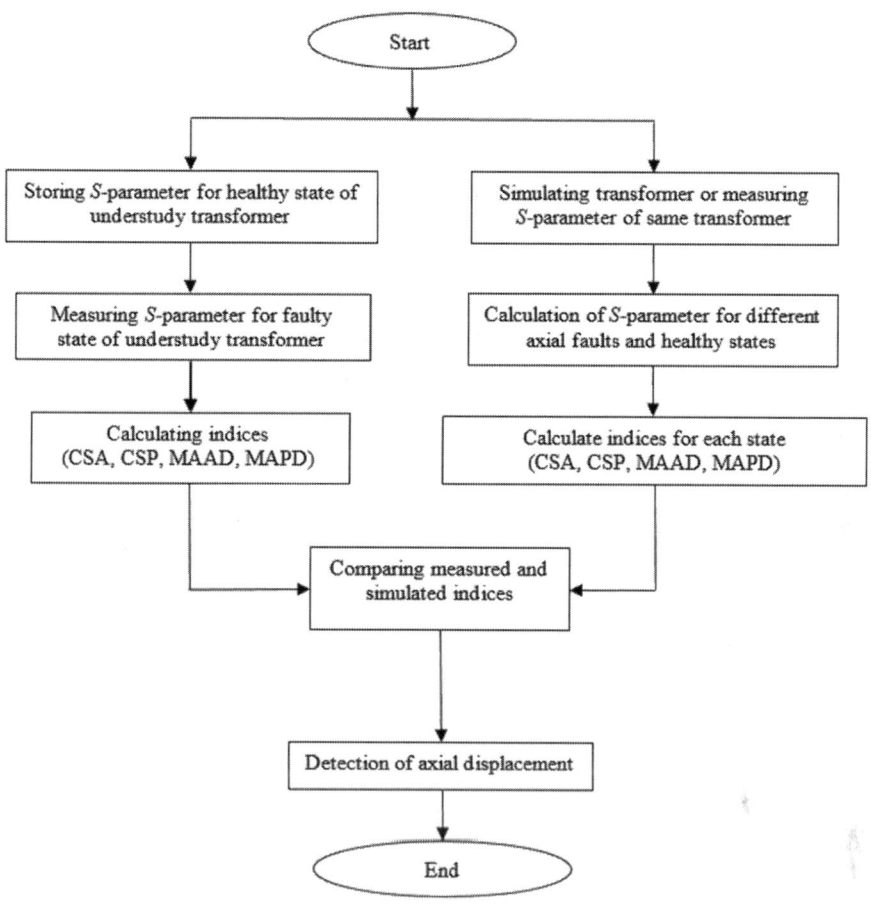

FIGURE 6.13

Procedure of axial displacement detection using *S*-parameter.

$$IC = \frac{Incorrect\ classified\ test\ data}{Whole\ test\ data} \times 100 \qquad (6.23)$$

The *IC* results for the different numbers k and E are summarized in Table 6.2. The results do not have acceptable accuracy. In addition, the results of this algorithm are discrete. So, in the next, the weighted k-NN regression method will be used to improve the accuracy and resolution.

Table 6.2 Accuracy of k-NN method for different values of k and E.

	$k = 3$	$k = 5$	$k = 7$
$E = 1$	19.149	27.66	34.043
$E = 0.5$	39.362	51.064	68.085

6.4.2.3.2 Weighted k-NN regression

Better results can be achieved by weighing in such a way that closer neighbors play a greater role than distant ones. Steps one to four of the proposed method are similar to the k-NN method described previously. Considering AD_j as the highest axial displacement value in class c_j, the average axial displacement value assigned to each class is calculated as follows:

$$AD_{c_j} = \frac{AD_j - AD_{j-1}}{2} \tag{6.24}$$

For class c_1, we have $AD_{c_1} = \frac{AD_1}{2}$. Each member of $l(S_q)$, $S_{q,t}$, is the delegate of AD_{c_j} axial displacement value if it is a member of class c_j, shown as $ADS_{q,t}$. Therefore, the estimated axial displacement value of S_q can be calculated as formulated by Eq. (6.25):

$$ADS_q = \frac{\sum_{i=1}^{k} AEDS_{S_q,S_{q,i}} \times ADS_{q,t}}{\sum_{i=1}^{k} AEDS_{S_q,S_{q,i}}} \tag{6.25}$$

The axial displacement value of an unknown S_q pattern is not necessarily limited to the database class values and can be changed continuously based on the weights, $AEDS_{S_q,S_{q,i}}$. As a result, this method can continuously estimate axial displacement values. Table 6.3 presents the *IC* results of the k-NN regression algorithm for estimating unknown axial displacement from *S*-parameters. Comparison of Table 6.2 with Table 6.3 shows a significant improvement in determining the accuracy of the axial displacement of the transformer winding. For example, for $k = 3$ and $E = 0.5$ mm, the *IC* value in Table 6.2 is 39.362%, whereas in Table 6.3 it is 4.26%. The minimum detectable axial displacement in the proposed method is 0.1 mm or 0.125% of the transformer winding length, while in the FRA method it is 0.5% of the transformer winding height (Nirgude et al., 2008). This method shows more sensitivity than the FRA method. The calculated estimates by k-NN in comparison with DT method will be shown in the next part.

Table 6.3 *IC* of weighted k-NN regression for different values of k and E.

	$k = 1$	$k = 3$	$k = 5$	$k = 7$	$k = 9$	$k = 11$
$E = 0.1$	38.30	46.81	55.32	57.45	61.70	63.83
$E = 0.2$	12.77	23.40	30.85	40.43	37.23	36.17
$E = 0.3$	6.38	11.70	17.02	21.28	24.47	25.53
$E = 0.4$	5.32	6.38	9.57	8.51	12.77	18.09
$E = 0.5$	4.26	4.26	6.38	6.38	8.51	17.02
$E = 0.6$	3.19	4.26	2.13	2.13	7.45	12.77
$E = 0.7$	2.13	1.06	2.13	1.06	4.26	10.64
$E = 0.8$	1.06	1.06	0.00	1.06	4.26	8.51
$E = 0.9$	1.06	1.06	0.00	1.06	2.13	8.51
$E = 1$	1.06	1.06	0.00	1.06	2.13	6.38

6.4.2.3.3 Decision tree method

DT is a nonparametric method that uses a recursive structure. Finding the best split point and performing splitting is the most important part of the DT algorithm (Hejazi, Gharehpetian, Moradi, Mohammadi, & Alehosseini, 2011). The DT developed in this study is a classification and regression tree (Breiman et al., 2017). Compared to ANN, DT has a lower computational complexity. This algorithm automatically selects features and at each decision stage, its output is an if-then test. The main two steps of DT are splitting index and feature selection.

The splitting index step evaluates the "goodness" of different branching options according to a particular characteristic. The division criterion is usually the Gini index described in the literature (Kingsford & Salzberg, 2008). The feature selection step means selecting a subset of all available features to represent the data before applying the learning algorithm, which is a common way to simplify and speed up computations (Chandrashekar & Sahin, 2014). The proposed algorithm for detecting axial displacement using the DT method is described in the following three steps:

Step 1: The S-parameters measured at different axial displacements of the model are stored in a database.

Step 2: The S-parameters are measured at unknown axial displacement.

Step 3: The axial displacement value is estimated using the DT method.

Amplitude, phase, or both can be used as feature vectors to classify S-parameters. The best results are obtained using both amplitude and phase as feature vectors. A part of the DT of 75 training data is shown in Fig. 6.14. In this figure, $M(f)$ and $P(f)$ show the amplitude and phase of the S-parameter at frequency f.

The most important phase characteristic is at $f = 8.925$ GHz, the first node of Fig. 6.14. At the second level, the amplitude is important at $f = 10.05$ GHz and $f - 7$ GHz. This DT has been tested using 19 samples. Table 4.4 compares the results of the DT method with the weighted k-NN regression method. Considering AD_r and AD_e as actual and estimated displacement values, respectively, the calculated results of the weighted k-NN regression are listed in Table 6.4. In this table, the *Error* is absolute of $(AD_r\text{-}AD_e)$ divided by winding height.

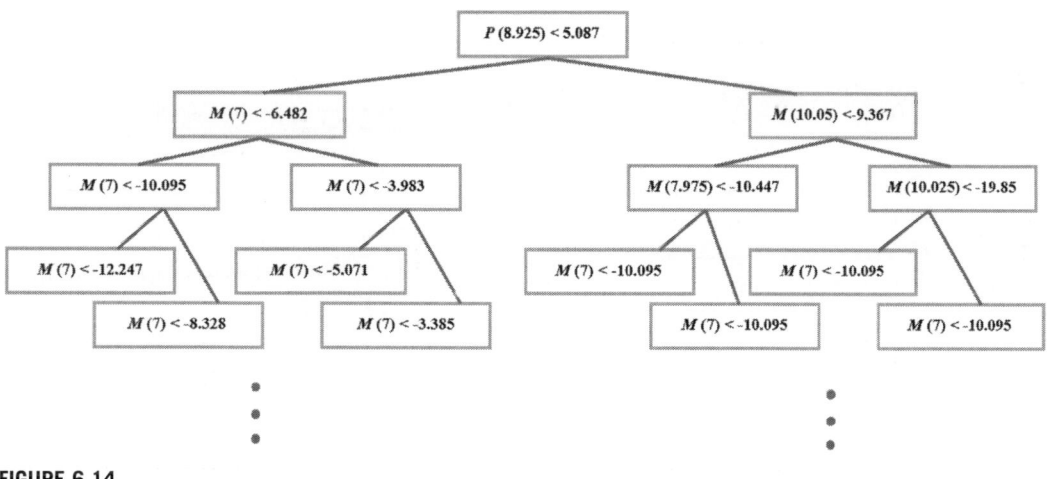

FIGURE 6.14

Part of DT

Table 6.4 Comparison of DT method with weighted k-NN regression results.

AD_r (mm)	DT method		k-NN regression									
			k = 3		k = 5		k = 7		k = 9		k = 11	
	AD_e (mm)	Error (%)	AD_e (mm)	Error (%)	AD_e (mm)	Error (%)	AD_e (mm)	Error (%)	AD_e (mm)	Error (%)	AD_e (mm)	Error (%)
−4.3	−3.8	0.625	−4.5	0.25	−4.4	0.125	−4.3	0	−4.2	0.125	−4.2	0.125
−3.9	−3.5	0.5	−4.0	0.125	−4.0	0.125	−4.1	0.25	−4.0	0.125	−4.0	0.125
−3.3	−2.5	1	−3.6	0.375	−3.6	0.375	−3.6	0.375	−3.5	0.25	−3.4	0.125
−2.9	−3	0.125	−2.9	0	−2.8	0.125	−2.7	0.25	−2.7	0.25	−2.7	0.25
−2.3	−2.8	0.625	−2.5	0.25	−2.4	0.125	−2.4	0.125	−2.3	0	−2.4	0.125
−1.9	−1.5	0.5	−1.9	0	−1.8	0.125	−1.8	0.125	−1.8	0.125	−1.9	0
−1.3	−1.2	0.125	−1.4	0.125	−1.5	0.25	−1.4	0.125	−1.5	0.25	−1.4	0.125
−0.9	−1	0.125	−1.1	0.25	−0.7	0.25	−0.4	0.625	−0.1	1	−0.2	0.875
−0.3	0	0.375	−0.4	0.125	−0.3	0	−0.2	0.125	−0.3	0	−0.4	0.125
0.1	0	0.125	−0.1	0.25	−0.1	0.25	−0.2	0.375	−0.3	0.5	−0.2	0.375
0.7	0.2	0.625	1.0	0.375	0.9	0.25	0.9	0.25	0.8	0.125	0.7	0
1.1	1.2	0.125	1.0	0.125	1.0	0.125	1.1	0	0.9	0.25	0.8	0.375
1.7	1.5	0.25	1.5	0.25	1.3	0.5	1.3	0.5	1.4	0.375	1.4	0.375
2.1	2.2	0.125	2.1	0	1.9	0.25	2.0	0.125	1.8	0.375	1.7	0.5
2.7	2.5	0.25	2.8	0.125	2.7	0	2.6	0.125	2.2	0.625	2.2	0.625
3.1	3	0.125	3.2	0.125	3.2	0.125	3.3	0.25	3.4	0.375	3.3	0.25
3.7	3.8	0.125	3.7	0	3.7	0	3.6	0.125	3.5	0.25	3.6	0.125
4.1	3.9	0.25	4.2	0.125	4.1	0	4.1	0	4.0	0.125	4.0	0.125
4.7	4.6	0.125	4.4	0.375	4.3	0.5	4.2	0.625	4.1	0.75	3.5	1.5
Average error (%)	0.322		0.171		0.184		0.230		0.309		0.322	

It is clear that DT can distinguish between upward and downward axial displacements. The winding is moved in 94 steps. Each step is equal to 0.1% of the whole winding height. The average percentage of errors is compared in the last row of Table 6.4. The results show that the k-NN regression method with $k = 3$ has the minimum error and is preferable to the DT method for detecting axial displacement.

The DT method has a maximum error of 1% (winding length) in the worst case, while in the k-NN algorithm, it is 1.5% for $k = 11$. Therefore, both methods are more accurate than the FRA method, with a maximum error of 4% (Karimifard et al., 2008). The minimum axial displacement detectable in the proposed method is 0.125% of the transformer winding height, while in the FRA method it is 0.5% (Nirgude et al., 2008).

6.4.2.3.4 ANN method

ANNs are widely used in identifying and modeling nonlinear systems, for which obtaining a mathematical model is difficult and tedious. The ANN's ability to learn and map input−output has led to its widespread use (Hoang et al., 2021; Nazir et al., 2020; Ramkumar et al., 2020). In the method of detecting the transformer winding axial displacement by S-parameters, it is not possible to find a definite relationship between the axial displacement value and the S-parameters. Different neural network topologies (number of layers and neurons in each hidden layer) have been tested for best performance. This study uses a multilayer perceptron network. The best performance is achieved with networks with 71 inputs, two hidden layers with 15 and 8 neurons, and 1 output. The forward/backward propagation algorithm is used for training. Fig. 6.15 shows the structure of this network.

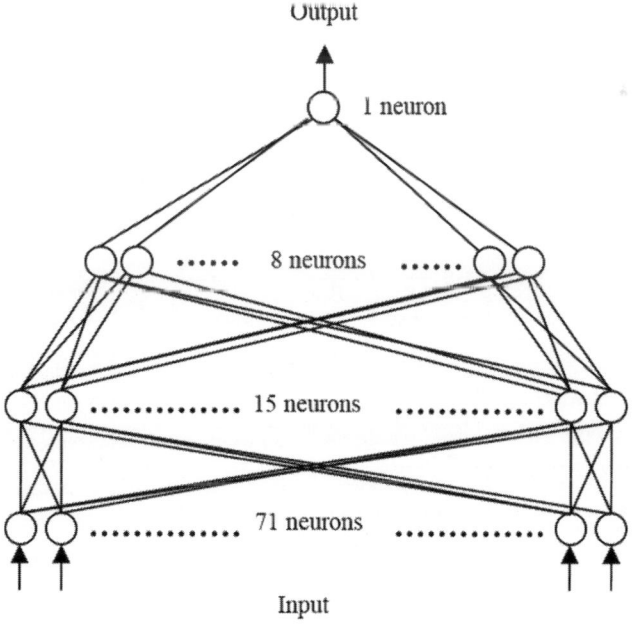

FIGURE 6.15

ANN structure for axial displacement detection.

The amplitude and phase of the S-parameters for each axial displacement have been measured by the network analyzer and categorized into training and test data. The procedure of determining displacement is summarized as follows:

Step 1: Measuring the S-parameters of the transformer.

Step 2: Creating a database for S-parameters of the axial displacements.

Step 3: Training the ANN.

Step 4: Using the trained ANN to estimate the extent of the axial displacement.

The results, listed in Table 6.5, show that the displacement by the ANN is predictable with a maximum error of 0.6%. The average error in this method is 0.24%. Therefore, by comparing Table 6.4, the k-NN regression method with $k = 3$ provides the best estimation among the studied methods.

Table 6.5 Actual and estimated axial displacement using ANN.

Actual displacement (mm)	Estimated displacement (mm)	Error (%)
−4.3	−4.297	0.00375
−4	−3.859	0.17641
−3.7	−3.689	0.01338
−3.5	−3.523	0.02851
−3.2	−2.772	0.53466
−2.9	−2.803	0.12104
−2.5	−2.574	0.09223
−2.2	−1.93	0.33804
−1.7	−1.658	0.052
−1.4	−1.413	0.01586
−1.1	−0.9	0.25042
−0.7	−0.681	0.02405
−0.5	−0.966	0.58301
−0.1	−0.243	0.17919
0.2	0.5436	0.42955
0.4	0.995	0.74375
0.7	0.8354	0.16921
1	0.9929	0.00887
1.3	1.5758	0.3447
1.6	1.8648	0.33103
2	2.1716	0.21449
2.6	2.538	0.07748
2.9	2.9251	0.03134
3.3	3.6926	0.49078
3.7	3.6013	0.12338
4	4.298	0.37249
4.3	4.0099	0.36269
4.7	4.2289	0.58883

6.4.3 Radial deformation study

As in RD, each faulty state is measured in different angles; the correlation index (*CI*) is calculated by the following equation:

$$CI(\theta) = \frac{\sum_{i=1}^{N}\left(\left(|S_{x,\theta}(i)| - |\overline{S}_{x,\theta}|\right)\left(|S_{0,\theta}(i)| - |\overline{S}_{0,\theta}|\right)\right) + \sum_{i=1}^{N}\left(\left(\angle S_{x,\theta}(i) - \angle\overline{S}_{x,\theta}\right)\left(\angle S_{0,\theta}(i) - \angle\overline{S}_{0,\theta}\right)\right)}{\sqrt{\left(\sum_{i=1}^{N}\left(\left(|S_{x,\theta}(i)| - |\overline{S}_{x,\theta}|\right)^2 + \left(\angle S_{x,\theta}(i) - \angle\overline{S}_{x,\theta}\right)^2\right)\right)\left(\sum_{i=1}^{N}\left(\left(|S_{0,\theta}(i)| - |\overline{S}_{0,\theta}|\right)^2 + \left(\angle S_{0,\theta}(i) - \angle\overline{S}_{0,\theta}\right)^2\right)\right)}} \quad (6.26)$$

In the previous indices, the amplitude and phase are included separately. Using the *CI*, the amplitude and phase information or the real and imaginary part of the *S*-parameter can be simultaneously used to calculate an index. Fig. 6.16 shows the *CI* variations for models one to four at angles 0 to 360 degrees. The purpose of this index is only to detect the presence of a fault.

According to the obtained data from the measurement, in models one and three, which have a symmetry axis, a symmetric variation is also observed in the *CI*. As can be seen in Figs. 6.6 and 6.8, the models two and four have two and four axes of symmetry, respectively. Therefore, two and four axes of symmetry are observed in the *CI* curve. Figs. 6.17 and 6.18 show the amplitude and phase values of the measured *S*-parameter data at 0, 90, and 270 degrees of the first model. The classification of RD models is studied in the next subsection.

6.4.4 Classification of fault type and extent/location

In the previous subsections, the detection of a fault and the estimation of the axial fault extent was done by assuming that type of fault is known. In this subsection, the type of fault, radial or axial, is unknown and should be determined first. Then, for the axial fault, the extent and for the radial fault, the location is determined.

To train and test the pattern recognition system, we use the real part of the *S*-parameter because it is more sensitive to the fault extent (Behkam et al., 2021). We randomly select 70% of the total data for training and 30% for test. Therefore, considering described states of axial deformation, we have 63 states for training which is a 63 × 201 matrix, and 31 states for test which is a 31 × 201 matrix. For RD, we have 49 states for training which is a 49 × 201 matrix, and 24 states for test which is a 24 × 201 matrix.

A designed system for classification should be able to solve the following three problems:

- Separation of the axial displacement fault and four different states of RD fault.
- Estimation of the axial displacement extent.
- Estimation of the RD location.

Given these three issues, we define the classes used in the pattern recognition methods as follows:

- For the first problem, five classes are defined, as given in Table 6.6.
- For the second problem (axial fault extent estimation), 15 classes are defined according to the data listed in Table 6.7.
- For the problem of RD location detection, three classes are defined according to Table 6.8.

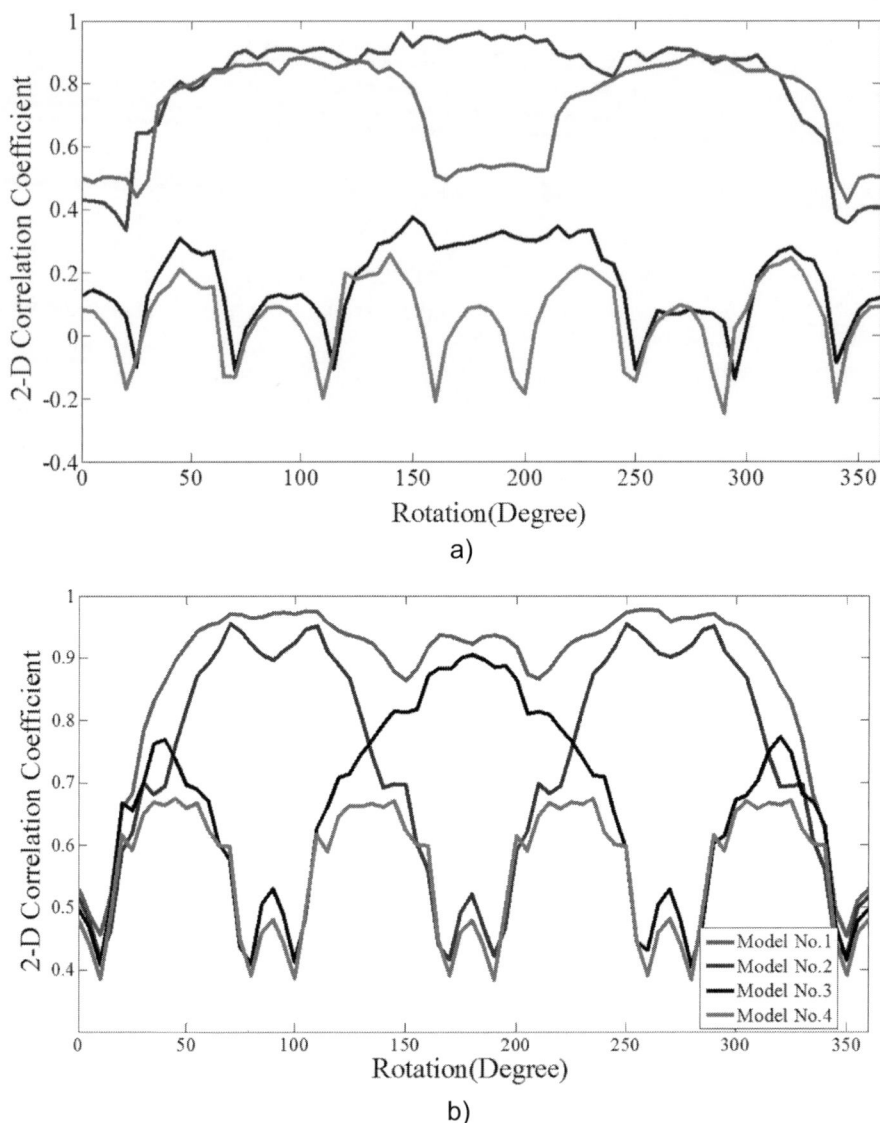

FIGURE 6.16

CI values vs. rotation degree for four models, (A) measured and (B) simulated values.

Using *S*-parameter data, the classification is carried out by Bayesian, Parzen window, k-NN, support vector machine (SVM), and ANN methods. The results are shown in Table 6.9. It can be seen that totally the SVM method has better performance than the others.

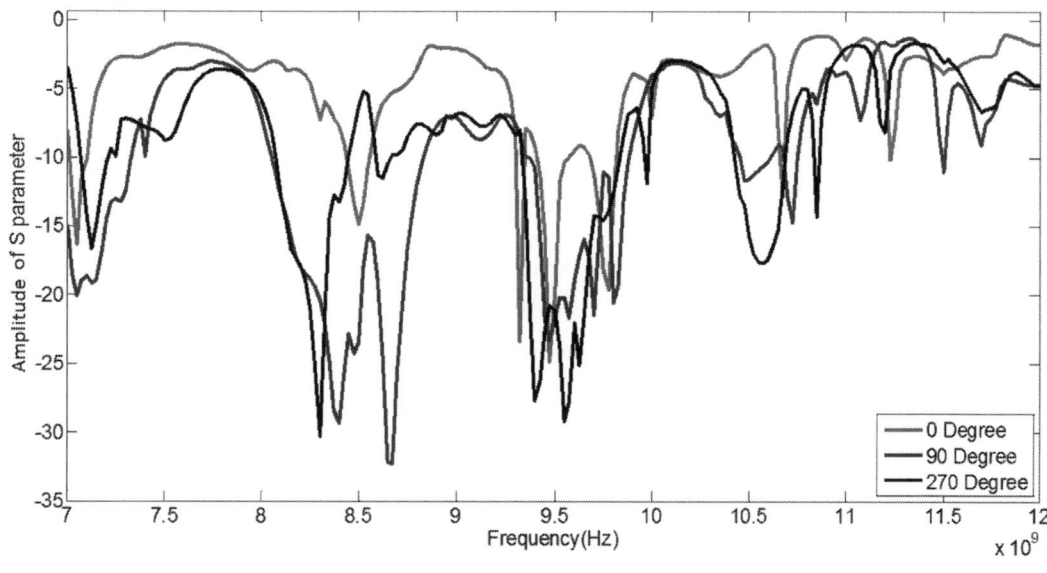

FIGURE 6.17

S-parameter amplitude for measured data of model one at 0, 90, and 270 degrees.

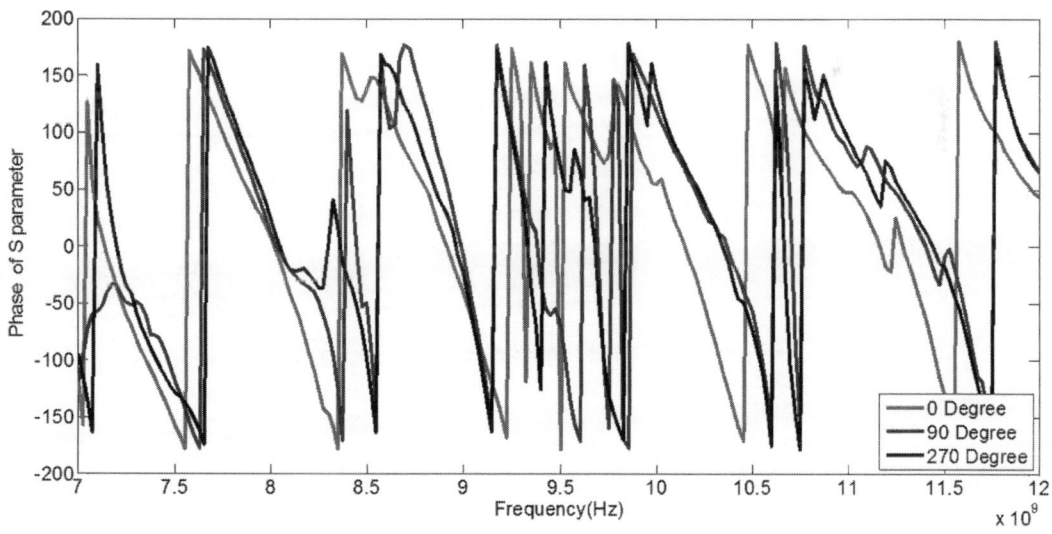

FIGURE 6.18

S-parameter phase for measured data of model one at 0, 90, and 270 degrees.

Table 6.6 Classes for axial and radial separation.

Class 1	RD model I
Class 2	RD model II
Class 3	RD model III
Class 4	RD model IV
Class 5	Axial

Table 6.7 Classes defined for second problem.

Class	Number of positions	Start of interval (mm)	End of interval (mm)
1	6	−4.7	−4.2
2	6	−4.1	−3.6
3	6	−3.5	−3.0
4	6	−2.9	−2.4
5	6	−2.3	−1.8
6	6	−1.7	−1.2
7	6	−1.1	−0.6
8	10	−0.5	0.4
9	6	0.5	1.0
10	6	1.1	1.6
11	6	1.7	2.2
12	6	2.3	2.8
13	6	2.9	3.4
14	6	3.5	4.0
15	6	4.1	4.6

Table 6.8 Radial deformation classes.

Class 1	Range of −60 (300) degrees to +60 degrees
Class 2	Range of +60 degrees to +120 degrees and range of −120 (240) degrees to −60 (300) degrees
Class 3	Range of +120 degrees to +240 degrees

6.5 Advantages and disadvantages

In this chapter, using the *S*-parameter and definition of indices, axial displacement and RD faults were detected. Then, by assuming the type of fault as axial, the extent of fault was estimated using DT method, k-NN, and ANN. In the last step, assuming that the type of fault is unknown, using classification methods such as Bayesian, ANN, k-NN, Parzen window, and SVM, the type, extent, and location of the fault were determined.

Table 6.9 Accuracy of classification results using *S*-parameter.

	Bayesian	Parzen window	k-NN	SVM	ANN
Separation of axial and radial faults	100 97.64	96.85 88.98	99.21 86.61	99.22 98.43	99.22 97.64
Axial fault extent estimation	96.77 93.55	93.55 93.55	83.871 80.65	93.54 89.1	93.54 89.1
Radial fault detection (model #1)	100 100	100 100	100 100	100 100	100 100
Radial fault detection (model #2)	100 75	100 70.83	100 70.83	100 95.83	100 95.83
Radial fault detection (model #3)	87.50 91.66	87.5 100	91.66 100	91.66 95.83	91.66 95.83
Radial fault detection (model #4)	66.67 50	62.5 41.66	62.5 54.1667	70.83 66.66	70.83 66.66

The advantages of the *S*-parameter method are as follows:

- Use of antennas that are installed for PD detection. Therefore, there is no need to add special hardware to the transformers that have PD antennas.
- The monitoring is online.
- Electromagnetic waves have no side effect on the transformer.
- The transformer condition can be easily checked at any time.
- It cannot be influenced by the transformer load due to being independent of the transformer electrical circuit.

The limitations of the proposed *S*-parameter method are as follows:

- Like the FRA method, this method depends on the position of the tap changer. Results must be measured at the same tap position to be comparable.
- Like the FRA method, this method is sensitive to the oil dielectric specification that changes as it ages.
- Compared to the FRA method, the most important limitation of this method is the impossibility of using it to detect radial faults of the winding inner layers. Because electromagnetic waves in the UHF band cannot pass through short distances between disks and that was the reason to model the transformer winding as a cylinder metal. These drawbacks indicate that the proposed method cannot completely replace the FRA method. Using the proposed method online along with the FRA method can be effective for merging information and making the best decision.
- Another limitation is the database which will no longer be useable by changing the type of transformer. In addition, transformer simulation, as a large and complex mechanical structure, in electromagnetic analysis software needs a supercomputer. It is almost impossible to match the simulation results with practical measurements for complex structures such as transformers.
- Another problem is the impossibility of extracting an analytical relation for changes in the *S*-parameter in terms of winding mechanical deformation.

References

Ahmadi, S.-A., & Sanaye-Pasand, M. (2021). A robust multi-layer framework for online condition assessment of power transformers. *IEEE Transactions on Power Delivery, 37*(2), 947–954.

Ahmadi, S.-A., Sanaye-Pasand, M., & Davarpanah, M. (2019). Preventing maloperation of distance protection due to CCVT transients. *IET Generation, Transmission & Distribution, 13*(13), 2828–2835.

Ahmadi, S.-A., Sanaye-Pasand, M., Jafarian, P., & Mehrjerdi, H. (2020). Adaptive single-phase auto-reclosing approach for shunt compensated transmission lines. *IEEE Transactions on Power Delivery, 36*(3), 1360–1369.

Akhavanhejazi, M., Gharehpetian, G. B., Faraji-Dana, R., Moradi, G. R., Mohammadi, M., & Alehoseini, H. A. (2011). A new on-line monitoring method of transformer winding axial displacement based on measurement of scattering parameters and decision tree. *Expert Systems with Applications, 38*(7), 8886–8893.

Azadifar, M., Karami, H., Wang, Z., Rubinstein, M., Rachidi, F., Karami, H., Ghasemi, A., & Gharehpetian, G. B. (2020). Partial discharge localization using electromagnetic time reversal: A performance analysis. *IEEE Access, 8*, 147507–147515.

Behkam, R., Karami, H., Naderi, M. S., & Gharehpetian, G. B. (2021). Generalized regression neural network application for fault type detection in distribution transformer windings considering statistical indices. *COMPEL-The International Journal for Computation and Mathematics in Electrical and Electronic Engineering, 41*(1), 381–409.

Breiman, L., Friedman, J. H., Olshen, R. A., & Stone, C. J. (2017). *Classification and regression trees*. Routledge.

Chai, H., Phung, B. T., & Mitchell, S. (2019). Application of UHF sensors in power system equipment for partial discharge detection: A review. *Sensors, 19*(5), 1029.

Chandrashekar, G., & Sahin, F. (2014). A survey on feature selection methods. *Computers & Electrical Engineering, 40*(1), 16–28.

Ghanizadeh, A. J., & Gharehpetian, G. B. (2013). Application of characteristic impedance and wavelet coherence technique to discriminate mechanical defects of transformer winding. *Electric Power Components and Systems, 41*(9), 868–878.

Hejazi, M. A., Gharehpetian, G. B., Moradi, G., Alehosseini, H. A., & Mohammadi, M. (2011). Online monitoring of transformer winding axial displacement and its extent using scattering parameters and k-nearest neighbour method. *IET Generation, Transmission & Distribution, 5*(8), 824–832.

Hejazi, M. A., Gharehpetian, G. B., Moradi, G. R., Mohammadi, M., & Alehoseini, H. A. (2011). Application of classifiers for on-line monitoring of transformer winding axial displacement by electromagnetic non-destructive testing. *Electric Power Components and Systems, 39*(4), 387–403.

Hoang, A. T., Nižetić, S., Ong, H. C., Tarelko, W., Le, T. H., Chau, M. Q., & Nguyen, X. P. (2021). A review on application of artificial neural network (ANN) for performance and emission characteristics of diesel engine fueled with biodiesel-based fuels. *Sustainable Energy Technologies and Assessments, 47*, 101416.

Idri, A., Benhar, H., Fernández-Alemán, J. L., & Kadi, I. (2018). A systematic map of medical data preprocessing in knowledge discovery. *Computer Methods and Programs in Biomedicine, 162*, 69–85.

Jawad, G. N., & Akbar, M. F. (2021). IFFT-Based microwave non-destructive testing for delamination detection and thickness estimation. *IEEE Access, 9*, 98561–98572.

Jia, D., & Zhuang, X. (2021). Learning-based algorithms for vessel tracking: A review. *Computerized Medical Imaging and Graphics, 89*, 101840.

Karami, H., Azadifar, M., Mostajabi, A., Rubinstein, M., Karami, H., Gharehpetian, G. B., & Rachidi, F. (2020). Partial discharge localization using time reversal: Application to power transformers. *Sensors, 20*(5), 1419.

Karami, H., Hejazi, M. S. A., Naderi, M. S., Gharehpetian, G. B., & Mortazavian, S. (2012). Three-dimensional simulation of PD source allocation through TDOA method. *The 4th Conference on Thermal Power Plants, 1–4.*

Karami, H., Gharehpetian, G. B., & Hejazi, M. S. A. (2013). Oil permittivity effect on PD source allocation through three-dimensional simulation. In *Int'l. Power System Conf.* (pp. 1–5). Tehran-Iran.

Karami, H., Gharehpetian, G. B., Norouzi, Y., & Akhavan-Hejazi, M. (2020). Simultaneous radial deformation and partial discharge detection of high-voltage winding of power transformer. *IET Electric Power Applications, 14*(3), 383–390.

Karami, H., Gharehpetian, G. B., Norouzi, Y., & Hejazi, M. A. (2016). GLRT-based mitigation of partial discharge effect on detection of radial deformation of transformer HV winding using SAR imaging method. *IEEE Sensors Journal, 16*(19), 7234–7241.

Karami, H., Tabarsa, H., Gharehpetian, G. B., Norouzi, Y., & Hejazi, M. A. (2019). Feasibility study on simultaneous detection of partial discharge and axial displacement of HV transformer winding using electromagnetic waves. *IEEE Transactions on Industrial Informatics, 16*(1), 67–76.

Karimifard, P., Gharehpetian, G. B., & Tenbohlen, S. (2008). Determination of axial displacement extent based on transformer winding transfer function estimation using vector-fitting method. *European Transactions on Electrical Power, 18*(4), 423–436.

Kingsford, C., & Salzberg, S. L. (2008). What are decision trees? *Nature Biotechnology, 26*(9), 1011–1013.

Maier, L. C., Jr., & Slater, J. C. (1952). Field strength measurements in resonant cavities. *Journal of Applied Physics, 23*(1), 68–77.

Malekizadeh, M., Karami, H., Karimi, M., Moshari, A., & Sanjari, M. J. (2020). Short-term load forecast using ensemble neuro-fuzzy model. *Energy, 196*, 117127.

McClean, S. I. (2003). Data mining and knowledge discovery. In R. A. Meyers (Ed.), *Encyclopedia of physical science and technology* (3rd ed., pp. 229–246). Academic Press. https://doi.org/10.1016/B0-12-227410-5/00845-0

Mirala, A., & Sarraf Shirazi, R. (2017). Detection of surface cracks in metals using time-domain microwave nondestructive testing technique. *IET Microwaves, Antennas & Propagation, 11*(4), 564–569.

Moradzadeh, A., Sadeghian, O., Pourhossein, K., Mohammadi-Ivatloo, B., & Anvari-Moghaddam, A. (2020). Improving residential load disaggregation for sustainable development of energy via principal component analysis. *Sustainability, 12*(8), 3158.

Nazir, M. S., Alturise, F., Alshmrany, S., Nazir, H., Bilal, M., Abdalla, A. N., Sanjeevikumar, P., & M Ali, Z. (2020). Wind generation forecasting methods and proliferation of artificial neural network: A review of five years research trend. *Sustainability, 12*(9), 3778.

Nirgude, P. M., Ashokraju, D., Rajkumar, A. D., & Singh, B. P. (2008). Application of numerical evaluation techniques for interpreting frequency response measurements in power transformers. *IET Science, Measurement & Technology, 2*(5), 275–285.

Pozar, D. M. (2011). *Microwave engineering*. John wiley & sons.

Rahbarimagham, H., Porzani, H. K., Hejazi, M. S. A., Naderi, M. S., & Gharehpetian, G. B. (2015). Determination of transformer winding radial deformation using UWB system and hyperboloid method. *IEEE Sensors Journal, 15*(8), 4194–4202.

Ramkumar, K. B., Rajkumar, P. K., Ahmmad, S. N., & Jegan, M. (2020). A review on performance of self-compacting concrete—use of mineral admixtures and steel fibres with artificial neural network application. *Construction and Building Materials, 261*, 120215.

Sadiku, M. N. (2000). *Numerical techniques in electromagnetics*. CRC press.

Sanjari, M. J., Karami, H., & Gooi, H. B. (2016). Micro-generation dispatch in a smart residential multi-carrier energy system considering demand forecast error. *Energy Conversion and Management, 120*, 90–99. https://doi.org/10.1016/j.enconman.2016.04.092

Sanjari, M. J., Karami, H., & Gooi, H. B. (2017). Analytical rule-based approach to online optimal control of smart residential energy system. *IEEE Transactions on Industrial Informatics, 13*(4), 1586–1597. https://doi.org/10.1109/TII.2017.2651879

Tarimoradi, H., Karami, H., Gharehpetian, G. B., & Tenbohlen, S. (2021). Sensitivity analysis of different components of transfer function for detection and classification of type, location and extent of transformer faults. *Measurement, 187,* 110292.

Wan, C. (2019). Feature selection paradigms. In C. Wan (Ed.), *Hierarchical feature selection for knowledge discovery: Application of data mining to the biology of ageing* (pp. 17−23). Springer International Publishing. https://doi.org/10.1007/978-3-319-97919-9_3

Zebari, R., Abdulazeez, A., Zeebaree, D., Zebari, D., & Saeed, J. (2020). A comprehensive review of dimensionality reduction techniques for feature selection and feature extraction. *Journal of Applied Science and Technology Trends, 1*(2), 56−70.

Time domain analysis of EMWs in transformers

7

Hossein Karami[1], Gevork B. Gharehpetian[2] and Seyed-Alireza Ahmadi[3]

[1]*Niroo Research Institute (NRI), High Voltage Studies Research Department, Tehran, Iran;* [2]*Amirkabir University of Technology (AUT), Electrical Engineering Department, Tehran, Iran;* [3]*School of Electrical and Computer Engineering, College of Engineering, University of Tehran, Tehran, Iran*

7.1 Simulation in CST software

To analyze and simulate the electromagnetic waves in the transformer, computer simulation technology (CST) software is used (3D EXPERIENCE Company, 2022). The CST Studio Suite has the capability of three-dimensional simulation and calculating different characteristics of electromagnetic waves (Karami et al., 2016, 2019). Its high-performance packages for analyzing, optimizing, and designing electromagnetic components use a calculation of the Maxwell equation for discretized space of the understudy environment. Using this software, time domain analysis of traveling electromagnetic waves in transformers is investigated (Karami et al., 2018). The first step is modeling a transformer in CST software, as explained in the next.

7.1.1 Simulated model

In this section, the time domain simulation of the laboratory transformer winding models, described in Chapter 4 will be presented in the CST software environment. These models are chosen to compare the simulation and experimental results. As mentioned, the used model has the following items:

- Transmitter (Tx) and receiver (Rx) antennas
- Transformer winding model
- Signal propagation environment model

7.1.1.1 Transmitter and receiver antennas

As mentioned in Section 4.5.2, since in the mechanical defect monitoring of a transformer winding, a narrow pattern with a high gain is required to prevent the effects of the environment, Vivaldi antennas are used in the understudy setup. The frequency bandwidth of the used antennas is between 1 and 10 GHz, but in the proposed configuration, it is used in the range of 3.1–6.3 GHz (Mokhtari et al., 2010). The structure and dimensions of the used antenna are shown in Fig. 7.1. In this antenna, the sublayer is made of Ro-4003 and has length, width, and thickness of 92, 76, and 0.508 mm, respectively. Also, the other two used layers are made of copper, which have a thickness of 0.1 mm. Fig. 7.2 shows the S_{11} parameter of the antenna in terms of amplitude and phase.

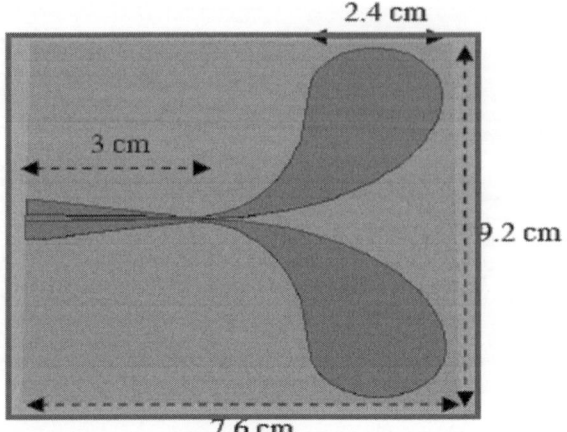

FIGURE 7.1

Dimensions of antenna.

7.1.1.2 Transformer winding model

The simulated winding is similar to the laboratory model because it is based on actual winding (Alehosseini et al., 2015), as described in Section 4.5.1, and therefore this leads to make the results of simulation comparable with the measurement study. Fig. 7.3 shows a schematic of the simulated design of the winding, which consists of five disks with the height of 2 cm and the diameter of 30 cm to model the transformer windings and five spacers to model the distance among the disks with the height of 0.44 cm and the diameter of 30 cm. Disk no. 5, as shown in Fig. 7.4, models the radial deformation of the winding with some sectors. This disk can be exchanged with other disks to model radial deformation at different winding heights. The specifications of different sectors are given in Table 7.1.

According to the measurement setup, shown in Fig. 7.5, for detecting axial displacement and radial deformation of the winding, this configuration is developed in the CST software (see Fig. 7.6).

As presented in Fig. 7.6 and 7.5, the winding model is located in the center of this setup, and the Tx and Rx antennas are located at a distance of 60 cm from the center of the winding.

7.1.2 Boundary conditions

Different types of boundaries are utilizable, each of which can be applied to the project's bounding box of any face (xmin/xmax/ymin/ymax/zmin/zmax). The description of them and their schematics are provided in Table 7.2.

The study of the transformer winding is carried out in two conditions: with and without considering the transformer tank. In the first case, the open space conditions are defined as boundary conditions. In the second one, the tank model is considered and $E_t = 0$, which is a model of the metal environment as boundary conditions.

7.1.3 Analysis of received signal

The frequency range of the ultra-wideband (UWB) transceiver is 3.1–6.3 GHz. Due to this frequency range, the Gaussian pulse shown in Fig. 7.7, is used as the excitation signal in the transmitter (Hejazi et al., 2011b).

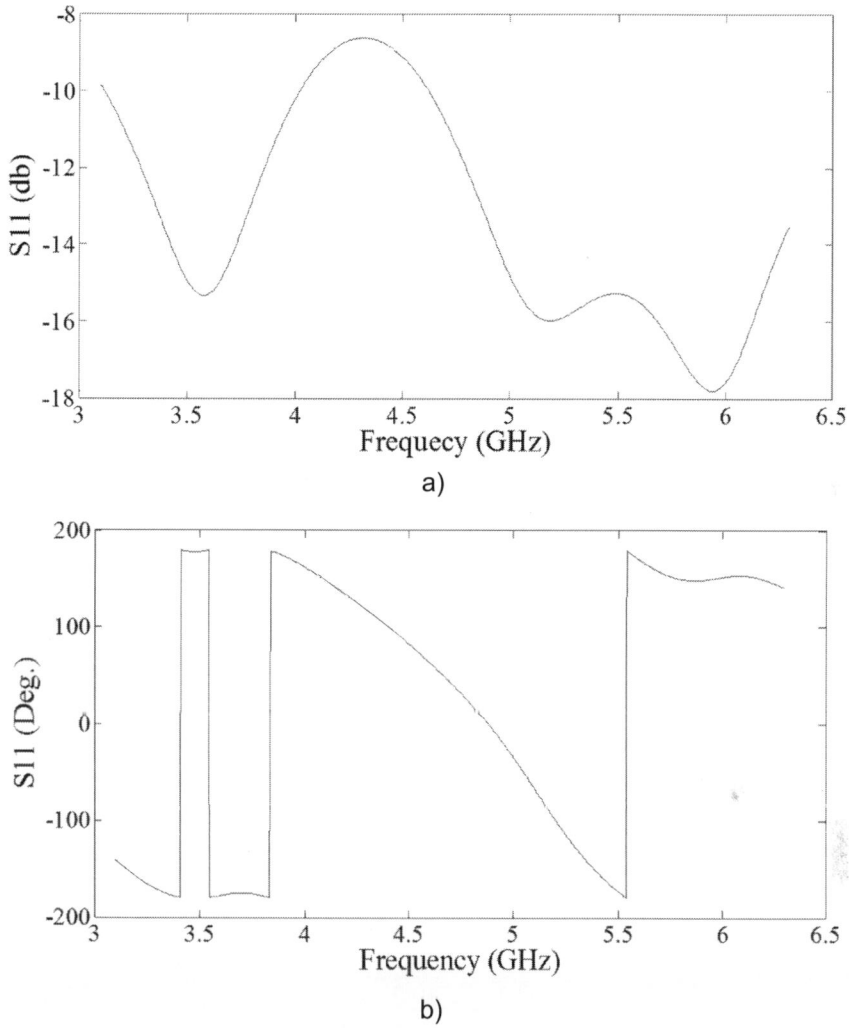

FIGURE 7.2

Amplitude and phase of S_{11} parameter for used vivaldi antenna.

The Tx antenna sends the signal and after passing the propagation environment and reflecting from the winding model, the Rx antenna receives it. Fig. 7.8 shows the simulated signal recorded by the Rx antenna in the healthy state. As can be seen, the received signal consists of three parts related to the different signal propagation paths shown in Fig. 7.9.

The first path, the red line in Fig. 7.9, includes the received signal from the direct path between two antennas, the line of sight, which travels less and reaches the receiving antenna sooner. The second signal path, the blue path in Fig. 7.9, creates the information that can be used to detect the winding

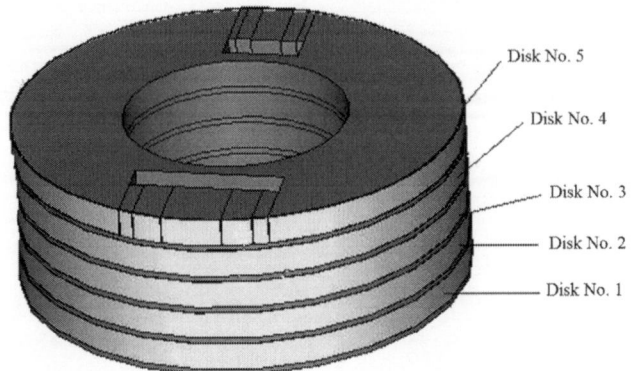

FIGURE 7.3

Simulated scheme of transformer winding model.

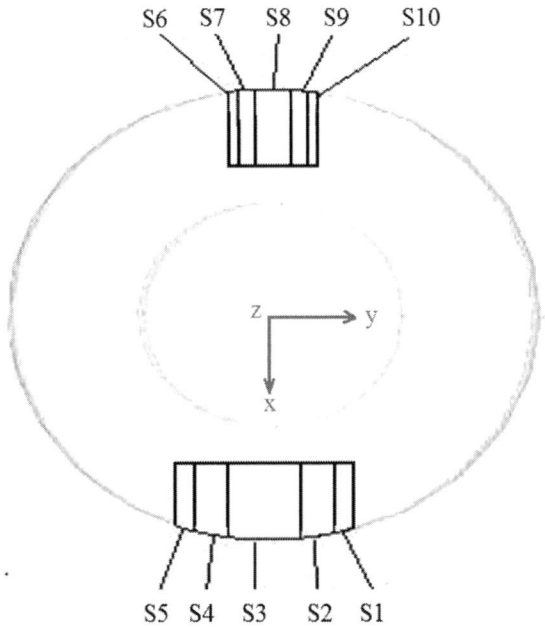

FIGURE 7.4

Designed disk to model radial deformations.

defect. The third signal path includes diffraction of the signal from the target and is reflected from other parts of the environment.

Since the received signal contains the information needed to detect mechanical faults, it is important to determine the arrival time of the desired signal part in the receiver. To do this, the distance

Table 7.1 Specifications of radial deformation disk segments.	
Segment	**Length**
S1	1 cm
S2	2 cm
S3	4 cm
S4	2 cm
S5	1 cm
S6	0.5 cm
S7	1 cm
S8	2 cm
S9	1 cm
S10	0.5 cm

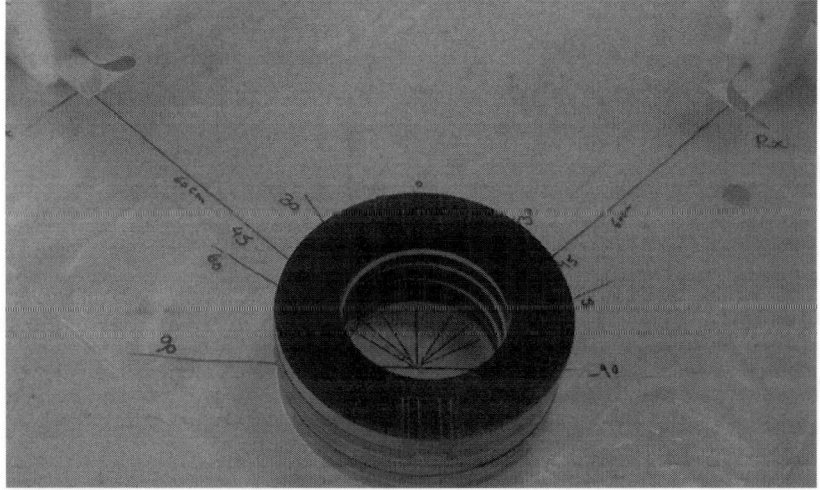

FIGURE 7.5

Antennas and winding model in measurement setup.

of the transformer winding model from the transmitter and receiver is calculated. Therefore the approximate arrival time of the desired multipath and the wave propagation speed in the environment are calculated. Determining a time window that encloses fault information and removing signals of other multipath such as direct path, is the most important part of the problem for this analysis. Assuming that the propagation velocity is equal to the velocity of light, i.e., c, the time for a wave from a path of length d to reach the receiver is calculated by Eq. (7.1).

$$\tau = \frac{d}{c} \qquad (7.1)$$

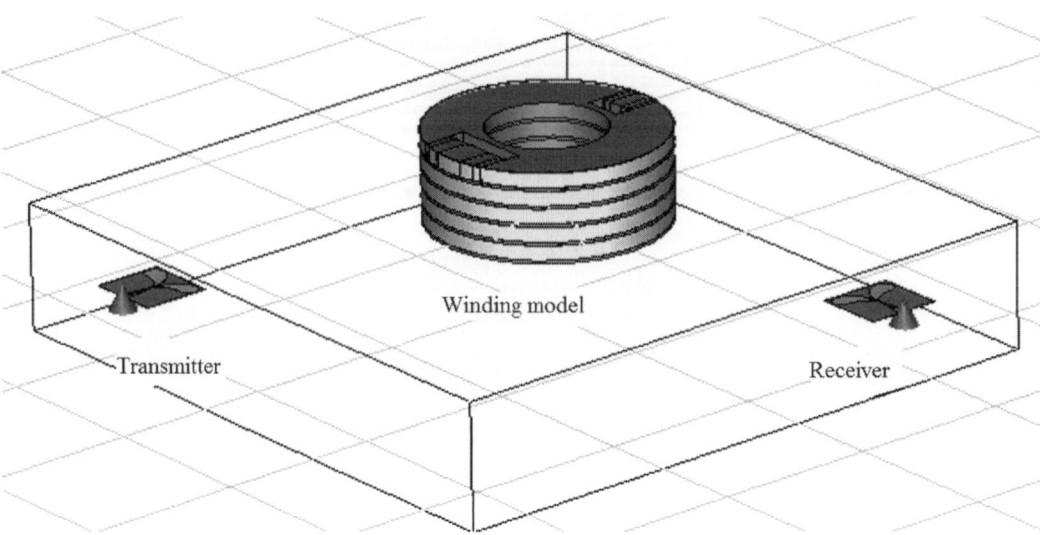

FIGURE 7.6

Modeling setup in simulation environment.

Table 7.2 Description of boundary conditions in CST software.	
Electric: Acts as a complete electric conductor. Whole regular magnetic fluxes and tangential electric fields are set to zero.	
Magnetic: Acts as a complete magnetic conductor. Whole regular electric fluxes and tangential magnetic fields are set to zero.	
Open (PML): Acts as free space. This border can be crossed by the waves via the least of reflections. It should be noted, in the case of a unit cell simulation with the general purpose frequency domain solver, open boundaries are came off by a Floquet port.	

Table 7.2 Description of boundary conditions in CST software.—cont'd

Open (add space): Like the open (PML), but combines some additional space for far-field calculation. For the antenna problems, this act is suggested.	
Periodic: Links two contradictory boundaries with an ascertainable phase shift in such a way that, in the matching direction, the computation domain is simulated to be periodically expanded. Therefore altering one boundary to intermittent always alters the contradictory boundary to intermittent as well. A phase shift may be entered for the mentioned boundary in the phase shift/scan angle property page by choosing an intermittent condition. By simulating antenna models with intermittent boundaries, examining the near-field effect among the individual antennas is enabled. However, an infinitely extended pattern for the antenna is shown for the resulting structure. Consequently, the related results of far-field have the highest accuracy along with the explanation of an antenna array (ideally large) in the postprocessing of far-field.	

Continued

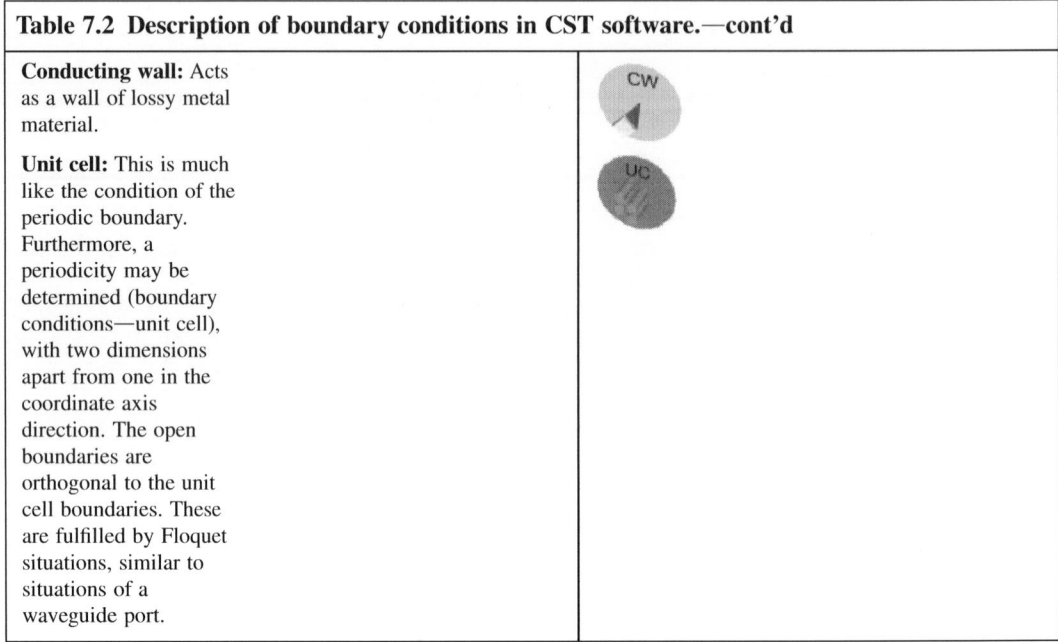

Table 7.2 Description of boundary conditions in CST software.—cont'd	
Conducting wall: Acts as a wall of lossy metal material.	
Unit cell: This is much like the condition of the periodic boundary. Furthermore, a periodicity may be determined (boundary conditions—unit cell), with two dimensions apart from one in the coordinate axis direction. The open boundaries are orthogonal to the unit cell boundaries. These are fulfilled by Floquet situations, similar to situations of a waveguide port.	

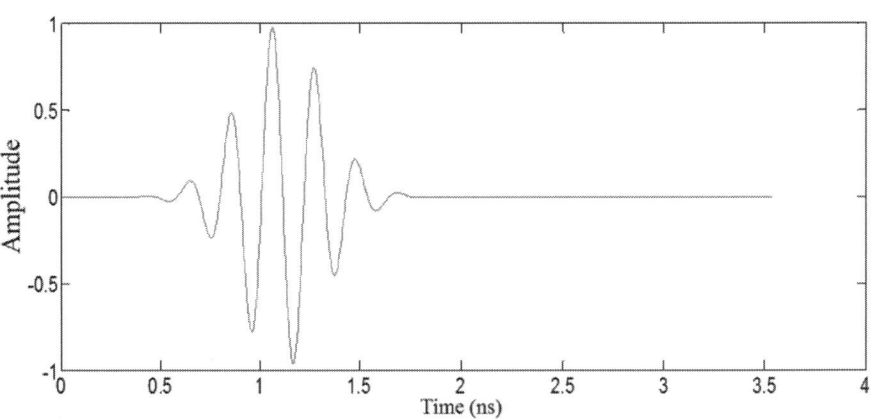

FIGURE 7.7

Gaussian pulse signal of transmitter.

According to Fig. 7.9, if the blue distance is d_W and the line of sight from the Tx to the Rx antennas is d_{AA}, given that the first received multipath by the receiver corresponds to the direct path between two antennas, the difference in arrival time of the multipath related to the winding model can be calculated using the difference between the direct path and the path related to the winding model, which is the difference of blue and red lines in Fig. 7.9. According to the configuration introduced in Chapter 3, Δd and its related time are obtained according to Eqs. (7.2) and (7.3), respectively.

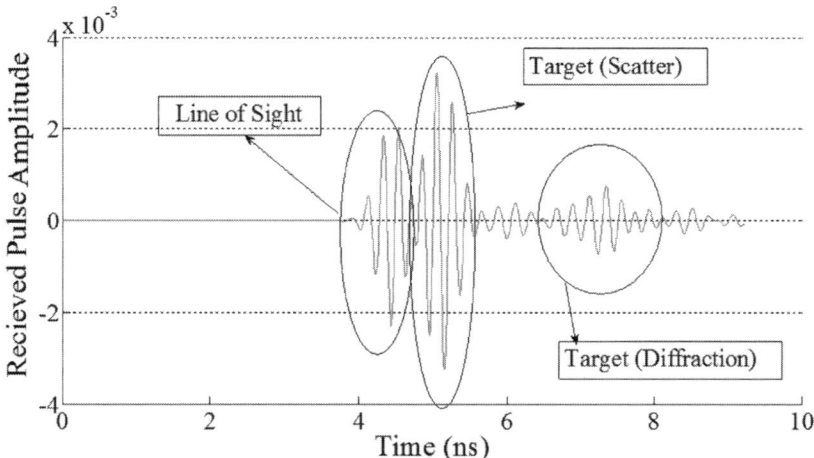

FIGURE 7.8

Simulated signal recorded by Rx antenna in healthy state.

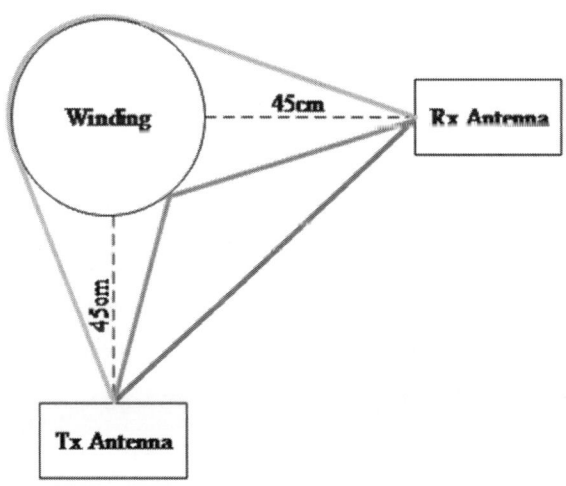

FIGURE 7.9

Multipaths from transmitter to receiver.

$$\Delta d = d_W - d_{AA} \tag{7.2}$$

$$\Delta \tau = \frac{\Delta d}{c} = \frac{d_W - d_{AA}}{c} \tag{7.3}$$

It means that after $\Delta \tau$ from receiving the first signal related to the line of sight, the waveform related to the desired path can be recorded. For the understudy setup, d_W and d_{AA} are 106.08 and

84.85 cm, respectively. Therefore the distance and related time difference are 21.21 cm and 0.7076 ns, respectively, calculated by Eqs. (7.4) and (7.5).

$$\Delta d = d_W - d_{AA} = 106.08 - 84.85 = 21.23 cm \tag{7.4}$$

$$\Delta \tau = \frac{0.2123}{3 \times 10^8} = 0.7076 ns \tag{7.5}$$

The third signal path (orange path) is related to the creeping wave around the winding due to the upper and lower edges causing diffraction and propagation of waves around the winding. To determine the time lag of this part of the signal, the related path difference should be determined. The third path is approximately 160.68 cm and so, we have a 75.83 cm path difference that leads to 2.5276 ns as calculated below.

$$\Delta d = 160.68 - 84.85 = 75.83 cm \tag{7.6}$$

$$\Delta \tau = \frac{0.7583}{3 \times 10^8} = 2.5276 ns \tag{7.7}$$

According to the calculated time differences, different parts of the received signal can be seen in Fig. 7.8.

7.1.4 Axial displacement detection

The healthy state signal is considered the reference signal. Then, different axial displacement states are created in the model and for each one, the received signal in the Rx antenna is stored for comparison with the reference signal. The automated external defibrillator (AED) of the received signals (Hejazi et al., 2011a), defined in the previous chapter, was measured for 90 displacement states from −4.5 cm to +4.5 cm with 1 mm steps relative to the reference position. It is assumed that the winding model has positive and negative displacements, upward, and downward, from the central position selected as the reference. The increase in this index is proportional to the increase in axial displacement and can be used to detect this displacement. However, this index cannot differentiate upward or downward displacement detection because it is almost an even function of the winding displacement.

7.1.5 Radial deformation detection

To model the winding radial deformation, the embedded components, described in 7.1.1.2, move radially outward while the location of the transmitter and receiver antennas and the disks is fixed. The transmitted signal, after passing through the propagation environment and reflection from the winding model, is recorded by the receiving antenna. The healthy state signal is used as the reference signal. Different radial deformations are applied to the model, and for each one, the received signal is stored to compare with the reference signal. Table 7.3 lists the specifications of the simulations, including the extent and segment of different deformations. The position of the deformed disk is exchanged with the positions of disks 1, 3, and 5 shown in Fig. 7.3.

Fig. 7.10 compares the received signals in the faulty states on disk 1 with the healthy state. As can be found out, the changes are only in a specific period of the received signal and not in the whole. Fig. 7.11 shows the AED index variations for different extents and segments of radial deformation on disk 5.

Table 7.3 Radial deformation specifications on selected disks in simulation.

Segment number	Extent value (mm)
S8	20− .15− .10− .5− .5+ .10+ .15+ .20+
S9	20− .15− .10− .5− .5+ .10+ .15+ .20+
S10	20− .15− .10− .5− .5+ .10+ .15+ .20+

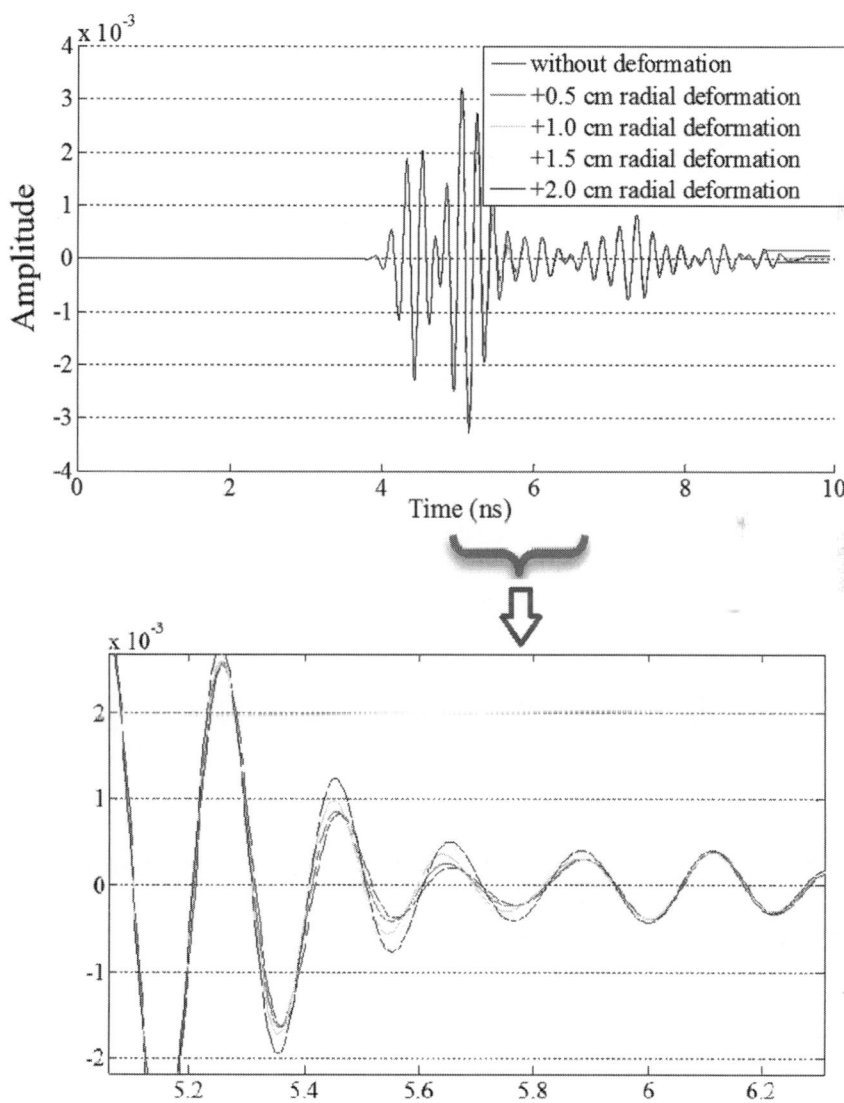

FIGURE 7.10

Received signals in the healthy and faulty states on disk 1.

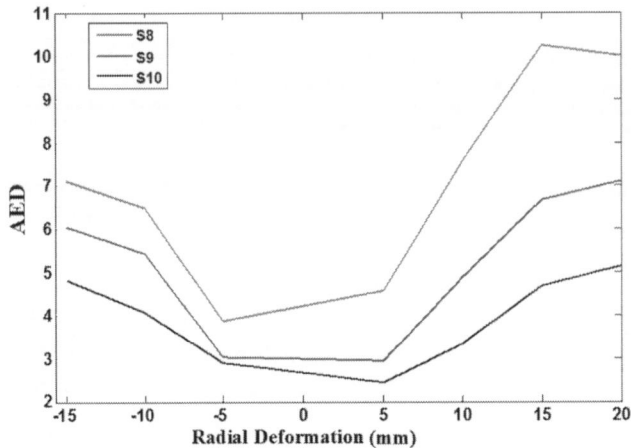

FIGURE 7.11

AED variations in terms of radial deformation on disk 5.

It is assumed that the segments compared with the healthy state, have positive/negative displacements (forward/backward). As shown, the increase in radial deformation is proportional to the increase in the index. Therefore the AED can be used as an indicator to detect radial deformation and estimate the extent of deformation. It can be seen that the proposed configuration is more sensitive to bulgy deformation than the concave one. Segment S_8 has higher variations because it has the biggest surface facing transmitted signal, and so, causes more changes in the recorded reflected signal. Similarly, this theory is also established in the case of segment S_9, whose radial deformation causes more surface area change than S_{10}.

Fig. 7.12 shows the AED variations for the radial deformation at different places, which are the top, middle, and bottom of the model. This variation is due to moving segment S_8, selected because of its

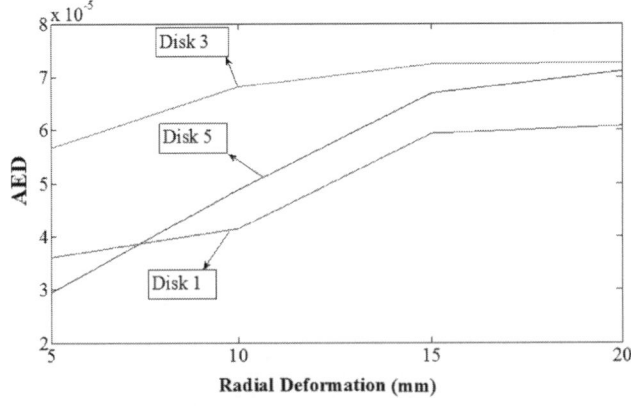

FIGURE 7.12

AED index variations with respect to radial deformation of S_8 segment on different places.

high effects on AED variations compared with other segments, as shown in Fig. 7.11. It can be concluded that radial deformation at the middle of the winding causes more variations in the AED index because disk 3 is located in the main lobe of the transmitter antenna, so its changes have a greater effect on the signal and cause more MED.

7.2 Experimental study
7.2.1 Procedure description
All the steps expressed in the previous subsection can be implemented in an experimental study. As mentioned, the obtained signals in the faulty cases of the transformer winding model are compared with the reference signal. A part of the wave that corresponds to the return waves from the transformer can be separated. Due to the constancy of other components in the experiment setup except the radial deformation, obtained waves from different faults, which are W'_2, W'_3 and, W'_n, are compared with W'_1, which is obtained wave from the healthy state using mean absolute distance criterion (MADC) calculated by Eq. (7.8).

$$C = \sum_{j=1}^{n} \lambda_j \times S \times t_{replacement} \times C_1 \times \left(\frac{P}{F}, i\%, n \right) \tag{7.8}$$

where, T_1 is the beginning of the time window and T_2 is its end. T_S is the sampling rate and N is the number of samples received in the window defined by Eq. (7.9).

$$N = \frac{T_2 - T_1}{T_s} \tag{7.9}$$

7.2.2 Axial displacement
The specifications of the laboratory setup, described in 4.5.1 and Fig. 4.16, are given in Table 7.4. Fig. 7.13 shows the received signal in different axial displacement states of the transformer winding model.

Table 7.5 shows the calculated MADC values for different axial displacement states. It seems that there is no straightforward relation between axial displacement value and MADC. So, the extent of the axial displacement cannot be detected, but this index can be used to detect the axial displacement existence.

7.2.3 Radial deformation
According to Fig. 4.16, the parameters considered for the laboratory setup are given in Table 7.6.

Table 7.4 Setup parameters of axial displacement experimental study.

Parameter	T_I	T_s	d_1	d_2	d_3	d_4
Value	100 ms	31.79 ns	130 cm	80 cm	103 cm	100 cm

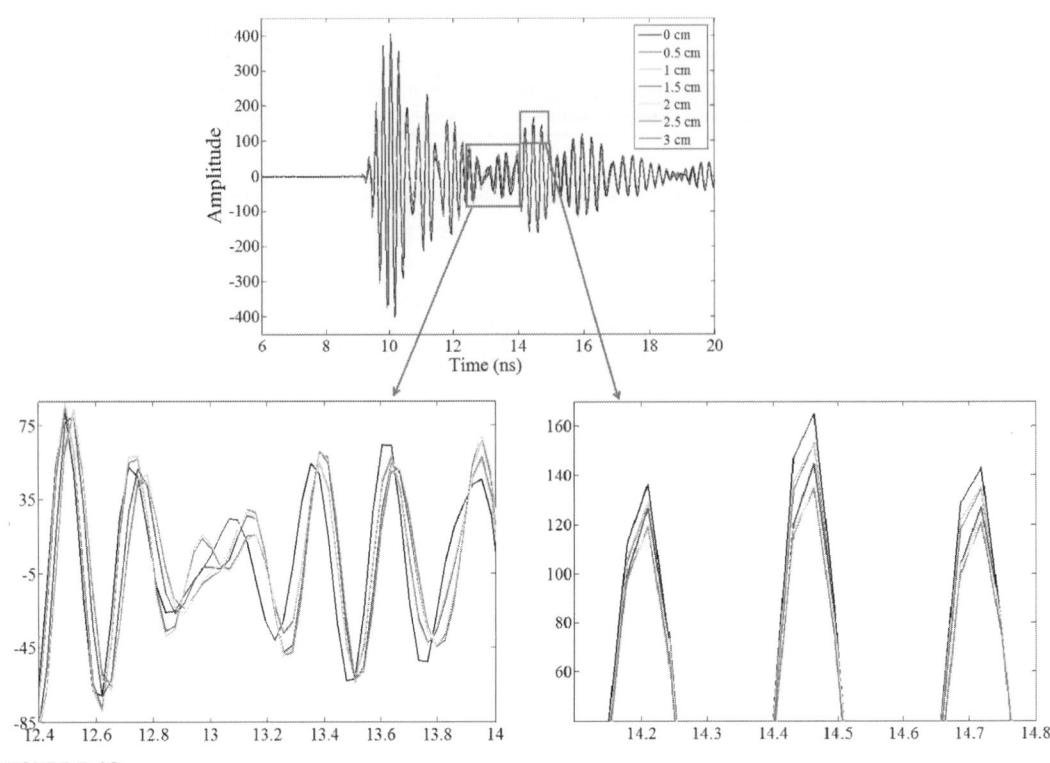

FIGURE 7.13

Received signals for different axial displacements.

Table 7.5 Calculated MADC values for different axial displacements.

Axial displacement value (cm)	3	2.5	2	1.5	1	0.5
MADC	1.89	1.69	1.54	1.82	1.33	1.41

Table 7.6 Experimental setup parameters of radial deformation study.

Parameter	T_I	T_s	d_1	d_2	d_3	d_4
Value	100 ms	31.79 ns	150 cm	100 cm	125 cm	100 cm

In this experiment, the segmented disk is in position two and the segment with 2 cm height and width is moved outward from 1 to 4 cm. Fig. 7.14 shows the received signals for radial deformations compared with the healthy state.

As expressed in Section 4 and shown in Fig. 7.15A, there are three segments in model no. 1 with the widths of 0.5, 1, and 2 cm. The height of all segments is 2 cm. Therefore we can have radial

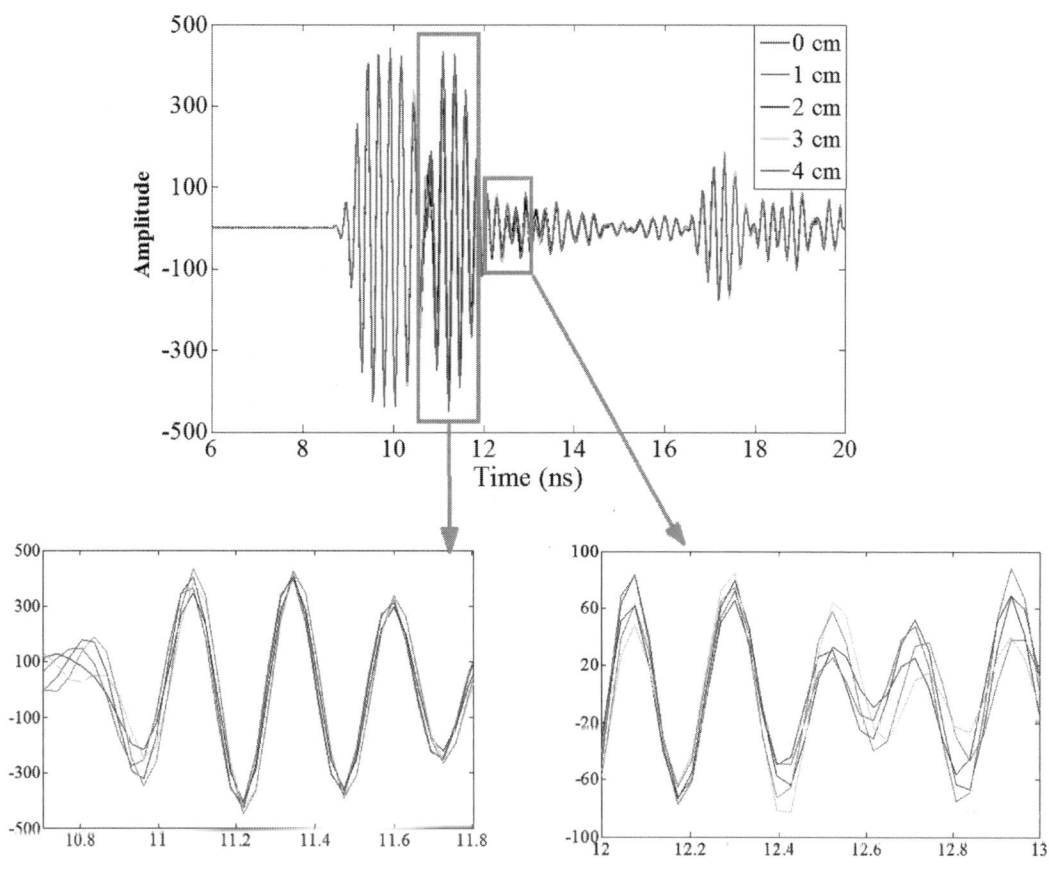

FIGURE 7.14

Received signals for different radial deformations.

deformation with areas of 1, 2, and 4 cm^2. Table 7.7 present the calculated MADC values for the radial deformations with different areas and bulgy sizes. The results in Table 7.5 show that the MADC variation for axial displacement is between 1.3 and 1.9. But according to Table 7.7, the variation of MADC for radial deformation is between 0.2 and 0.8. Therefore the value of the MADC can be used as a measure for the separation of axial displacement and radial deformation faults.

More investigations of radial deformation can be done by different combinations of segments 1, 2, and 3, shown in Fig. 7.15B, and C. The dimensions details of the segments are listed in Tables 7.7 and 7.8.

The radial bulgy size is changed from zero to 40 mm with 2 mm steps. It means that we have 120 different states of radial deformation with different volumes, as listed in Tables 7.7 and 7.8. The MADC is calculated for different states arranged in ascending order, and the results for the first and second windows, shown in Fig. 7.14, are presented in Fig. 7.16.

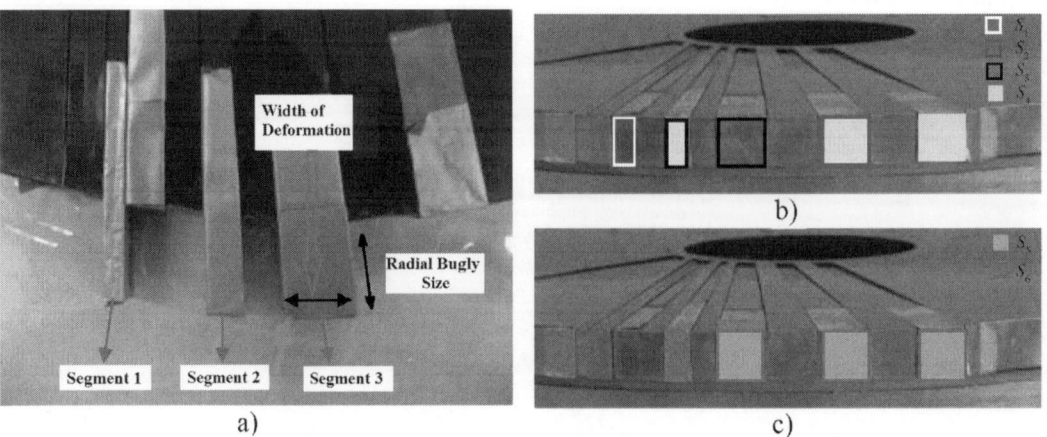

FIGURE 7.15

(A) configuration of radial deformation segments in model no. 1, (B) defined deformations for S_1 to S_4, and (C) defined deformations for S_5 and S_6.

Table 7.7 MADC value for different radial deformations.		
Radially bulgy size (cm)	**Displaced area (cm²)**	**MADC value**
1	4	0.3371
2	4	0.4396
3	4	0.4002
4	4	0.4605
1	2	0.3060
2	2	0.6137
3	2	0.4497
4	2	0.1940
1	1	0.4069
2	1	0.8008
3	1	0.2030
4	1	0.2901

Fitted curves for the first and second windows are calculated as follows, which can be used to estimate the radial deformation volume:

$$\text{First window}: f(x) = 4.239 \times 10^{-6}x^5 - 0.0005416x^4 + 0.02353x^3 - 0.3995x^2 + 3.584x + 6.43$$

$$\text{Second window}: f(x) = 9.006 \times 10^{-6}x^4 - 0.001053x^3 + 0.03569x^2 + 0.06971x + 4.293$$

Table 7.8 Dimensions of different segments combinations.

Segment combination	Height (cm) S width (cm)	Area (cm²)
S_1	(1) $t_{replacement}$ 2	2
S_2	(2) C_1 2	4
S_3	(1) $\left(\frac{P}{F}, i\%, n\right) 2 + 2C = \sum_{j=1}^{n} \lambda_j \times S \times t_{replacement} \times (C_1 \times (1+k)^n) \times \left(\frac{P}{F}, i\%, n\right) 2$	6
S_4	(2) $2 + 2\tau = \frac{d}{c} 2 + 1\ \Delta d = d_w - d_{AA}$ 2	10
S_5	(2) $\Delta\tau = \frac{\Delta d}{c} = \frac{d_w - d_{AA}}{c} 2 + 2 \frac{A(z^{-1})}{B(z^{-1})} 2 + 2x_t$ 2	12
S_6	(2) $y_t\ 2 + 2y_t\ 2 + 2\frac{C(z^{-1})}{D(z^{-1})} 2 + 1v_t$ 2	14

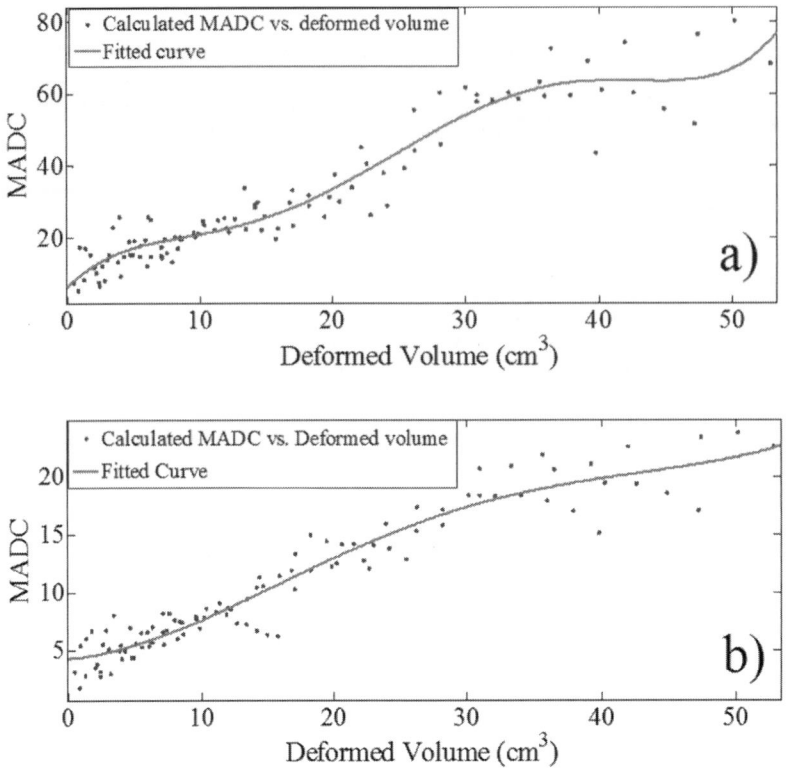

FIGURE 7.16

MADC calculated for different deformation volumes (cm³), (A) first window, and (B) second window.

7.2.4 Comparison of simulation results and practical experiments

Fig. 7.17 shows the received signal in the experiment study at the receiver in the healthy state (without any mechanical changes). As shown in this figure, the signal consists of three distinct parts: the line of sight, the scattered signal by the winding, and the multipath signal of the environment.

The healthy state signal in the simulation environment includes the line of sight, the scattered signal by the winding, and the creeping signal around the winding, as shown in Fig. 7.8. As can be seen, the signals of the simulation and measurement environments are different in the third part of the signal. In justifying these differences, the following descriptions can be enumerated:

1. In the measurement environment, the effect of the surrounding environment, which creates a multipath, cannot be eliminated. Therefore the third part of the measurement signal includes the multi-paths signal of the surrounding environment. However, in the simulation configuration, this environment is not modeled, so the multi-paths signal of the surrounding environment is not recorded in the Rx antenna.

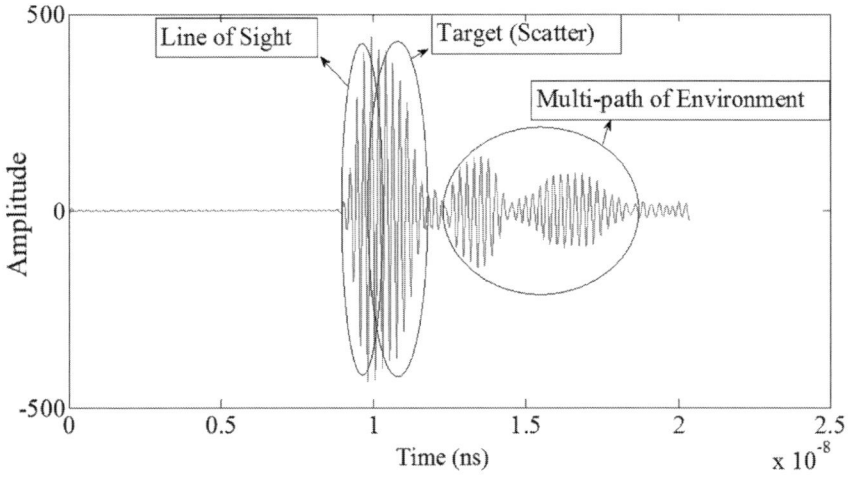

FIGURE 7.17

Received signal at the receiver in healthy state of experiment study.

2. In the simulation environment, the third part of the signal is due to the rotation of the wave around the winding. This signal is severely attenuated in the measuring medium and is not received in the Rx antenna.

To make the comparison possible, the signal sent in the actual transmitter and the signal used in the software must first be the same. This signal was first generated in the software using the device datasheet. After several experiments with the transceiver in the measuring medium, it was found that the frequency range for the device does not match the datasheet. Two experiments are performed to determine the characteristics of the signal sent by the transmitter. At first, as shown in Fig. 7.18, the transmitter was connected directly to the receiver (back-to-back) using a 30 dB attenuator to receive the transmitted signal with attenuation at the receiver.

Fig. 7.19 shows the signal recorded by the receiver. As it turns out, there is a huge difference between the signal in the range of 3.1–6.4 GHz, shown in Fig. 7.7, with this signal. Therefore it was found that the device frequency range is different from the value defined in the device datasheet.

In the second experiment, to obtain the transmitter's frequency range, the transmitter is directly connected to a spectrum analyzer to determine the frequency range of the signal. Figs. 7.20–7.22 shows the output of the spectrum analyzer. According to 10 dB frequency, the frequency range of the transmitted signal is between 3.4 and 5.2 GHz. Fig. 7.23 is created for the transmitted signal in the

FIGURE 7.18

Back-to-back experiment.

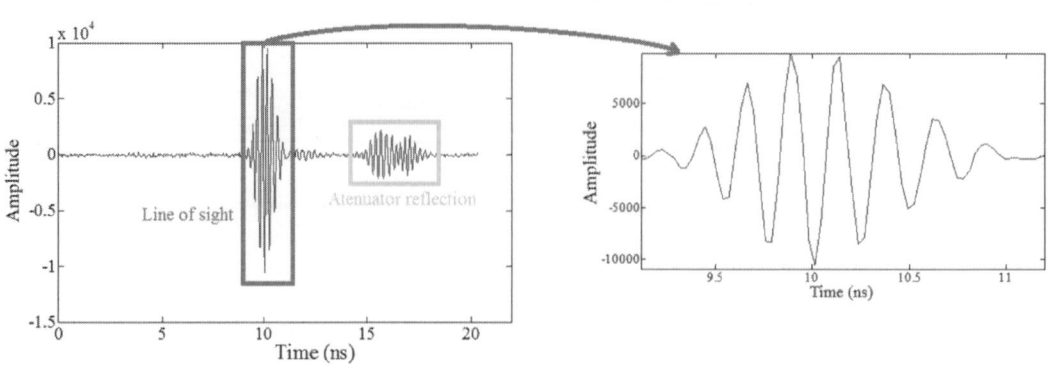

FIGURE 7.19

Received signal in back-to-back mode.

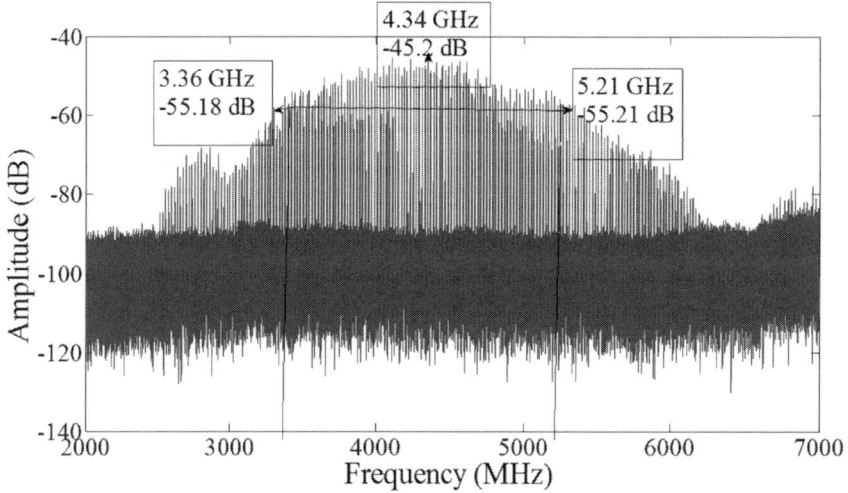

FIGURE 7.20

Output of spectrum analyzer for transmitter signal.

simulation environment by defining this frequency range in the CST software environment. This signal shape is similar to the transmitted signal in the measurement.

As the transmitted signal is changed, simulation in the CST software should be repeated again. Given that the experiments are performed on a wooden surface, this effect should also be considered in the simulation configuration. To make the results comparable with measurement, the following steps must be applied to the recorded signals:

Step 1: After receiving the simulated and measured signal in each faulty state, these signals are normalized based on their maximum.

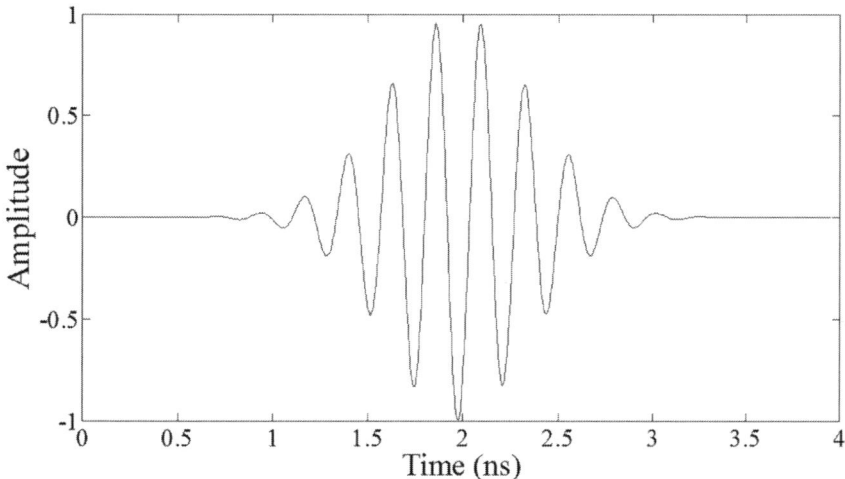

FIGURE 7.21

Transmitter signal defined in CST software with frequency range of 3.4—5.2 GHz.

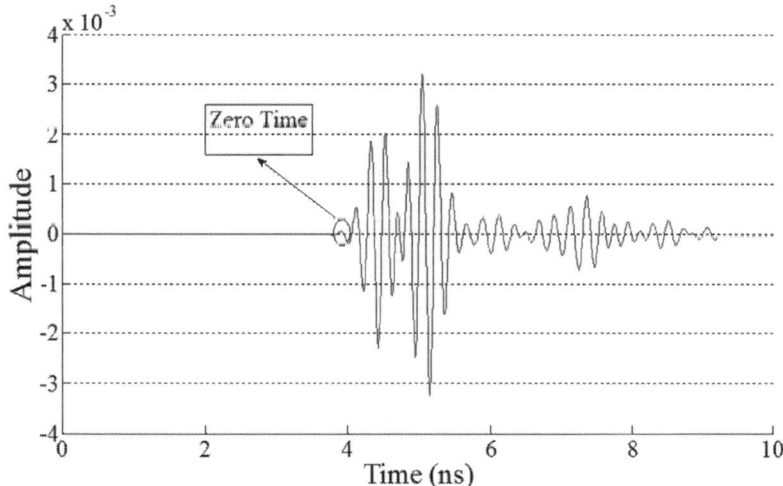

FIGURE 7.22

Zero time criterion selection for synchronizing signals.

Step 2: As the zero time of signals is important for comparison, we consider the first peak in signals of simulation and measurement results as the start time for comparing these shapes, as shown in Fig. 7.22.

Step 3: In the simulation environment, other signal reception paths are not modeled in the receiver. The multipath signals of the surrounding environment in the receiver are filtered in the measurement signal according to their arrival time.

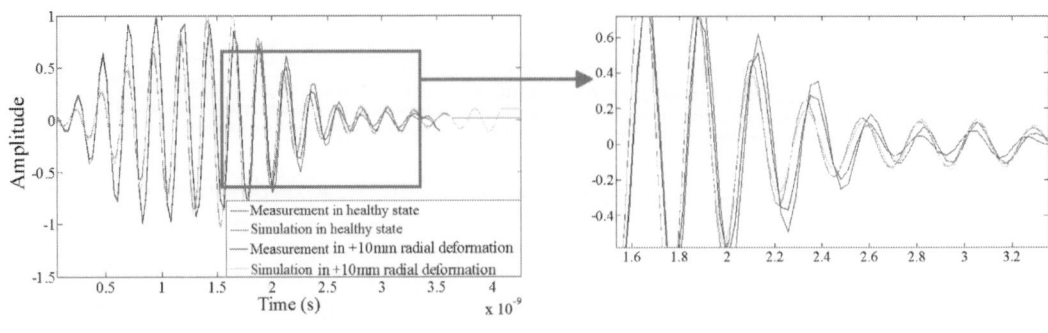

FIGURE 7.23

Comparison of received signal in measurement and simulation after simulation modifications.

By applying the above steps, the results are ready for comparison. Fig. 7.23 shows the measured and simulated signals for $+10$ mm deformation of segment S_8, shown in Fig. 7.4, compared with the healthy state. As can be seen, there is an acceptable similarity between the simulated signal and the measured one.

7.2.5 Sensitivity analysis

Given the proposed configuration in this chapter, it should also be considered whether all possible faults for the winding are within sight of the transceiver system. In the case of axial displacement, this does not pose a problem for the proposed configuration because any axial displacement is in line with the sight of the antennas and can be detected by the proposed approach. Some additional simulations are carried out to investigate the proposed configuration capability regarding the radial deformation.

As shown in Fig. 7.24, the transformer winding is rotated by 40 degree steps, and in each step, segment S_8 is moved from 0 to 20 mm with 5 mm steps. This will change the location of the defective

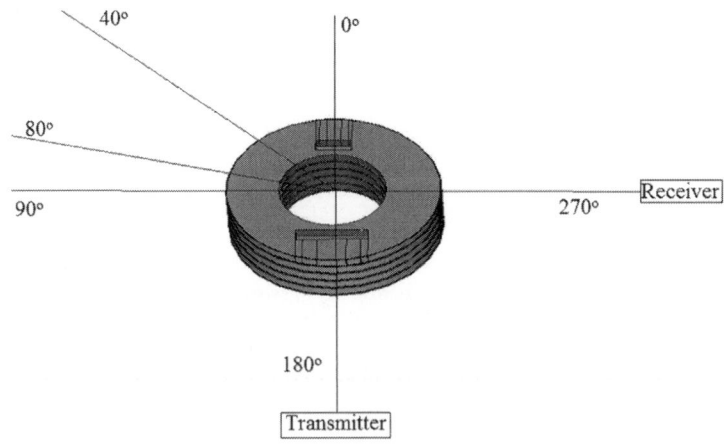

FIGURE 7.24

Direction of winding rotation.

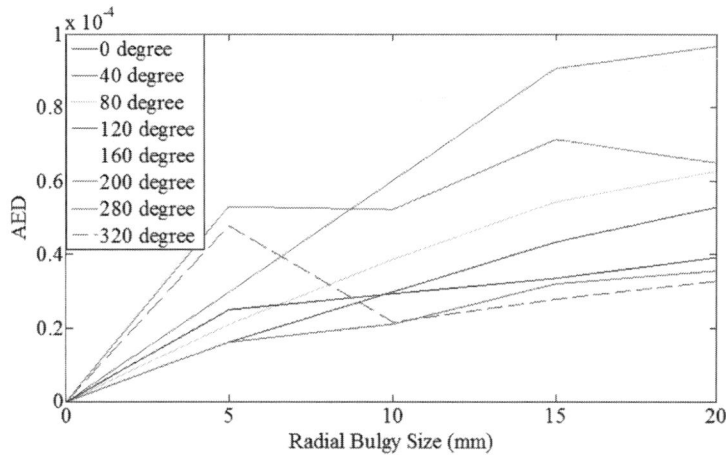

FIGURE 7.25

AED variations for different angles and deformation sizes.

part around the surrounding of the winding to see whether these faults can be detected by the antennas. It should be noted that only the winding rotates, and the location of the antennas is fixed.

Fig. 7.25 shows the AED variations for different angles in terms of deformation values. As shown in the figure at 0, 40, 80, 120, and 320 degrees, the AED value is lower than its values at 160, 200, 240, and 280 degrees. Because the third part of the signal (creeping part) has changed and so, it has a different amplitude and effect in comparison with the healthy state.

7.2.6 Tank effect

As the metallic tank is a part of a transformer, investigating its effect on the proposed approach is necessary. Due to the metallic nature of this tank, its effects on the recorded signal at the Rx antenna are predictable. Different situations of antennas and winding with a metallic wall should be analyzed. Then, their aggregation can introduce the effects of the tank on the results of the proposed monitoring approach.

7.2.6.1 Metallic wall near Tx antenna

In the first step in the simulation environment, a metallic wall is placed on the transmitter side by determining $E_t = 0$ in the boundary conditions of the CST software. Fig. 7.26 shows the location of the wall in the configuration and Fig. 7.27 shows the signal recorded by the Rx antenna in the healthy winding state. As it is specified, the received signal consists of four parts related to four signal paths.

The following signal paths in the understudy configuration, shown in Fig. 7.28, affect the recorded signal:

- Path 1, shown by the red line in Fig. 7.28, is related to the line of sight, the direct path between Tx and Rx antennas.

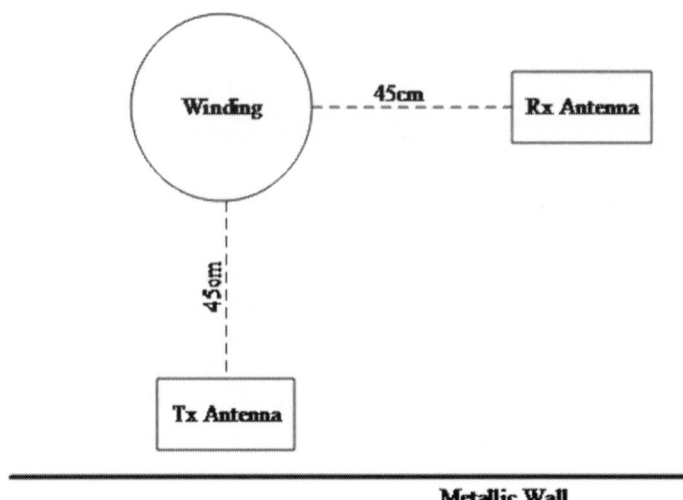

Metallic wall near Tx antenna.

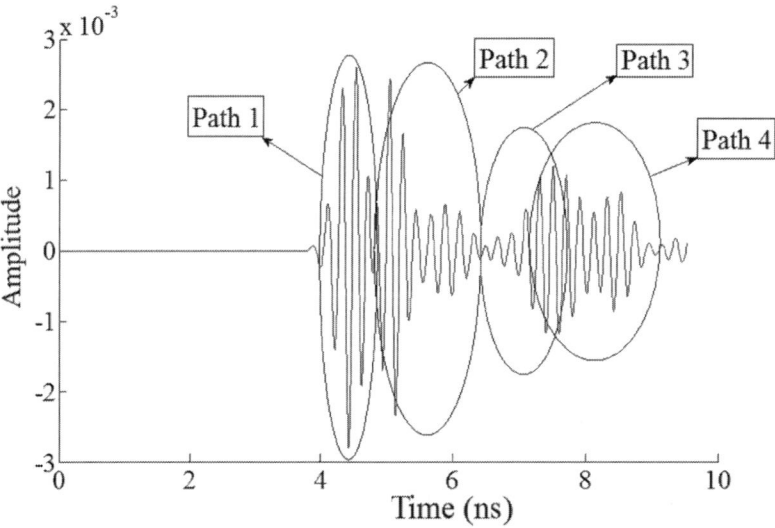

Recorded signal for metallic wall near Tx antenna.

- Path 2, shown by the blue line in Fig. 7.28, is related to the desired path between antennas and the winding model.
- Path 3, shown by the yellow line in Fig. 7.28, is related to the creeping wave around the winding.

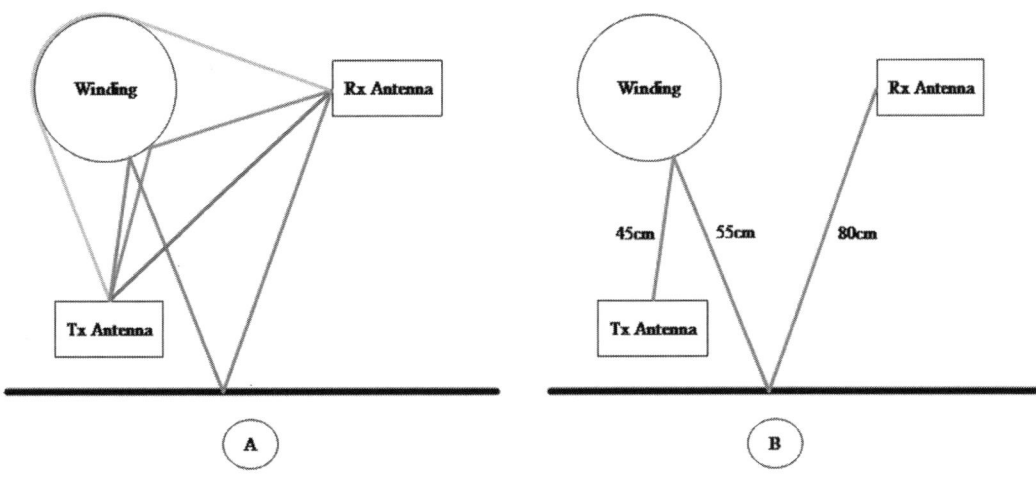

FIGURE 7.28

(A) Signal paths for wall near Tx antenna, and (B) distance values of fourth path.

- Path 4, shown by the green line in Fig. 7.28, is related to the multipath of the receiver from the winding and the metallic wall. It can be said that the transmitter side causes the signal to reach the receiver with more reflections.

The time-domain analysis of paths one to three has been described before. The analysis of all parts of the signal in this configuration can be finalized by calculating the time difference of path four regarding the main path, as follows:

$$\Delta t = \frac{\Delta d}{c} = \frac{1.80 \quad 0.84}{3 \times 10^8} = \frac{0.96}{3 \times 10^8} = 3.2 ns \tag{7.10}$$

7.2.6.2 Metallic wall near Rx antenna

If the metallic wall is placed on the receiver side, the received signal is similar to Fig. 7.27.

In justifying this phenomenon, according to the symmetric conditions drawn in Fig. 7.29, the signal travels the same distance for the fourth path in two setups. In this figure, the signal traveling path for the metallic wall near Tx and Rx antennas are shown with solid and dotted green lines, respectively. The symmetric axis is shown by the red dashed line.

7.2.6.3 Metallic wall near winding sides

We place the metallic walls on the remaining two sides of the setup. In Fig. 7.30, the setups and the fourth signal paths are plotted. It can be seen that these two setups have a symmetric axis similar to the previous and so, the same received signals are similar. The analysis of the arrival times of paths is the same as before, so we avoid repeating it. However, the fourth signal path, the green line in Fig. 7.30, is shorter in these two cases than the previous ones, so the related signal should reach the receiver sooner. In addition, the distance of the third and fourth signal paths are approximately near,

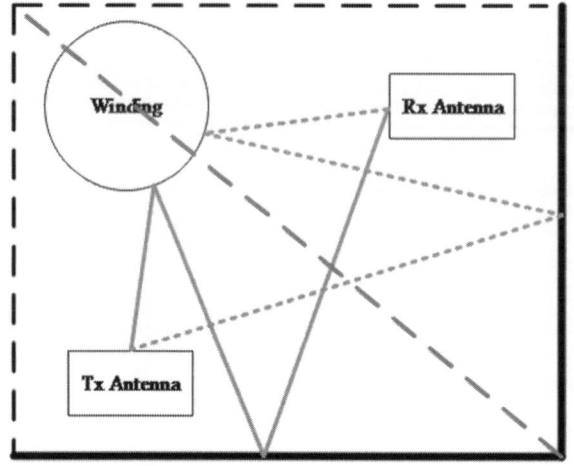

FIGURE 7.29

Symmetric conditions for metallic wall near Tx and Rx antennas.

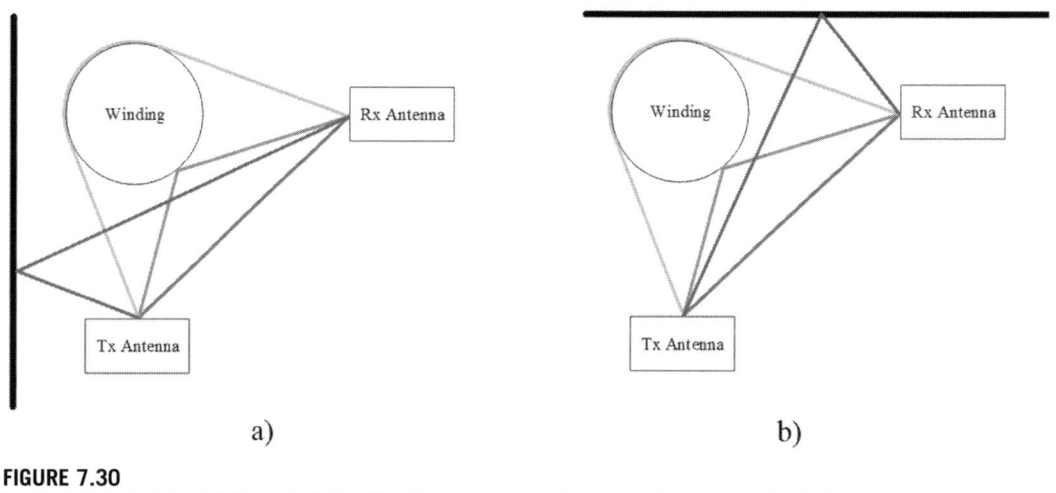

a)   b)

FIGURE 7.30

Setups of metallic wall near winding.

and so the received signals of these paths are combined more. Fig. 7.31 shows the signal in the receiver for these two steps.

7.2.6.4 Surrounded by metallic tank walls
The previous subsections show that the effect of the walls is receiving more reflections of the transmitted signal due to the signal multipath. Considering all the walls together leads to receiving a total of signal reflections of winding and different walls with different time delays, which their separation is

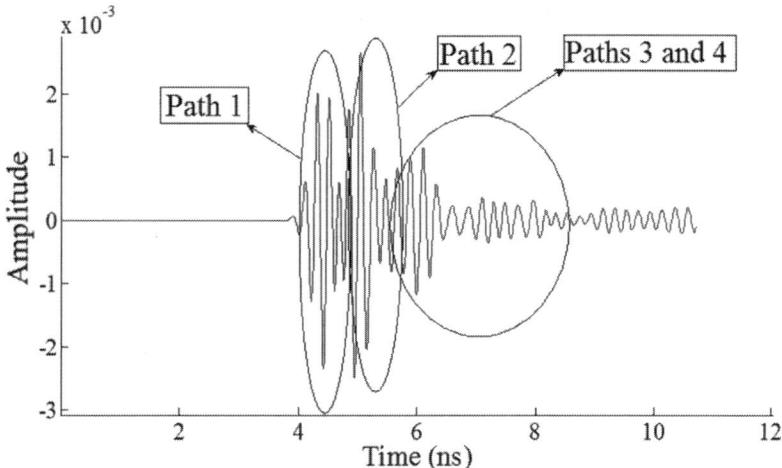

FIGURE 7.31

Received signal for metallic wall near winding.

difficult. According to Fig. 7.32, the received signal in the Rx antenna is due to the following signal paths:

- Line of sight between two antennas, shown by the red line.
- Reflection from the winding model, shown by the blue line.
- Creeping wave of winding, shown by the yellow line.
- Reflection from the tank wall near the antennas, shown by the green solid and dotted lines.
- Reflection from the tank wall near the winding model, shown by the violet solid and dotted lines.

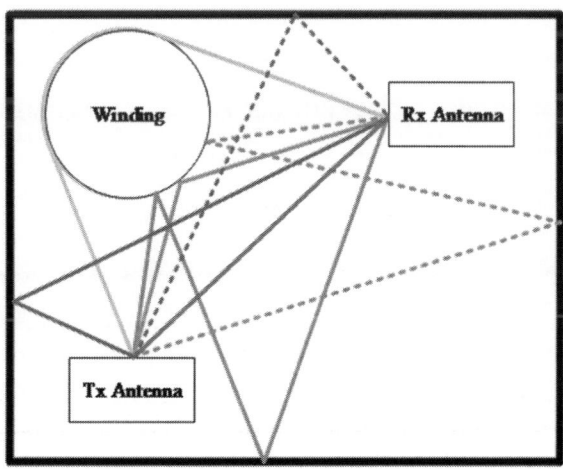

FIGURE 7.32

Different signal paths from Tx to Rx antennas in presence of metal tank.

Generally, the received signal has several types of multipath with different time delays. In the simulation environment, it is sufficient to define all the boundary conditions as $E_t = 0$. Fig. 7.33 compares the received signal for the reference position with the 10 mm axial displacement state due to the existence of the metallic tank. It is obvious that the signals are the same in the line of sight part. In addition, the signal changes are remarkable in the part of winding reflection, which makes visible the detection of the axial displacement.

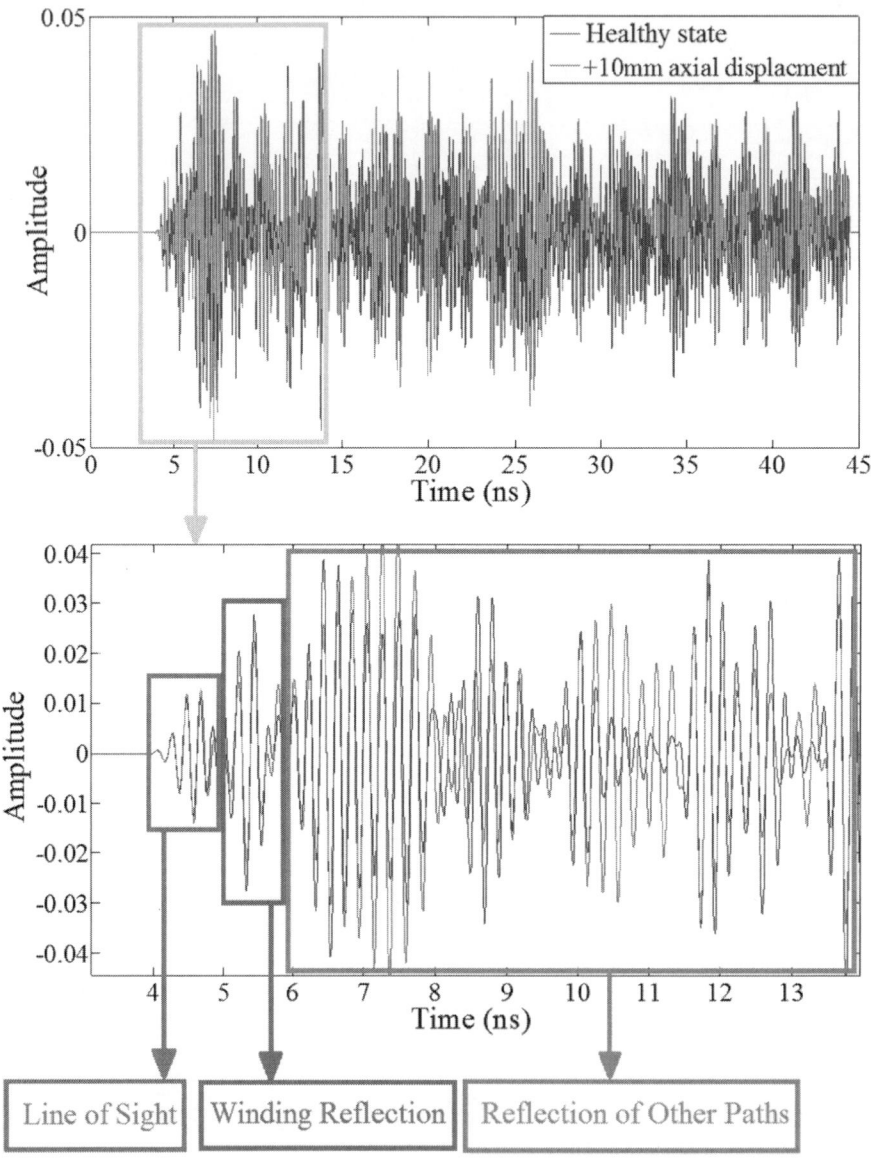

FIGURE 7.33

Comparison of received signals in healthy and +10 mm axial displacement states considering tank.

The axial displacement has been applied from -20 mm to $+20$ mm in 2 mm steps and the AED index variations are depicted in Fig. 7.34. The variation trend is similar to the system without the tank, even more linear, and AED fits better with the displacement value. Therefore the axial displacement detection using AED faces no problem considering the tank existence because the used method is comparative and the tank addition has occurred in both healthy and faulty states.

Fig. 7.35 shows the received signals in the case of applying a radial deformation of $+10$ mm on segment S_8 on disk three and in a healthy state. Similar to Fig. 7.33, the recorded signal is a combination of the line of sight, reflection from the winding and other reflections. Due to similarity with Fig. 7.33, they are not separated and shown again.

To show the effect of different radial deformations on the AED index variations, Fig. 7.36 is prepared. The value of this index for the radial deformation of the segment S_8 from -20 mm to $+20$ mm is calculated. As it turns out, changes in the bulgy mode have caused a further change in AED. In the case of bulgy radial deformation, the receiving path due to the winding reflection, because of the protruded segment is reduced and the variations in the received signal are more noticeable. The minimum wavelength according to the highest frequency in the transmitter waves is equal to 4.76 cm. It means that the dimensions of the concave deformation created on segment S_8 will be comparable to the length of waves, and so the defect is detectable. However, the signals see it as approximately more similar to a flat surface than a concave defect and so, changing the depth of this hole makes small changes in the received signal.

According to the simulations performed in this subsection, we conclude that a metallic wall increases the number of multipath, but does not affect the comparison method, and the index can be used to diagnose faults because the detection is based on comparison, and in both healthy and faulty cases tank effect exists. The investigation of the tank effect is carried out by simulations. As the experimental results have similar behavior, repetition of results with the experimental study is avoided.

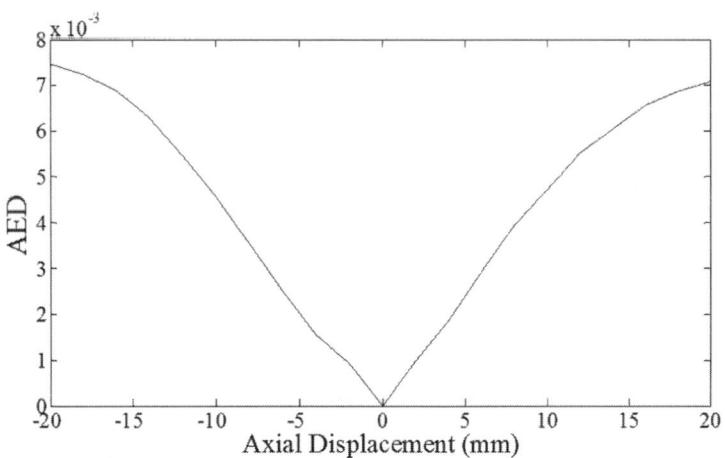

FIGURE 7.34

AED variations versus axial displacement values.

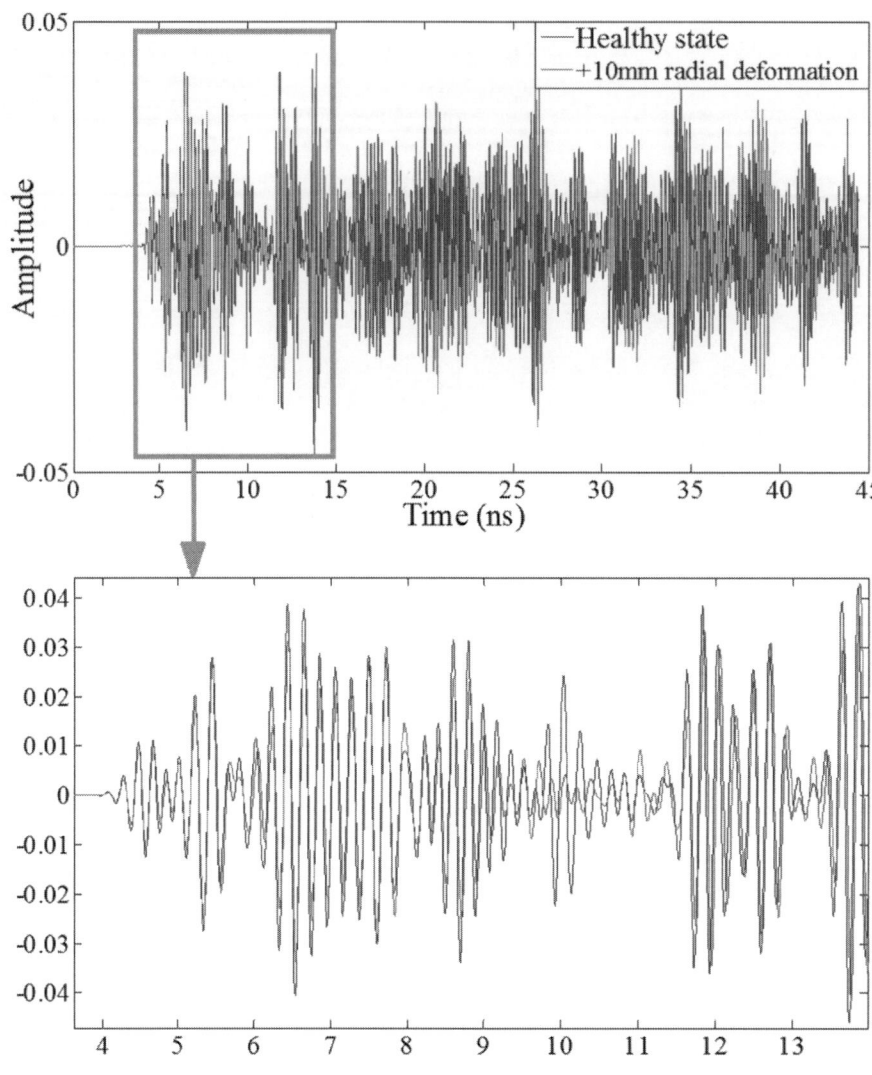

FIGURE 7.35

Comparison of received signals in healthy and radial deformation of $+10$ mm on segment S_8 of disk no. 3.

7.2.7 Classification of type, location and extent

The classification of type, location, and extent has been described and carried out in Section 6.4. The procedure is similar, and this subsection shows only an example for extent estimation of transformer winding axial displacement using an artificial neural network (ANN). The ANN is easy to use and widespread in engineering problems (Behkam et al., 2022; Sanjari et al., 2016). The total number of data for the axial displacement includes different 91 states, which indicate the axial displacements

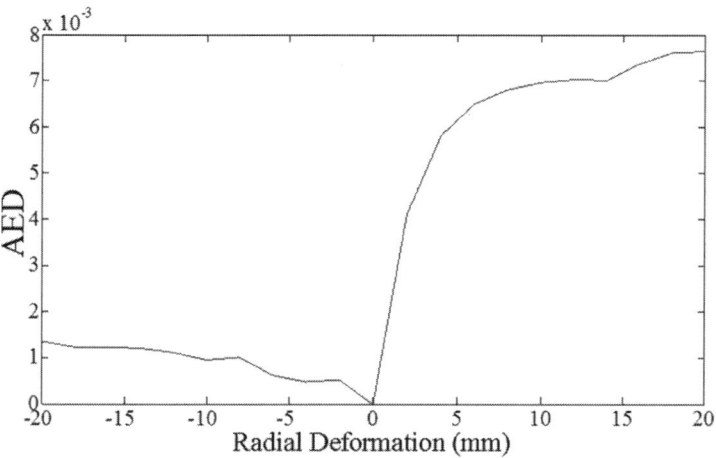

FIGURE 7.36

AED variations versus radial deformation values for segment S_8 of faulty disk no. 3.

from the position of -4.5 cm to $+4.5$ cm in 1 mm steps. To estimate the extent of axial displacement, a four-layer neural network with two hidden layers is used. The selection of neural network structure was based on using the trial-and-error method. According to the 91 simulated cases, 75% of the data is randomly used for training and the rest for performance testing.

The neural network must be able to detect the displacement value of the winding from the healthy state. If the displacement was greater than the specified limit value, we conclude that an error has occurred. After training the neural network, any case can be applied to the trained network according to its simulation data and then calculate its output, representing the estimated extent of displacement or the distance relative to the healthy state. As mentioned, 25% of cases have been used to test the trained neural network. The neural network output for each of the test cases is listed in Table 7.9.

Table 7.9 Estimation of axial displacement extent using trained neural network.

Actual axial displacement (mm)	Estimated displacement value (mm)	Difference (mm)
-43	-43.0521	0.0521
-39	-39.2531	0.2531
-35	-31.9015	1.9015
-25	-24.9316	0.0015
-27	-24.7507	2.2493
-23	-22.0467	0.9533
-19	-17.6789	1.3211
-15	-13.1485	1.8515
-11	-10.2080	0.8020

Continued

Table 7.9 Estimation of axial displacement extent using trained neural network.—cont'd

Actual axial displacement (mm)	Estimated displacement value (mm)	Difference (mm)
−7	−7.8786	0.8786
−2	−4.4891	−2.4891
0	0.989	0.9897
2	3.7970	1.7970
6	6.6829	0.6829
10	10.1422	0.1422
14	14.202	0.2029
18	17.6542	0.3458
22	22.6974	0.6974
26	29.8779	3.8779
30	29.6038	0.3962
34	32.6749	1.3251
38	39.3486	1.3486
42	43.012	1.0126

It turns out that the trained network specifies the distance well. In order to show the accuracy of the used method, the accuracy index is defined as the ratio of the number of correctly detected cases to the total number of test cases. This index is equal to 87% for the trained network.

References

3DEXPERIENCE Company. (2022). *CST studio. Electromagnetic field simulation software*. https://www.3ds.com/products-services/simulia/products/cst-studio-suite/.

Alehosseini, A., Hejazi, M. A., Mokhtari, G., Gharehpetian, G. B., & Mohammadi, M. (2015). Detection and classification of transformer winding mechanical faults using UWB sensors and Bayesian classifier. *International Journal of Emerging Electric Power Systems, 16*(3), 207−215.

Hejazi, M. A., Gharehpetian, G. B., Moradi, G. R., Mohammadi, M., & Alehoseini, H. A. (2011a). Application of classifiers for on-line monitoring of transformer winding axial displacement by electromagnetic non-destructive testing. *Electric Power Components and System, 39*(4), 387−403.

Behkam, R., Karami, H., Naderi, M. S., & Gharehpetian, G. B. (2022). Generalized regression neural network application for fault type detection in distribution transformer windings considering statistical indices. *COMPEL-The International Journal for Computation and Mathematics in Electrical and Electronic Engineering, 41*(1), 381−409. https://doi.org/10.1108/COMPEL-06-2021-0199

Hejazi, M. S. A., Ebrahimi, J., Gharehpetian, G. B., Mohammadi, M., Faraji-Dana, R., & Moradi, G. (2011b). Application of ultra-wideband sensors for on-line monitoring of transformer winding radial deformations−A feasibility study. *IEEE Sensors Journal, 12*(6), 1649−1659.

Karami, H., Gharehpetian, G. B., Norouzi, Y., & Hejazi, M. A. (2016). GLRT-based mitigation of partial discharge effect on detection of radial deformation of transformer HV winding using SAR imaging method. *IEEE Sensors Journal, 16*(19), 7234−7241.

Karami, H., Gharehpetian, G. B., Norouzi, Y., & Hejazi, M. A. (2018). Experimental study on elimination of partial discharge effect on detection of radial deformation of high voltage transformer winding using electromagnetic waves. In *2018 IEEE international conference on environment and electrical engineering and 2018 IEEE industrial and commercial power systems Europe* (pp. 1−5). EEEIC/I&CPS Europe).

Karami, H., Tabarsa, H., Gharehpetian, G. B., Norouzi, Y., & Hejazi, M. A. (2019). Feasibility study on simultaneous detection of partial discharge and axial displacement of HV transformer winding using electromagnetic waves. *IEEE Transactions on Industrial Informatics, 16*(1), 67−76.

Mokhtari, G., Gharehpetian, G. B., Faraji-Dana, R., & Hejazi, M. A. (2010). On-line monitoring of transformer winding axial displacement using UWB sensors and neural network. *International Review of Electrical Engineering (IREE), 5*(5).

Sanjari, M. J., Karami, H., & Gooi, H. B. (2016). Micro-generation dispatch in a smart residential multi-carrier energy system considering demand forecast error. *Energy Conversion and Management, 120*, 90−99.

Syntactic aperture radar imaging

Gevork B. Gharehpetian[1], Hossein Karami[2] and Seyed-Alireza Ahmadi[3]

[1]*Electrical Engineering Department, Amirkabir University of Technology (AUT), Tehran, Iran;* [2]*High Voltage Studies Research Department, Niroo Research Institute (NRI), Tehran, Iran;* [3]*School of Electrical and Computer Engineering, College of Engineering, University of Tehran, Tehran, Iran*

8.1 Concept of syntactic aperture radar

The syntactic aperture radar (SAR) imaging method is used in many applications such as topography, geology, military, environmental monitoring, civil infrastructure monitoring, etc. What is important in this method is the procedure of imaging and processing data to create an image from the data, which are described in the following.

8.1.1 Procedure of imaging

Fig. 8.1A illustrates two-dimensional (2D) imaging using the SAR method. The transmitter and receiver antennas simultaneously move on the X-axis while the electromagnetic waves (EMWs) are sent to the target and received at different points with equal intervals. The transmitted waves hit the target and are uniformly reflected in all directions.

The received signal in each antenna position is called a scan, a time signal obtained by sampling the received waves at a specific position. Different scans are obtained by changing the position of the antennas along the X-axis. The datasets collected during the scanning can be considered as a function, $f(x,t)$, related to the antenna position (x) and time (t), as shown in Fig. 8.1B. It is worth noting that the signals received at the receiver are sampled by the analog-to-digital converter and considered discrete-time signals. In addition, the number of scanned points on the X-axis is limited. Therefore, both time (t) and position (x) variables are considered discrete variables, leading to a discrete function represented as an M^*N vector, where N is the number of observed samples in each of M different positions of target scanning.

Suppose there is a target point at (x_T, z_T), scanned by moving the antennas along the X-axis with the SAR method. Each scan includes only one pulse, which is the reflection of the transmitted pulse by the Tx antenna from the target. The delay in each pulse equals the duration of the wave propagation from the transmitter to the target and its reflection in the opposite direction, i.e., twice the wave velocity in the environment multiplied by the distance between the target and antennas. Thus, each scan has a

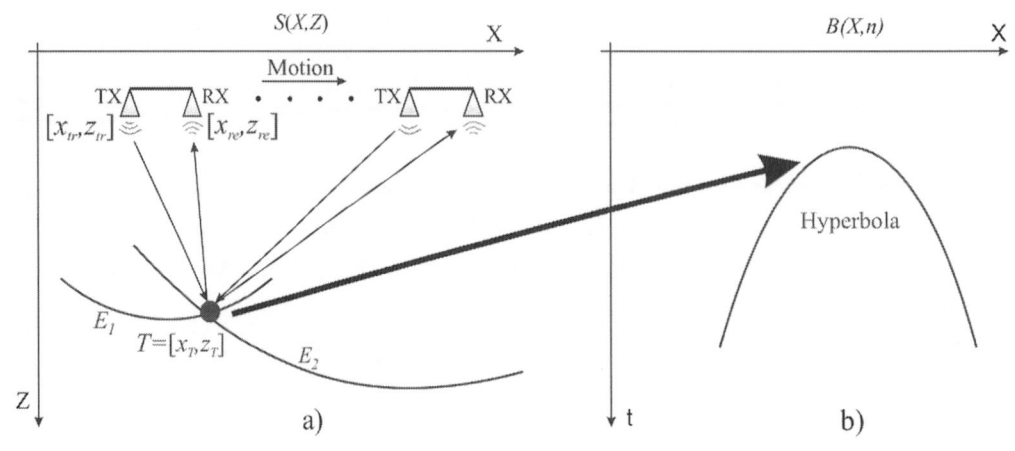

FIGURE 8.1

SAR 2D model illustration, (A) procedure of imaging, and (B) time-position function.

maximum value after the aforementioned delay. If x is the position of the antennas and $\tau(x)$ indicates the delay of the received pulse in each scan, we have:

$$\tau(x) = \frac{2r}{c} = \frac{2}{c}\sqrt{(x - x_T)^2 + z_T^2} \tag{8.1}$$

In other words,

$$\left(\frac{c}{2}\right)^2 \tau(x)^2 - (x - x_T)^2 = z_T^2 \tag{8.2}$$

Equation (8.2) represents the relationship between the propagation delay or the maximum point in each scan and the position of the antennas in that scan. A hyperbola is obtained by drawing the equation mentioned above on the X-τ plane, the vertex of which is located at the point (x_T, τ_T), where τ_T indicates the wave propagation delay corresponding to the z_T distance. Therefore, SAR imaging converts each target point on the X-Z plane into a hyperbola on X-t one. However, we have the f function, in which a hyperbola is achieved when its maximum points are drawn on the X-t plane. To obtain the coordination of the target, the hyperbola should be returned to the point where the target is located in the X-Z plane, described in the next section (Aftanas, 2009).

8.1.2 Data processing

As described in the previous chapter, the ultra wideband (UWB) pulse sent by the transmitter hits various objects, the reflection of which reaches the receiver. Thus the transmitted wave travels several different paths from the transmitter to the receiver, called multipath. Fig. 8.2 illustrates an example related to the recorded signal by the Rx antenna with three main parts. The first part (cross talk), which has the least delay, takes the shortest path and directly reaches the receiver from the transmitter. The second part is considered as the reflected wave from the target, the delay of which corresponds to the distance from the target to the antenna. The third part, which has the longest delay, is regarded as the

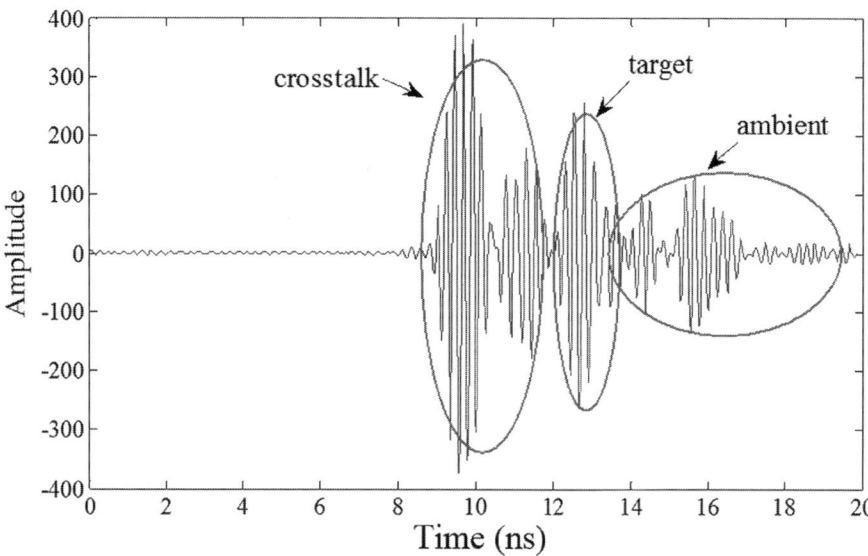

FIGURE 8.2

Sample received signal affected by multipaths.

reflection of the waves hitting the surrounding ambient, such as a wall, table, and the like. The cross talk part and the reflection of the ambient create a lot of clutter and distortion in creating the image of the target and affect its quality significantly. To obtain the high-quality target image, only the part of the signal reflected by the target is considered and other parts are eliminated (Rahbarimagham et al., 2015).

Several steps of calibration and processing are performed on the signals. First, the time base (zero time) in each received signal is determined in different antenna positions, and the signals are shifted to the base. In other words, all of the scans are synchronized with each other, which plays a significant role in the image quality because the image recognition criterion is based on the time delay of the received pulses. The first peak in the received signal is used to identify the time base. It means that the delay is calculated by comparing the first received signal peak, the start of the cross talk part, and the first transmitted signal peak, i.e., the delay equals the distance between the two antennas divided by the wave velocity. The time base is calculated by knowing the abovementioned delay and time of the first peak in the received signal.

Then, the unwanted parts of the received signal, including the cross talk and the effect of the ambient, are eliminated and the reflected signal from the target is separated. To this aim, the appropriate time window containing the target is selected, part of the signal placed in the window is saved, and the rest is eliminated. The maximum and minimum paths traveled by the signal from the antenna to the target are used to determine the time window containing the reflected pulse from the target. Such operation is repeated for each antenna position on the X-axis. Fig. 8.3 demonstrates the signal obtained after the abovementioned procedure, which is utilized in the next steps to obtain the image (Rahbarimagham et al., 2017).

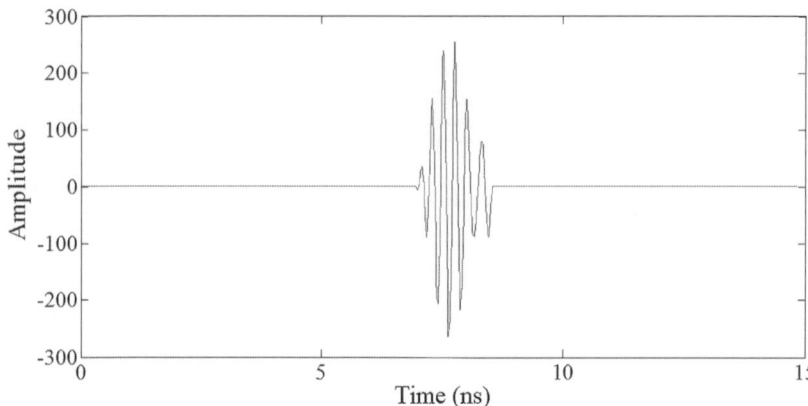

FIGURE 8.3

Signal obtained after two processing steps.

8.2 Imaging using SAR methods

Assume that the data measured through the SAR method is stored in matrix $B(x,t)$. As indicated, each target with the appropriate shape can be considered as a set of points, which can be retrieved by returning the hyperbole corresponding to each point to that target, resulting in creating the target image. Retrieving an image from the measured data is called migration.

Migration algorithms convert time-domain signals into depth-domain ones. Depth means the distance between the target and the X-axis (z_p). In other words, such algorithms convert the measured data stored in the vector $B(x,t)$ into an image $I(x,z)$ (Mortazavian et al., 2015). Most migration algorithms operate based on wave propagation linearization, meaning that the reflections inside and among objects are ignored, known as the optics or Born approximation (Scales, 1995).

8.2.1 Back projection algorithm

As the simplest migration algorithm, back-projection regards waves as a beam of direct motion without considering the wave equation. Then the target image is obtained by calculating the delay between sending and receiving the signal (Ulander et al., 2003). Calculating the wave propagation delay in the ambient is regarded as the first step in back-projection imaging. The depth is simply calculated through multiplying the wave velocity by the time when the velocity is constant. Otherwise, the latency is calculated as follows:

$$T_d = \int_0^z \frac{dz}{V(z)} \qquad (8.3)$$

The following equation calculates the wave propagation delay from the transmitter to the target and its return to the receiver when the wave propagation velocity is assumed to be half ($V = c/2$).

$$T_d = \int_0^R \frac{2dr}{c(r)} \tag{8.4}$$

The equation in constant velocity is as follows.

$$T_d = \frac{2R}{c} \tag{8.5}$$

where, R indicates the distance between the target and the antenna.

Fig. 8.4 displays the schematic for achieving an image by applying UWB waves. First, the transmitter and receiver antennas are simultaneously moved on the X-axis and the target is scanned at different equal interval points (SAR imaging). If the distance between two measuring points equals L, the length of each measured point, is as follows.

$$x_k = L \times k \tag{8.6}$$

The scan at the k-th point of view, x_k, is demonstrated with $b(x_k)$. As shown in Fig. 8.4, the entire ambient, in which the target is located, is divided into many square parts, i.e., pixels, each of which can be considered as a hypothetical target. Consider the intended pixel P with a coordination of (x_i, z_j), where the target is located. The transmitted waves hit the target, and their reflection returns to the Rx antenna. When the antenna is located at the x_k point, the delay in the received pulse of the k-th scan is calculated, as follows.

$$T_{dP,k} = \frac{2R}{c} = \frac{2}{c}\sqrt{(x_k - x_i)^2 + z_j^2} \tag{8.7}$$

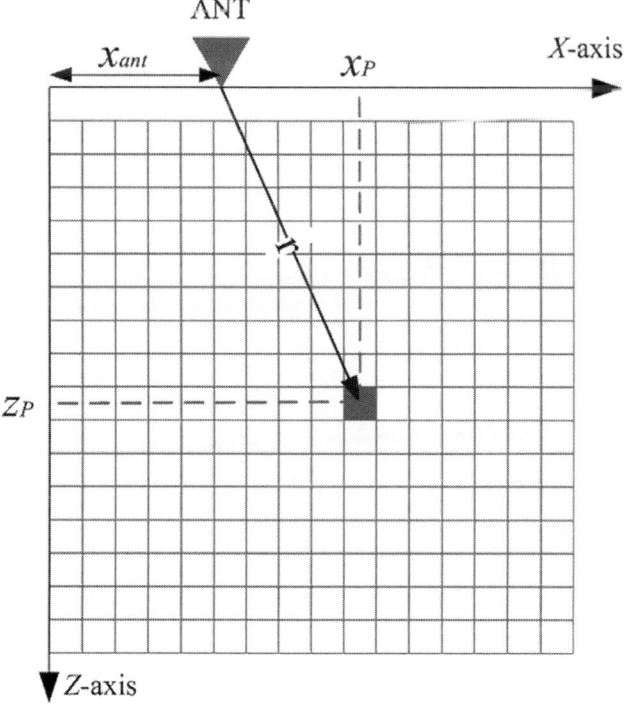

FIGURE 8.4

SAR 2D imaging.

As illustrated in Fig. 8.5, the value of the received signal at the time base equals the peak of the received pulse when it is shifted as $T_{dP,k}$. Calculating the delay and creating the appropriate time shift for each scan on the pixel P is repeated. Then, the sum of the shifted signals is calculated as (Eq. 8.8).

$$I(x_p, z_p) = \sum_k b_k(t + T_{dP,k})\Big|_{t=0} \qquad (8.8)$$

The value of the time-shift corresponds to the received pulse delay, and so, the amplitude of the summed signal will be large at the time base in all of the scans when the target is really located at point P. Otherwise, the value of the summed signal at the time base will be small and close to zero. Therefore the value of the summed signal at the time base indicates the amount of wave reflection from point P.

The aforementioned steps are repeated for all the pixels to obtain the target image. The summed value demonstrated in Eq. (8.8) is considered a discrete two-variable function $I(x,z)$, whose value at each point represents the reflection intensity of the waves from that point. The 2D image of the target is obtained by drawing the abovementioned function, which is known as the back-projection (Herman, 1995).

Note that the signal obtained in the aforementioned step is regarded as a UWB pulse. However, a narrow pulse is required to obtain a high-quality image utilizing the back-projection algorithm. So, only a small neighborhood around the signal peak should be considered. Thus, the signal from the abovementioned processing steps becomes a narrow pulse, as shown in Fig. 8.6.

Briefly, the steps in the back-projection migration algorithm can be described as follow:

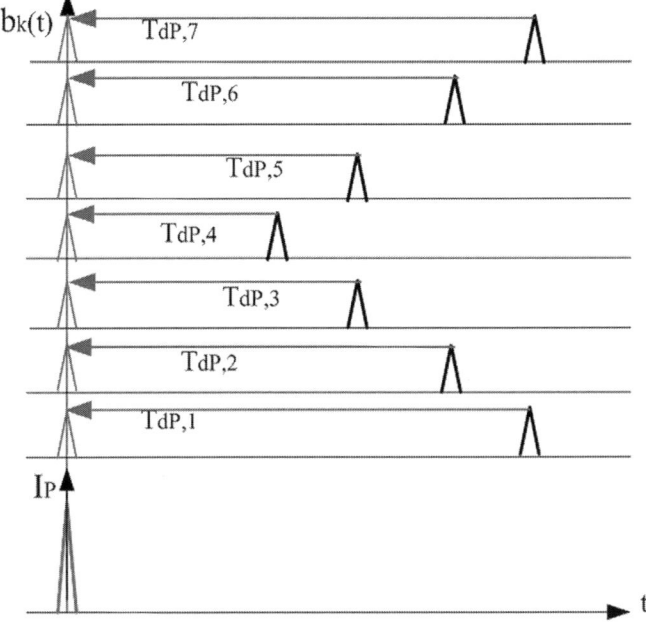

FIGURE 8.5

Appropriate time shift in received signal.

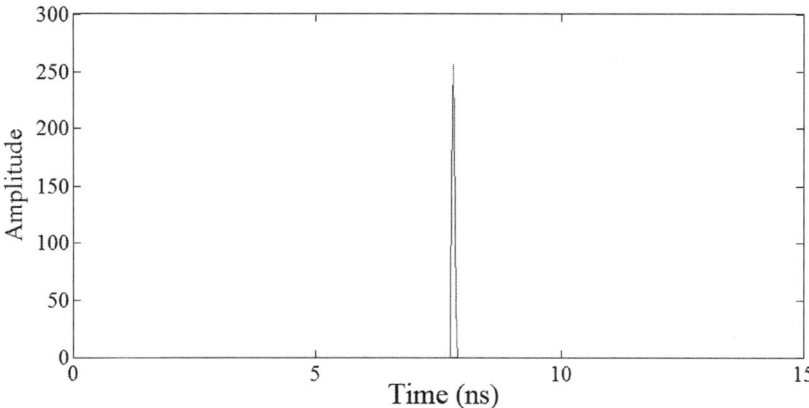

FIGURE 8.6

Pulse peak reflected from target.

Step 1. Reflected waves are received at different points on the X-axis (x_i) and stored as a time signal ($b_k(t)$).

Step 2. Necessary processing is performed on the received signal, and the peak of the signal reflected from the target is extracted.

Step 3. The X- and Z-axes are divided into M and N parts, respectively, resulting in having $M \times N$ pixels. The coordinates in each pixel are displayed by (x_i, z_j), where i and j indicate the pixel number on the X- and Z-axes, respectively.

Step 4. Initialize the parameter values: $i = 1, j = 1, k = 1, K, I = [0]_{M^*N}$, where K represents the number of measuring points on the X-axis.

Step 5. Calculate the wave propagation delay using Eq. (8.7).

Step 6. Obtain the wave reflection from the corresponding pixel by applying the following recursive equation.

$$I(x_i, z_j) = I(x_i, z_j) + b_k(t + T_{dP,k})|_{t=0} \tag{8.9}$$

Step 7. Set $k = k+1$. Go to Step 5, when $k \le K$. Otherwise, go to the next step.

Step 8. Set $i = i+1$. Go to Step 5, when $i \le M$. Otherwise, go to the next step.

Step 9. Set $j = j+1$. Go to Step 5, when $j \le N$. Otherwise, go to the next step.

Step 10. Draw the reflection intensity function, I, to obtain the image.

8.2.2 Phase difference

To improve the obtained image, the received signals are moved to the frequency domain using the Fourier transform (FT), and the time delay is modeled as phase change, known as the phase difference algorithm (Ji et al., 2022). This is the back-projection algorithm when FT transfers the received signals to the frequency domain and replaces the time shift with phase. By taking FT from Eq. (8.8), the time delay, $-T_d$, is considered as the phase shift, $e^{+j\omega Td}$, and we have:

$$I_f(x, y) = \sum_{x_k} A_k(f).\exp(j2\pi f T_{dP,k}) \tag{8.10}$$

where $A_k(f)$ is regarded as the discrete FT of b_k. However, $A_k(f)$ has a large bandwidth and all of its frequency components should contribute to creating the image. Therefore, Eq. (8.10) is rewritten as follows.

$$I(x,z) = \sum_l \sum_{x_k} A_k(f).\exp\left(j2\pi l f_0 T_{dP,k}\right) \qquad (8.11)$$

where, f_0 represents the frequency corresponding to the sampling time of the received signal (T_s), calculated by Eq. (8.12).

$$f_0 = \frac{1}{T_s} \qquad (8.12)$$

It is clear that l is changed, so $l f_0$ covers the entire frequency range of the received signal.

Briefly, the steps of the phase difference migration algorithm can be described as follows:

Step 1. Reflected waves are received at different points of the X-axis (x_i) and stored as a time signal ($b_k(t)$).

Step 2. Two processing steps, including synchronizing the time base and selecting the appropriate time window, are performed on the received signals.

Step 3. FT for each received signal is calculated.

$$A_k(f) = F\{b_k(t)\} \; k = 1, 2, ..., K \qquad (8.13)$$

Step 4. The X- and Z-axes are divided into M and N parts, respectively, resulting in $M \times N$ pixels. The coordinates in each pixel are demonstrated by (x_i, z_j), where i and j indicate the part number on the X- and Z-axes, respectively.

Step 5. Initialize the value of the parameters.

$$i = 1, j = 1, k = 1, l = 1, J = [0]_{M*N*L}, I = [0]_{M*N}, K, N, L$$

where, K and L are considered as the number of measuring points on the X-axis and the number of frequencies for the FT output of the received signal, respectively.

Step 6. Calculate the wave propagation delay by Eq. (8.8).

Step 7. Obtain the amount of reflection at l-th frequency component of the received signal for related pixels utilizing the following recursive equation.

$$J\left(x_i, z_j, l f_0\right) = J\left(x_i, z_j, l f_0\right) + A_k(l f_0).\exp\left(j2\pi l f_0 T_{dP,k}\right) \qquad (8.14)$$

Step 8. Set $k = k+1$. Go to Step 6 when $k \leq K$. Otherwise, go to the next step.

Step 9. Obtain the amount of reflection in the pixel by summing the reflection related to each frequency component in the received signal, as follows:

$$I\left(x_i, z_j\right) = I\left(x_i, z_j\right) + \left|J\left(x_i, z_j, n f_0\right)\right| \qquad (8.15)$$

Step 10. Set $l = l+1$. Go to Step 6 when $l \leq L$. Otherwise go to the next step.

Step 11. Set $i = i+1$. Go to Step 5 when $i \leq M$. Otherwise go to the next step.

Step 12. Set $j = j+1$. Go to Step 5 when $j \leq N$. Otherwise go to the next step.

Step 13. Draw the reflection intensity function, I, to obtain the intended image.

8.2.3 Kirchhoff migration algorithm

The advanced method, which is widely used in deep-field imaging, is the Kirchhoff migration algorithm (KMM) (Mortazavian et al., 2012). The wave equation is solved utilizing Green's function, and an integral equation is achieved for imaging (Golsorkhi et al., 2012). The wave propagation velocity is assumed constant during the first step of the Kirchhoff imaging.

The field direction can be neglected and assume that the wave propagation ambient is homogeneous. The scalar wave equation in a sourceless ambient is shown in the following.

$$\nabla^2 \Psi - \frac{1}{c^2}\partial_t^2 \Psi = 0 \tag{8.16}$$

where, the scalar function $\Psi(r, t)$ is considered as the amplitude of the field at different points in space. Fig. 8.7 illustrates the geometry of the problem to find the response of the scalar wave equation within a closed volume, Ω, by knowing the boundary conditions on the continuous boundary, $\partial \Omega$. To solve the wave equation, one of the following boundary conditions should be specified:

- The initial value of the field $(\Psi(r, t)|_{t=0})$ and its vertical derivative $(\frac{\partial \Psi(r,t)}{\partial n}\big|_{t=0})$ in the whole volume Ω, as well as the field value on the boundary $(\Psi(r, t)|_{r \in \partial\Omega})$.

- The initial value of the field $(\Psi(r, t)|_{t=0})$ and its vertical derivative $(\frac{\partial \Psi(r,t)}{\partial n}\big|_{t=0})$ in the whole volume Ω, as well as the vertical derivative of field value on the boundary $(\frac{\partial \Psi(r,t)}{\partial n}\big|_{r \in \partial\Omega})$.

The clarity of the field and its vertical derivative at the boundary are called the Dirichlet and Neumann problems, respectively. Unlike the wave equation, in some differential equations, both boundary conditions may be known (Cauchy's theorem). Suppose the source is located in an empty

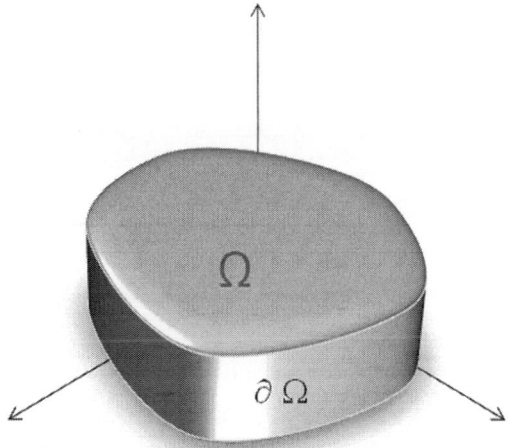

FIGURE 8.7

Geometry of volume Ω surrounded by $\partial \Omega$.

space on a portion of the boundary $\partial \Omega$. Thus the wave equation is solved in a homogeneous space. The solution to the wave equation can be calculated by summing the responses of the source point at infinity through the Huygens' principle. Such an approach can be expressed mathematically by defining the impulse response, or Green's function, as the solution to the wave equation with a point source. A Green's function ($\Gamma(r, t, r', t')$), which depends on the coordinates of the source and point of view, is defined as the solution to Eq. (8.17) regardless of the problem boundary conditions.

$$\nabla^2 \Gamma - \frac{1}{c^2}\partial_t^2 \, \Gamma = -4\pi\delta(r - r')\delta(t - t') \tag{8.17}$$

where, r' indicates the position of the point source, which creates the Green's function of the wave equation and t' represents the time of transmitting the Dirac delta pulse from the source. Note that the above equation is considered as a wave propagation equation derived from a point source located at the coordinates $r = r'$, emitting a Dirac delta pulse (δ) at the moment $t = t'$. The Green's function is the field resulting from the point source in space. After some calculations, we have:

$$4\pi\delta(r - r')\delta(t - t')\Psi = \Gamma \, \nabla^2\Psi - \Psi \, \nabla^2\Gamma + \frac{1}{c^2}\left(\Gamma\partial_t^2\Psi - \Psi\partial_t^2 \, \Gamma\right) \tag{8.18}$$

In other words,

$$4\pi\delta(r - r')\delta(t - t')\Psi = \nabla.[\Gamma \, \nabla\Psi - \Psi \, \nabla\Gamma] + \frac{1}{c^2}\partial_t(\Gamma\partial_t\Psi - \Psi\partial_t \, \Gamma) \tag{8.19}$$

The space-time integral is taken from Eq. (8.19) to eliminate the δ function as shown in Eq. (8.20). The spatial and time ranges of the integral are defined as Ω and $(-\infty, \infty)$, respectively.

$$\varepsilon(r)\Psi(r, t) = \int\limits_{-\infty}^{\infty} \int\limits_{\Omega} \nabla.[\Gamma \, \nabla\Psi - \Psi \, \nabla\Gamma]dv'dt' + \frac{1}{c^2} \int\limits_{-\infty}^{\infty} \int\limits_{\Omega} \partial_{t'}(\Gamma\partial_{t'}\Psi - \Psi\partial_{t'} \, \Gamma)dv'dt' \tag{8.20}$$

The function $\varepsilon(r)$ is obtained by a finite argument and equals 4π inside of Ω, 2π on the boundary and zero for others. The first term on the right side of Eq. (8.20), considered as an integral on the volume Ω, can be converted to a double integral on $\partial \Omega$ utilizing the divergence theorem, as follows:

$$\int\limits_{-\infty}^{\infty} \int\limits_{\Omega} \nabla.[\Gamma \, \nabla\Psi - \Psi \, \nabla\Gamma]dv'dt' = \int\limits_{-\infty}^{\infty} \int\limits_{\partial\Omega} [\Gamma \, \nabla\Psi - \Psi \, \nabla\Gamma].nda'dt' \tag{8.21}$$

After integrating the second term of Eq. (8.20) in terms of time, we have:

$$\int\limits_{\Omega} (\Gamma\partial_t\Psi - \Psi\partial_t \, \Gamma)dv' \Bigg|_{t=-\infty}^{t=+\infty} \tag{8.22}$$

Obviously, $\partial_t\Psi$ and Ψ can be assumed to be zero as long as the source starts working ($t = 0$). Therefore, as Green's function is a causal function and assuming Sommerfeld radiation conditions concerning Ψ, the lower and upper limits equal zero. Thus, the second term in Eq. (8.20) equals to zero and is summarized as Eq. (8.23).

$$\forall r \in R^3 : \ \varepsilon(r)\Psi(r,t) = \int\limits_0^\infty \int\limits_{\partial\Omega} [\Gamma \, \nabla\Psi - \Psi \, \nabla\Gamma].n \, da' \, dt' \tag{8.23}$$

The equation mentioned above is known as the Kirchhoff integral theorem, in which n is the unit vector perpendicular to $\partial\Omega$. The functions f and g are defined as the value of the wave function on the boundary and the vertical derivative of the wave function on the boundary, respectively, as follows:

$$f = \Psi|_{\partial\Omega}, \ g = \nabla\Psi = \left.\frac{\partial\Psi}{\partial n}\right|_{\partial\Omega} \tag{8.24}$$

By substituting Eq. (8.24) in Eq. (8.23), we have:

$$\forall r \in R^3 : \ \varepsilon(r)\Psi(r,t) = \int\limits_0^\infty \int\limits_{\partial\Omega} [\Gamma \, g - f \, \nabla\Gamma].n \, da' \, dt' \tag{8.25}$$

According to the problem and boundary conditions, the integral on the right side of Eq. (8.25) can be solved. The solution of the wave equation can be obtained in the whole space by knowing the functions f and g and their vertical derivatives on the boundary, due to the clarity of Green's function (Γ) and its vertical derivative ($\nabla\Gamma$) on the boundary. However, as both these functions cannot be simultaneously known, one is calculated in terms of another (Cauchy conditions). Therefore, the term limit on the left side of Eq. (8.25) is calculated as Eq. (8.26), where r leans toward the $\partial\Omega$ boundary.

$$\lim_{r \to \partial\Omega} \varepsilon(r)\Psi(r,t) = 2\pi f(r,t) \ \forall r \in \partial\Omega \tag{8.26}$$

Thus, Eq. (8.26) can be summarized as Eq. (8.27).

$$\forall r \in \partial\Omega : \ f(r,t) = \frac{1}{2\pi} \int\limits_0^\infty \int\limits_{\partial\Omega} [\Gamma \, g - f \, \nabla\Gamma].n \, da' \, dt' \tag{8.27}$$

The equation mentioned above is based on the functions f and g, and regarded as a Neumann problem. For example, Eq. (8.27) is considered as an integral equation for f when the vertical derivative Ψ at the boundary g is determined. The solution to the wave equation is achieved by placing it in the Kirchhoff integral when f is calculated from Eq. (8.27).

There is no specific assumption for the Green's function's boundary conditions, so the afore-mentioned equations are true for any Green's function or any function that holds in Eq. (8.17). Suppose, in a specific problem, the Green's function or its vertical derivative equals zero in all or parts of the boundaries. In that case, the Kirchhoff integral can solve the problem directly since the term containing the vertical derivative of the wave function, g, is eliminated. In addition, the Kirchhoff integral can be calculated in Eq. (8.25), and the solution to the wave equation can be obtained at all of the points in space only by knowing the wave function on the boundary, f. In other words, when Green's function is zero at the boundary, Eq. (8.25) can be summarized, as follows.

$$\forall r \in \Omega : \ \Psi(r,t) = -\frac{1}{4\pi} \int\limits_0^\infty \int\limits_{\partial\Omega} f(r',t') \partial_n \Gamma(r,t,r',t') da' \, dt' \tag{8.28}$$

In addition, $\varepsilon(r)$ is replaced by 4π (inside the boundary), and the vertical derivative of Green's function on the boundary ($\nabla\Gamma$) is displayed by $\partial_n\Gamma(r,t,r',t')$.

It is worth noting that Eq. (8.28) is not considered an integral one because the value of f is known. Similarly, the Neumann problem can be solved immediately due to the elimination of the unknown term containing the boundary condition when the vertical derivative of Green's function is zero at the boundary. However, having a Green's function with appropriate boundary conditions eliminates the need to solve the integral. When such a function does not exist, Eq. (8.28) should be solved numerically or by assuming an approximate Green's function, as implemented in the following sections by generalizing the results to non constant velocity ambient (Margrave & Lamoureux, 2019).

As mentioned, electromagnetic imaging is based on sending waves to an object and measuring the received wave. During the aforementioned process, the transmitter and receiver antennas are fixed relative to each other and move at different points in a line or a plane (Golsorkhi et al., 2021). The transmitter sends a UWB pulse, and the receiver records the reflected wave. Accordingly, the solution to the wave equation is available at different points in the Z = 0 plane as the problem data. Then, the data are converted into an image, called data migration. We have a recorded wave at each point of the measuring line. $f_i(t)$ is the data measured at i-th point. The whole data can be hypothesized as a 2D function, $B(x, t)$. The migration algorithm should convert time-length data (x-t), to length-width information (x-y), or a 2D image of an object. Similarly, the migration algorithm converts the data to a 3D image when they are recorded at different points on a plane (Mosayebi et al., 2014).

Assume the data is received on the Z = 0 plane and the ambient around the object has a constant velocity without any boundary. As demonstrated in Fig. 8.8, the volume Ω is considered a large hemisphere extending downward and bounded from above by the Z = 0 plane. The radius of the hemisphere is hypothesized to tend toward infinity due to the lack of boundary. Thus the surfaces containing the volume Ω, include a Z = 0 plane and a hemisphere with an extremely large radius. As displayed in Eq. (8.25), the Kirchhoff integral tends toward zero on its surface by increasing the radius of the hemisphere, resulting in reducing Eq. (8.25) to an integral on Z = 0 plane. However, showing the integral tendency of the hemisphere toward zero is complex. One can refer to (Goodman, 1996) to find the full proof of this theorem.

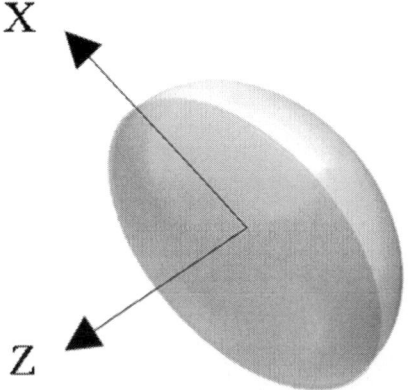

FIGURE 8.8

Volume Ω as large hemisphere bounded by Z = 0 plane.

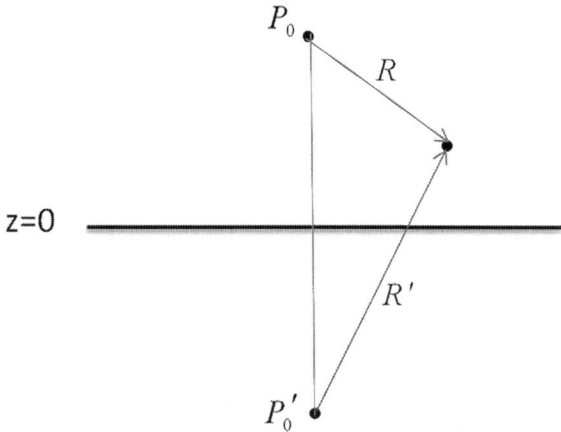

FIGURE 8.9

Schematic of selected green's functions.

As indicated, to simplify the Kirchhoff integral into Eq. (8.28), a Green's function or its vertical derivative should be zero at all of the points on the boundary. Being zero on the $Z = 0$ plane suffices for Green's function because the field on the hemisphere plane equals zero. The imaging method can be applied to find such a function due to the symmetry of the problem. As demonstrated in Fig. 8.9, two-point sources are placed at $P_0 = (x', y', z')$ and $P_0' = (x', y', -z')$, symmetric to the $Z = 0$ plane. The abovementioned sources emit a Dirac delta pulse at the moment $t = t'$ with amplitudes $+1$ and -1. Each source creates a spherical wave that propagates through space over time. The waves emitted by the sources on the $Z = 0$ plane neutralize each other and make the field resulting from Green's function on the plane zero due to their symmetry to the $Z = 0$ plane.

Green's function derived from the aforementioned sources is determined using Eq. (8.17), as follows:

$$\Gamma_r(r, t, r', t') = \frac{\delta\left(t - t' - R/c\right)}{R} - \frac{\delta\left(t - t' - R'/c\right)}{R'} \tag{8.29}$$

$$\Gamma_a(r, t, r', t') = \frac{\delta\left(t - t' + R/c\right)}{R} - \frac{\delta\left(t - t' + R'/c\right)}{R'} \tag{8.30}$$

where, $R = (x - x', y - y', z - z')$ and $R' = (x - x', y - y', z + z')$. Γ_r is a causal wave Green's function propagated through space over time and moved away from sources. In Eq. (8.30), the wave is propagated in the opposite direction over time and concentrated toward the points P_0 and P_0' due to the positive sign of R and R'. Therefore, Γ_a is considered as a noncausal Green's function.

One of the above mentioned Green's functions should be placed in the Kirchhoff integral. Since the migration algorithm aims to return the received waves to the point source, Γ_a is utilized, resulting in presenting Eq. (8.31).

$$\Psi(r,t) = -\frac{1}{4\pi} \int_0^\infty \int_{z'=0} f(r',t') \frac{\partial}{\partial z'} \left[\frac{\delta\left(t - t' + R/c\right)}{R} - \frac{\delta\left(t - t' + R'/c\right)}{R'} \right] da' \, dt' \qquad (8.31)$$

In addition, the following equation can be written since $R'(z') = R(-z')$.

$$\frac{\partial}{\partial z'}(R'(z')) = \frac{\partial}{\partial z'}(R(-z')) = -\frac{\partial}{\partial z'}(R(z')) \qquad (8.32)$$

Therefore, the integral is simplified, as follows:

$$\Psi(r,t) = -\frac{1}{2\pi} \int_0^\infty \int_{z'=0} f(r',t') \frac{\partial}{\partial z'} \left[\frac{\delta\left(t - t' + R/c\right)}{R} \right] da' \, dt' \qquad (8.33)$$

In addition, Eq. (8.34) can be written since the variable z' in R only appears in the term $(z-z')$.

$$\frac{\partial}{\partial z'} \left[\frac{\delta\left(t - t' + R/c\right)}{R} \right] = -\frac{\partial}{\partial z} \left[\frac{\delta\left(t - t' + R/c\right)}{R} \right] \qquad (8.34)$$

Thus, Eq. (8.33) is simplified, as follows:

$$\Psi(r,t) = \frac{1}{2\pi} \frac{\partial}{\partial z} \int_0^\infty \int_{z'=0} f(r',t') \frac{\delta\left(t - t' + R/c\right)}{R} da' \, dt' \qquad (8.35)$$

Finally, the integral in the time domain is eliminated due to the properties of the δ function as shown in Eq. (8.36).

$$\Psi(r,t) = \frac{1}{2\pi} \frac{\partial}{\partial z} \int_{z'=0} \frac{f\left(r', t + R/c\right)}{R} da' \qquad (8.36)$$

The abovementioned equation is the Schneider formula for migration in the Kirchhoff method. The question here is related to the boundary conditions at z' = 0. The assumed conditions are not in an actual experiment, so the solved problem is not related to a visible field. However, in Eq. (8.36) the boundary value of the field, f, is replaced with the recorded data to obtain the image. The wave is sent to and reflected from the object during the data recording. Therefore, the procedure is considered a reflection experiment, while the wave is hypothesized to propagate from the object to obtain Eq. (8.36). Thus the wave propagation time during the data recording is doubled. To adapt the data collection process to the real problem, the wave propagation velocity c is halved, along with setting t = 0, resulting in achieving Eq. (8.36), as follows.

$$\Psi(r) = \frac{1}{2\pi} \frac{\partial}{\partial z} \int\limits_{z'=0} \frac{f\left(r', 2R/c\right)}{R} \, da' \tag{8.37}$$

where R indicates the distance between any point on the $Z = 0$ plane and the point on $Z' = 0$ plane.

In addition, Eq. (8.37) can be considered as a convolution integral that propagates the data recorded at a level to a lower one. Mathematically, the $Z = 0$ plane has no special features and is only a plane in different parts of which the data are recorded. Therefore when the ambient includes several layers with different materials and the wave velocity is constant in each layer, the received data can be propagated from one layer to the lower one using the aforementioned integral. Based on the appendix to Schneider's study (Schneider, 1978), the Fourier transform in Eq. (8.37) can be written, as follows:

$$\Psi\left(k_x, k_y, z + \Delta z, \omega\right) = \Psi\left(k_x, k_y, z, \omega\right) . H\left(k_x, k_y, \Delta z, \omega\right) \tag{8.38}$$

where,

$$H = \exp\left(\pm j\Delta z \sqrt{\left(\omega/c\right)^2 - k_x^2 - k_y^2} \right) \tag{8.39}$$

This equation means that the downward propagation is regarded as a phase shift operation for a constant velocity ambient. Another form of the migration process is obtained by deriving from the right side of Eq. (8.37) based on Z. By utilizing the chain rule, we have:

$$\frac{\partial}{\partial z}\left(\frac{f\left(r', 2R/c\right)}{R}\right) = \frac{\partial R}{\partial z}\frac{\partial}{\partial R}\left(\frac{f\left(r', 2R/c\right)}{R}\right) = \cos\theta\left(\frac{1}{Rc} * \frac{\partial}{\partial R}f\left(r', 2R/c\right) - \frac{1}{R^2}f\left(r', 2R/c\right)\right) \tag{8.40}$$

After simplification and defining the virtual variable $t' = 2R/c$, we have:

$$\frac{\partial}{\partial z}\left(\frac{f\left(r', 2R/c\right)}{R}\right) = \frac{\cos\theta}{Rc}\left(\frac{\partial}{\partial t'}f(r', t') - \frac{c}{R}f(r', t')\right)\Bigg|_{t'=2R/c} \tag{8.41}$$

By substituting Eq. (8.41) in Eq. (8.40), $\Psi(r)$ is calculated as the intensity of the wave reflected from the point P located in the coordinates $r(x_p, y_p, z_p)$, as follows:

$$\Psi(r) = \frac{1}{2\pi} \int\limits_{z'=0} \frac{\cos\theta}{R.c}\left[\frac{\partial}{\partial t'}f(r', t') + \frac{c}{R}f(r', t')\right]\Bigg|_{t'=2R/c} da' \tag{8.42}$$

where, θ indicates the angle between the line perpendicular to the $Z = 0$ plane and the line connecting the point P and point of view, and $f(r', t')$ is the solution to the wave equation on the $Z = 0$ plane, which is considered as a temporal-spatial signal. The second term inside the bracket is insignificant compared to the first one and is usually eliminated due to being divided by R (Scales, 1995). Therefore, we have:

$$\Psi(r) = \frac{1}{2\pi} \int_{z'=0} \frac{\cos\theta}{R.\,c} \left[\frac{\partial}{\partial t'} f(r',t') \right] \Bigg|_{t'=2R/c} da' \tag{8.43}$$

Thus the intensity of the wave reflected from each point in space can be obtained by knowing the solution to the wave equation at all the points on the Z = 0 plane (X–Y plane) and applying Eq. (8.43). Note that the wave reflection intensity at any point reflects its electromagnetic properties. The wave reflection intensity at a point is considered as significant when a metal target is located there.

Note that the solution to the wave equation can only be measured at a limited number of points on the plane due to the limited number of antennas, which may reduce the image quality. The distance between the two adjacent antennas should not be more than half the transmitted pulse wavelength to achieve the appropriate image quality. The solution to the wave equation is measured only in a limited part of the Z = 0 plane, in addition to the discrete points of the measurement. Such limitations affect the image quality and should be larger than the target dimensions if possible.

Measuring the solution to the wave equation on the line Z = 0 (X-axis) suffices for achieving a 2D image of the target. The basis relation of Kirchhoff's migration, i.e., Eq. (8.43), can be summarized in two dimensions as written in Eq. (8.44) (Karami et al., 2019).

$$\Psi(r) = \frac{1}{2\pi} \int_{x'} \frac{\cos\theta}{R.\,c} \left[\frac{\partial}{\partial t'} f(r',t') \right] \Bigg|_{t'=2R/c} dl' \tag{8.44}$$

Fig. 8.10 illustrates obtaining a 2D image using the Kirchhoff algorithm (Karami et al., 2016). The solution to the wave equation is measured and stored at different points on the X-axis at the equivalent

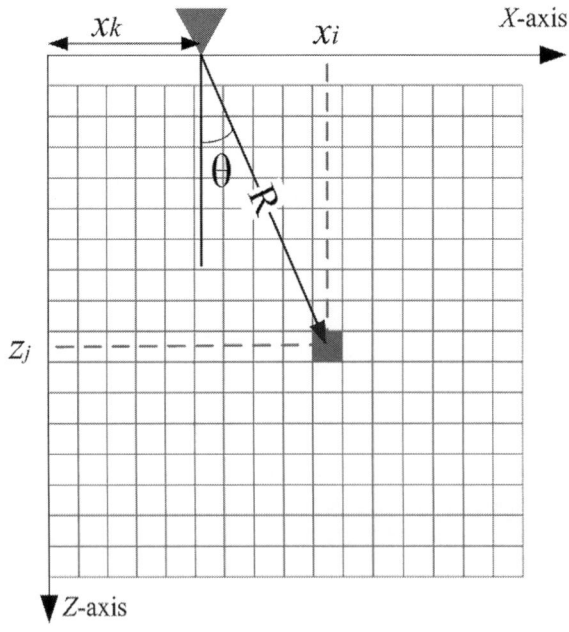

FIGURE 8.10

2D imaging utilizing Kirchhoff method.

distances by moving the antenna on the X-axis. The signal measured at the point x_k is demonstrated by $f(x_k, t')$. The surface is divided into small square elements. Then, each element is specified with an index (i, j), where i and j indicate the number of corresponding parts on the X- and Y-axes, respectively. The center point coordinates in each element are displayed with $r = (x_i, z_j)$.

Any intended element can be assumed as a potential wave propagator. A wave propagator is hypothesized to be located at the center point of the element with index (i, j), in which the emitted waves reach the boundary and are received by the antenna. Therefore, the Kirchhoff integral can calculate the initial emitted wave, i.e., Eq. (8.44). The Kirchhoff integral has a value at a given point when an object is located there. Otherwise, the Kirchhoff integral at that point equals almost zero. A 2D image of the target is obtained by repeating the abovementioned process for all of the points of the X-Z plane.

Note that the signals stored in each antenna position are considered as discrete-time functions. Thus, the time derivative in Eq. (8.44) is replaced by the difference. In addition, the number of points on which the data are recorded is limited on the X-axis, and the integral is converted to a summation as shown in Eq. (8.45).

$$\Psi(r) = \frac{1}{2\pi} \sum_{xk} \frac{\cos\theta}{R.c} [f(x_k, t+1) - f(x_k, t)]|_{t=2R/c} \tag{8.45}$$

Therefore, the image brightness in each point of the X-Z plane can be calculated by the following steps:

Step 1. The transmitter and receiver antennas are moved simultaneously on the X-axis. The reflected waves are received at different points of the X-axis $(x = x_k)$ and stored as a time signal. The received datasets are stored as a discrete bivariate function, $f(x_k, t)$.

Step 2. Two processing steps, including standardizing the time base and selecting the appropriate time window, are performed on the received signals.

Step 3. The X- and Z-axes are divided into M and N parts, respectively, resulting in having $M \times N$ pixels. The coordinates in each pixel are displayed by (x_i, z_j), where i and j indicate the part number on the X- and Z-axes, respectively. The coordinates in each element are observed as $P_{j,k} = (x_j, z_k)$.

Step 4. Two processing steps, including standardizing the time base and selecting the appropriate time window, are performed on the received signals.

Step 5. Value the parameters initially, such as $i = 1, j = 1, k = 1, \Psi = [0]_{M*N}$.

Step 6. Calculate the time difference of (x_k, t) and call it h.

$$\forall t: \ h(x_k, t) = f(x_k, t+1) - f(x_k, t) \tag{8.46}$$

Step 7. Calculate the distance between the element $P_{i,j}$ and the point of view (x_k).

$$R = \sqrt{(x_i - x_k)^2 + z_j^2} \tag{8.47}$$

Step 8. $\cos\theta$ is calculated by applying the following ratio.

$$\cos\theta = \frac{z_j}{R} \tag{8.48}$$

Step 9. $\Psi(x_i, z_j)$ is calculated using the following feedback equation.

$$\Psi(x_i, z_j) = \Psi(x_i, z_j) + \frac{1}{2\pi} \left(\frac{\cos\theta}{R \cdot c} \cdot h(x_k, t) \right)|_{t=2R/c} \tag{8.49}$$

Step 10. Set $k = k+1$. Go to Step 6 when $k \leq K$. Otherwise, go to the next step.

Step 11. Set $i = i+1$. Go to Step 6 when $i \leq M$. Otherwise, go to the next step.

Step 12. Set $j = j+1$. Go to Step 6 when $j \le N$. Otherwise, go to the next step.

Step 13. At each point of the X-Z plane, the value of the two-variable function $\Psi(x_i, z_j)$, indicates the amount of wave reflection. Draw the function as a 2D image whose color at each point represents the absolute value of the function value (Karami, Gharehpetian, Norouzi, et al., 2020). It should be noted that the data can be recorded in different parts of the X−Y plane, resulting in forming a 3D image.

8.3 Defects detection using SAR imaging method
8.3.1 Setup description

During all tests of this section, the imaging procedure uses the same layout consisting of two main parts: the PulsON transceiver and Vivaldi antennas, which were introduced in previous chapters. Fig. 8.11 shows the setup front view. As observed, the path of the antennas is alongside the Y-axis. According to the presented method, the final image is on the X-Z plane, which represents the top view of the winding. A lifting system has been designed and built to move the antennas in the above-mentioned direction. In addition, a calibrated tape has been mounted on one of the lift arms along the vertical movement path to identify the measuring points at a distance of 2 cm. At each point, 100 scans are taken, and the transformer image is obtained after averaging and processing the data. The number of measuring points equals 24 points in the aforementioned system, which cannot be increased due to the lift height limit; however, it covers the transformer height well to achieve the final image.

An actual single-phase winding of a power transformer with the characteristics specified in Table 8.1 is utilized in this chapter. Fig. 8.12 illustrates the image related to the high-voltage (HV) and low-voltage (LV) windings. The EMWs cannot penetrate inside the HV winding. Therefore, all of the obtained information from the received signals is related to the HV winding conditions (Rahbar-imagham et al., 2019).

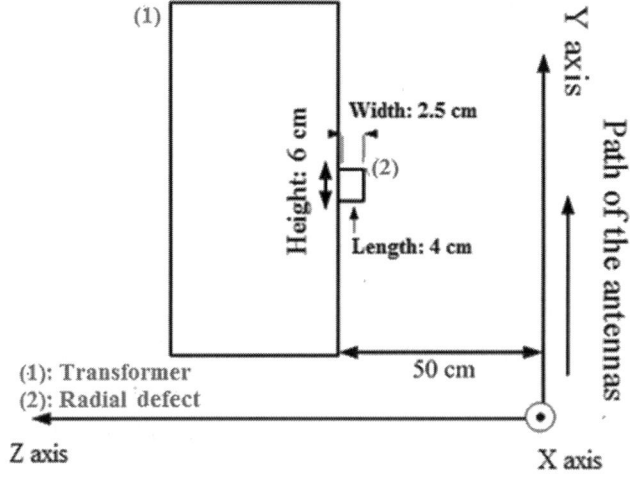

FIGURE 8.11

Imaging set-up front view.

Table 8.1 Specifications of understudy transformer winding used in tests.	
Nominal power	**80 kVA**
Winding height	536 mm
HV winding	38 disks
	External diameter: 353 mm
	Internal diameter: 237 mm
LV winding	Two layers with 27 turns and three output conductors
	External diameter: 212 mm
	Internal diameter: 186 mm

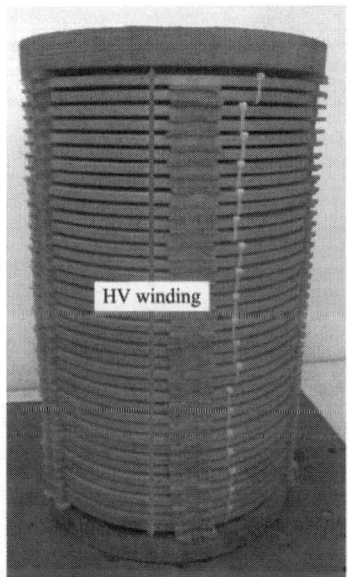
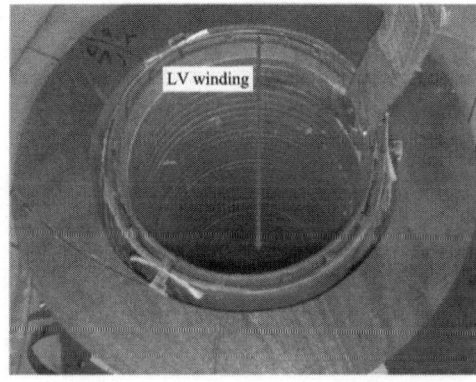

FIGURE 8.12

Single-phase power transformer winding utilized in tests.

8.3.2 Axial displacement

8.3.2.1 Detecting bulgy radial defect

To model the bulgy radial defect, a laboratory defect sample, which can be displaced along the vertical axis of the transformer, should be built. To this aim, as demonstrated in Fig. 8.13, the radial defect (RD) includes stacking three copper strips with total length, width, and height of 4, 2.5, and 6 cm, respectively, which are selected considering the width of the bulgy RD in the reported cases (Karami et al., 2016), which is equal to about 15% of the transformer height covers about 4-5 disks.

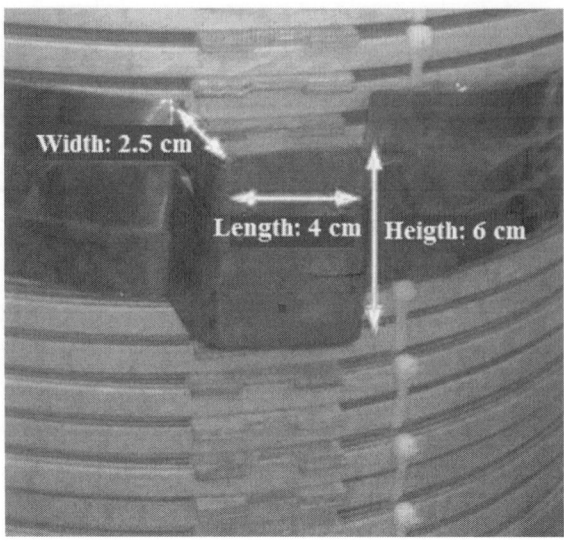

FIGURE 8.13

RD model.

During the tests, the RD is located at the bottom, middle, and top of the transformer winding with a distance of 7, 25, and 40 cm from the lowest part, respectively. The image of sound state (without RDs) should be achieved to be used as a reference image and be compared with any of the defected states for detecting the RDs. Then, the relative location of the RD is detected by comparing the images related to the different parts of winding, which is explained in the next. The image of the sound state is shown in Fig. 8.14.

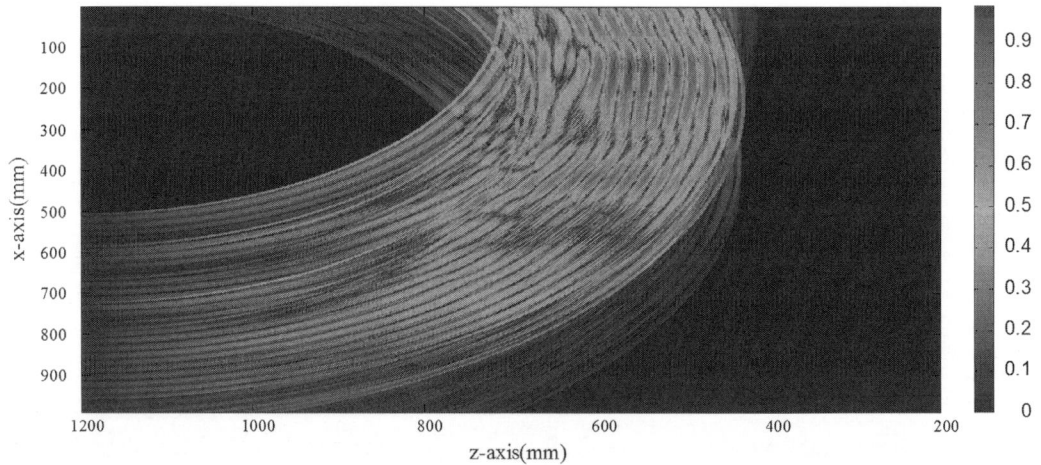

FIGURE 8.14

Winding sound state image.

As observed, the semicircles related to the transformer windings are detected well and more warm colors are perceived in the central area of the image as the closest point to the antennas. In addition, the colors in the aforementioned area are arranged in a regular and almost symmetrical manner, indicating the sound state of the windings.

Fig. 8.15 displays the images related to the three different RD locations indicated along the height of the winding. The warm color is increased compared to the sound state for all the RD locations. Further, the amount of distortion and discontinuity is increased compared to Fig. 8.14. Thus the ability to detect the RD in the vertical imaging method is proven. Furthermore, Fig. 8.15 indicates that the color turbulence is more increased for RD on the top and bottom of the winding compared with the RD located in the middle of the winding. Therefore the imaging method can be utilized to detect the relative location of the RD on the transformer.

8.3.2.2 Detecting concave radial defect

The concave deformations appear in the form of buckling or a combination of bumps and depressions. The RD created on the understudy winding at a distance of 2 cm from the upper part of the winding with a length, width, and height of 6, 1, and 3.5 cm, respectively, is displayed in Fig. 8.16. The defect covers four disks of the HV winding.

The imaging is performed based on the steps described in Section 8.2.3 and the generated image is shown in Fig. 8.17. Comparing Fig. 8.17 with Fig. 8.14, which is related to the sound state of the transformer winding, indicates that the presence of RD is well detected since the warm colors and turbulence in Fig. 8.17 have clearly reduced.

8.3.2.3 Locating radial defect

This section seeks to identify two main features in images obtained from SAR imaging, including the number of warm colors and turbulence to locate RD. Therefore, a quantitative definition of the terms should be provided.

Fig. 8.14 illustrates the image related to the sound state of the winding. As displayed in the color bar of the image, each of the colors applied in the images equals a color number between zero and one, starting with cold colors and ending with warm colors, respectively. Thus, the number of warm colors in an image can be determined by calculating the number of points with a number related to warm colors. After assessing the numerous images obtained from SAR imaging, the number of 0.45 is selected as the threshold of warm colors. Therefore, the high/low number of points with a color number greater than 0.45, N_{warm}, in an image compared with the reference image means a defect as one of the comparative parameters.

The amount of warm color scattering in an image obtained in the presence of an RD compared with the reference image is considered as the next comparative parameter. To this aim, the color number values of warm colors are summed and called T_{warm}. This value is used for comparison of images with reference to show how much whole of an image is warmed or cooled. Any difference of N_{warm} and T_{warm} can show the defect occurrence compared with the reference state.

8.3.2.3.1 Detection and locating RD along height

Single-section method: As discussed before, a visual comparison of sound and defected images can be useful to detect the occurrence of a defect. This funding is proved now by calculating the difference values of N_{warm} and T_{warm} between the image obtained from the defected state of the winding and that of the sound state, named the single-section method. The results are listed in Table 8.2.

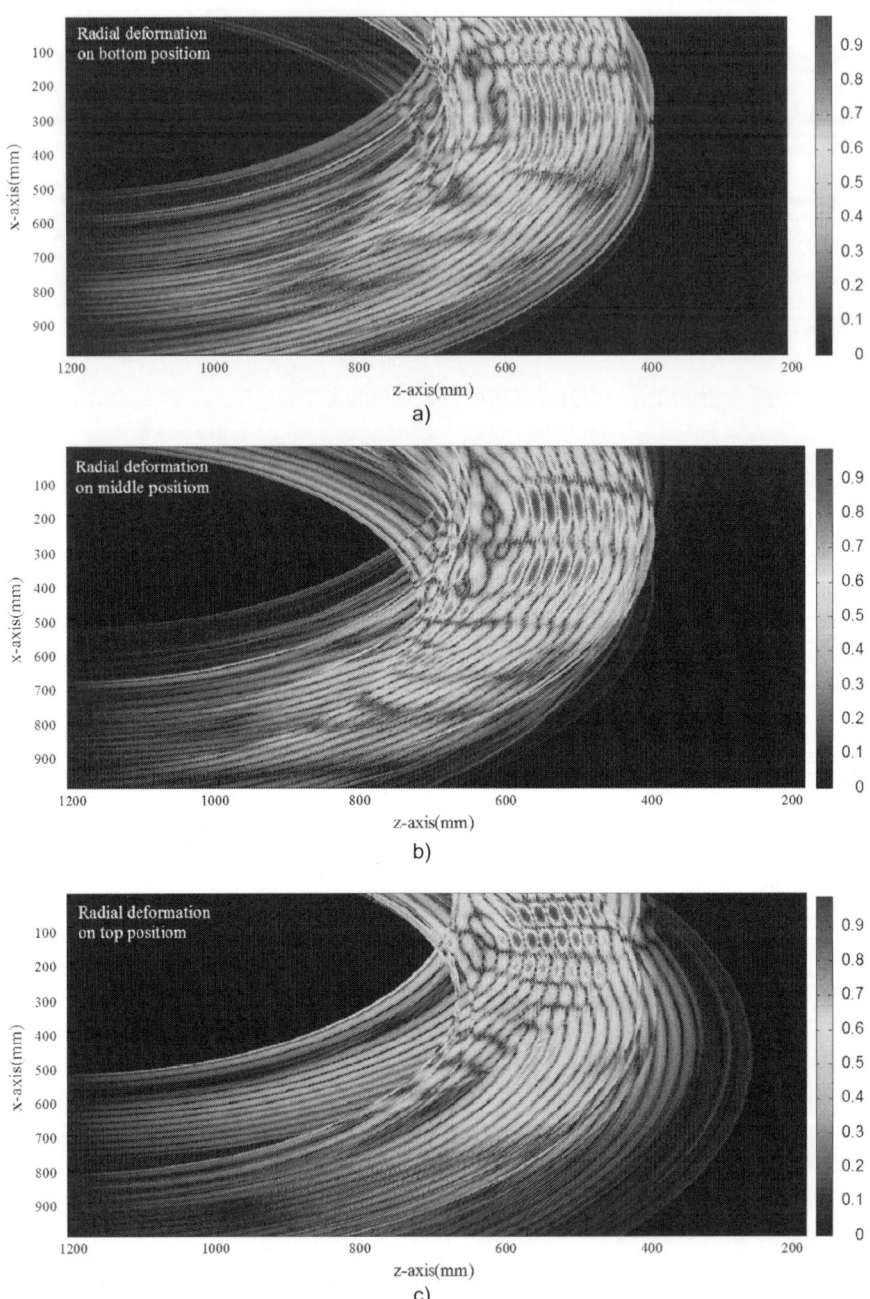

FIGURE 8.15

Images from defected state of transformer in imaging, (A) RD in lower part, (B) RD in middle part, and (C) RD in upper part.

FIGURE 8.16

Actual RD created on understudy winding.

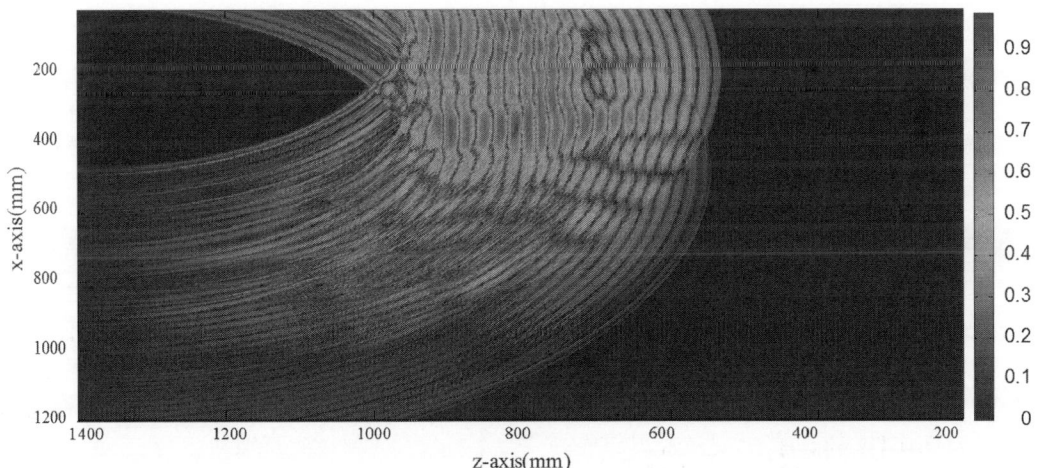

FIGURE 8.17

Image of actual concave RD on winding.

Table 8.2 Results of applying "single-section" method on images of winding for different RD locations.

RD at lower part of winding	N_{warm}	13905
	T_{warm}	6.6988e+03
RD in middle of winding	N_{warm}	17426
	T_{warm}	8.4919e+3
RD at upper part of winding	N_{warm}	19169
	T_{warm}	9.2055e+3

As represented in Table 8.2, the results achieved in the visual form are quantitatively obtained here, indicating that the presence of warm colors in the image of the defect in the lower part of the winding (Fig. 8.15A) is less than the image of the defect in the middle part (Fig. 8.15B) and similarly, less than the presence of the defect in the upper part (Fig. 8.15C). The single-section method only detects the presence of a defect and has no practical and useable information about its location. Next, a practical approach for RD location is suggested by increasing the sections.

Two-section method: In this part, the method accuracy in identifying the defect location is examined. In the "two-section" method, the antennas moving path length is divided into two sections. For example, the sound state image obtained from SAR imaging at 24 points is divided into two 12-point sections, one for the lower and the other one for the upper half of the winding. The same division is made on the images achieved from the defected states of the winding. Then, the values of N_{warm} and T_{warm} are calculated and compared between sound and faulty states. This method should be carried out after applying the single-section method, which means that the defect in the winding has been detected, and the question is whether the RD location is at the upper half of the winding height or the lower. So, there is no need to calculate differences like the single-section method. The results are listed in Table 8.3. The values of N_{warm} and T_{warm} in the image related to the upper half of the transformer are higher where an RD is at the upper part of the winding. The same argument is true for the presence of an RD at the lower part of the transformer. Thus, the location of the RD on the transformer winding can be determined by comparing defined parameters in two sections. It should be noted that by using the two-section method, only the presence of RD in the upper or lower part of the winding is evaluated since the images are divided into upper and lower parts. Increasing locating accuracy by increasing the number of sections is evaluated in the next section.

Table 8.3 Results of applying "two-section" method on images of winding for different RD locations.

RD at lower part of winding	N_{warm}	Image of upper part: 27178
		Image of lower part: 28344
	T_{warm}	Image of upper part: 1.2427 e+04
		Image of lower part: 1.4599 e+04
RD at upper part of winding	N_{warm}	Image of upper part: 28171
		Image of lower part: 22384
	T_{warm}	Image of upper part: 1.1074 e+04
		Image of lower part: 9.8388 e+03

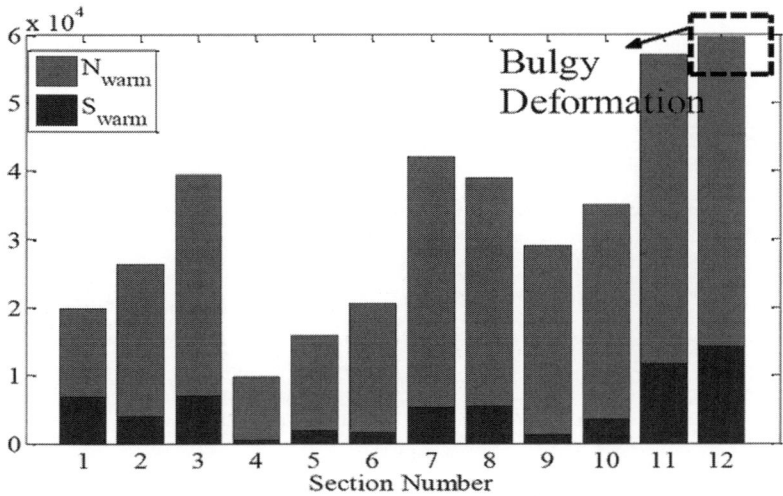

FIGURE 8.18

Bulgy RD locating with 12-section method for defect at bottom of winding.

12-section method: The mentioned procedure is conducted by dividing the winding height into 12-sections and generating an image of each section, meaning that the location of the RD can be achieved with an accuracy of $\frac{1}{12} \times 53$ or 4.41 cm. Fig. 8.18 shows the results implemented with the 12-section method where the RD is at the bottom of the winding. A similar procedure can be applied for bulgy or concave RD at any winding location. For example, Fig. 8.19 shows the results of applying the 12-section method for locating a concave RD. The accuracy of locating can be improved by increasing the number of sections.

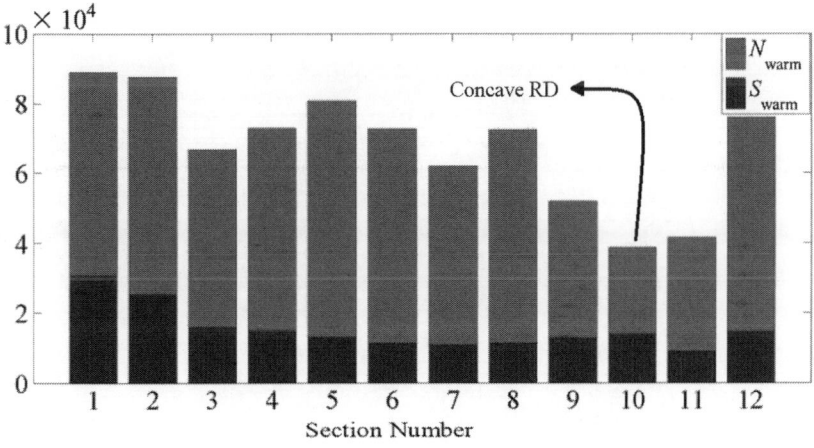

FIGURE 8.19

Concave RD locating with 12-section method.

8.3.2.3.2 Locating RD on circumscribed circle

Based on the previous results, SAR imaging can be used as an efficient method for detecting the winding RD. To indicate the efficiency of the abovementioned method in detecting the location of an RD on circumscribed circle, specific tests are performed, in which an RD is created in a winding model in different degree locations, and SAR imaging is performed. Then, using the KMM, 2D images for each state are obtained and compared visually. To this aim, four different cases are designed. In the first case, the winding is regarded as to be sound without any RD. In the second to fourth cases, an RD occurs in three different positions of the circumscribed circle.

Fig. 8.20 illustrates the resulting image in the first (sound) case, in which a schematic of the winding top view is placed on the image as a red circle. In the front of the winding, several arcs have been appeared in the resulting image. The demonstrated arcs are considered as uniform and have a slight color change, indicating that the winding is regarded as intact and there is no RD on its surface.

Fig. 8.21 displays the images obtained in the second to fourth cases, in which the winding top view is presented on the image as an arc in black. As observed in Fig. 8.21A and B, by comparing the generated image and location of the winding defect, it is concluded that the defect location is well detectable and corresponds to the actual condition. Where the defect is at the center, as shown in Fig. 8.21C, there is a uniform color distribution with increased warm colors, compared with Fig. 8.20, which shows the defect occurrence. Therefore, an approximate location of the defect on the circumscribed circle can be detected by visual comparison.

8.3.3 Detecting and estimating axial displacement

The understudy winding is hard to be displaced for axial displacement studies due to its weight and available facilities. So, a light-winding model should be utilized, which must be close to an actual transformer winding. Fig. 8.22 demonstrates this model. The axial displacement of the winding is studied by moving the entire winding model along its axis. Fig. 8.23 and Fig. 8.24 display the schematic and laboratory setups, respectively.

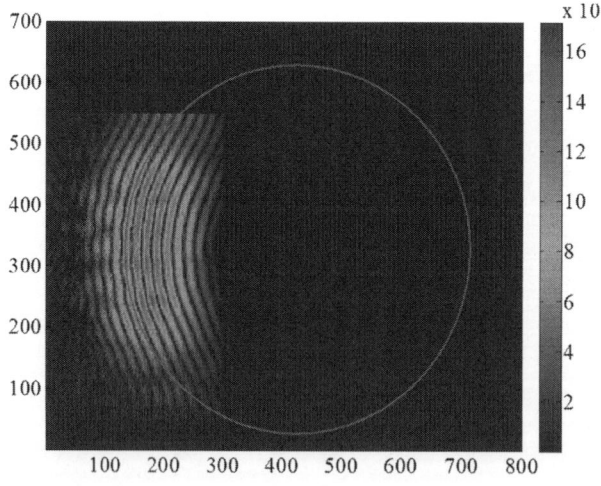

FIGURE 8.20

Image of sound state and schematic of winding top view.

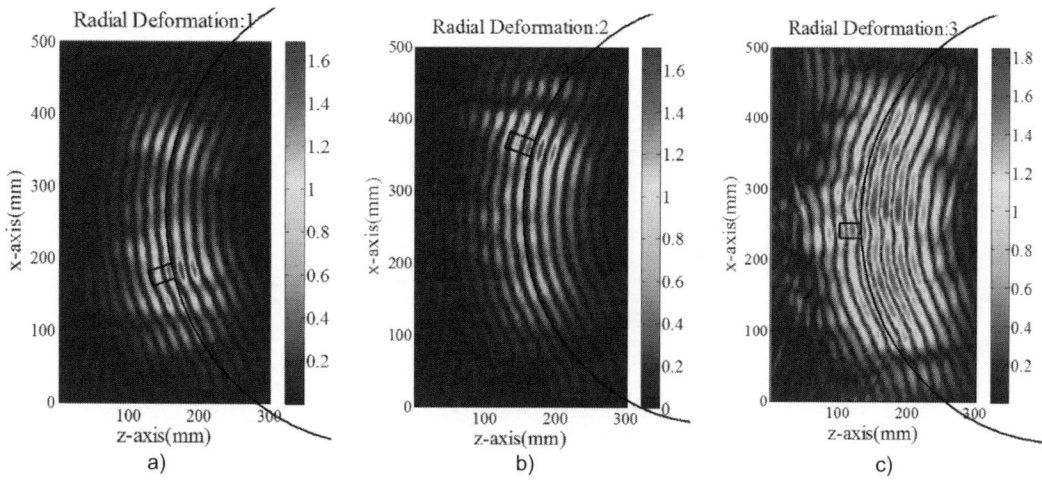

FIGURE 8.21

Images of defected states for different locations of RD on circumscribed circle.

FIGURE 8.22

View of light transformer winding model for axial displacement studies.

The axial displacement study is performed in three different states. The winding model is placed in a specific position in the first state, which is assumed to be sound with no axial displacement. In the second and third states, the model is moved by 1 and 2 cm along the cylinder axis, equal to 1.5% and 3% of the model height, respectively. It means that we have three states: no axial displacement, 1 cm displacement, and 2 cm displacement. The procedure of SAR imaging and generating images is similar to the RD study.

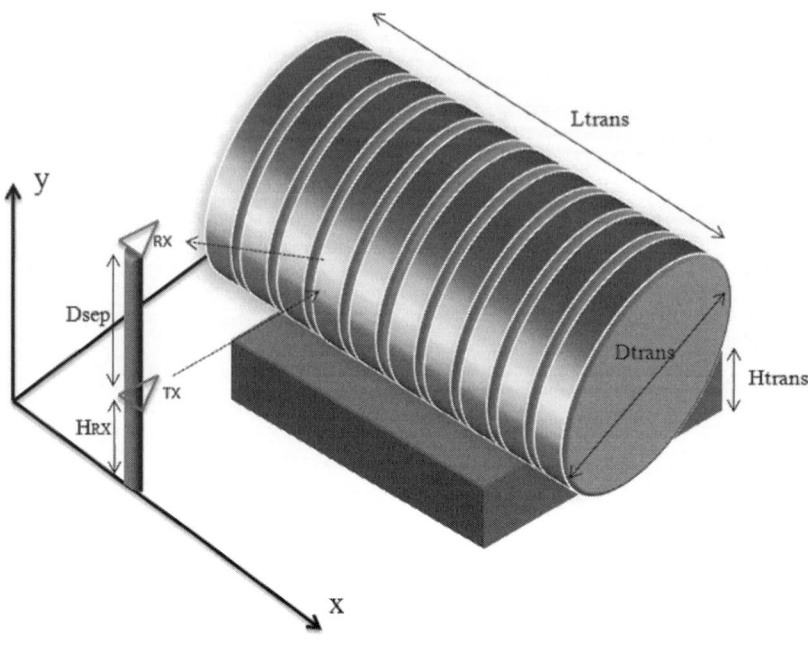

FIGURE 8.23

Schematic of axial displacement test setup.

FIGURE 8.24

Laboratory setup of axial displacement studies.

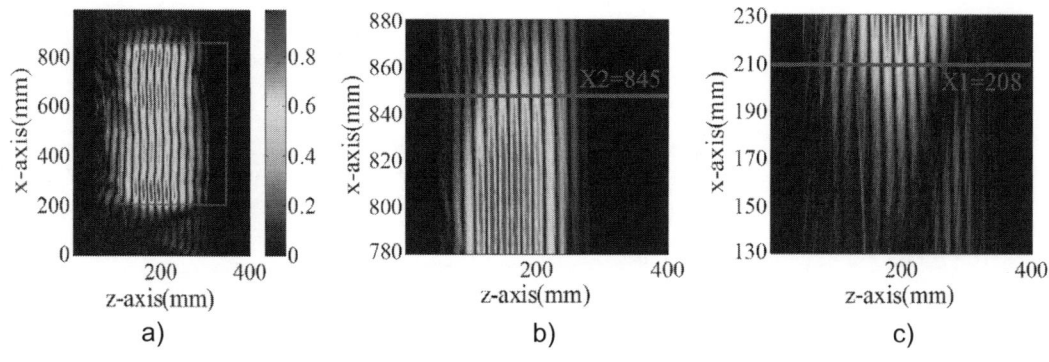

a)   b)   c)

FIGURE 8.25

Results of sound state for axial displacement study, (A) main image, (B) upper part zoom, (C) lower part zoom.

Fig. 8.25A shows the generated image of the sound state of the winding model. In order to study axial displacement using SAR imaging, the approximate location of the top and bottom of the winding in the generated images should be defined. To this aim, the first point with a color number greater than the specified limit, such as 0.4, is found from the top/bottom of the image to demonstrate the start/end position of winding in the image. The actual location of the model is illustrated with a red rectangle in Fig. 8.25A. The image related to the winding model appears in several vertical lines, the start and end points of which indicate the model position along the X-axis. The upper and lower parts of the image are magnified to determine the exact position of the winding (Fig. 8.25B and Fig. 8.25C).

In Fig. 8.25B, the end point, marked by a red horizontal line and displayed by X_2, indicates the position of the top of the model cylinder. This line shows that there is no point with a color number greater than 0.4 above it. Similarly, the end point of the lower part of the image is marked by a red line, and its location is indicated by X_1. This line determines that the first point from the bottom of the image with a colored number greater than the specified limit is found, and this horizontal line is drawn through it.

The images achieved from the second and third states are displayed in Fig. 8.26 and Fig. 8.27, respectively. The position of the model in each of the images is indicated by X_1 and X_2 lines. The position of the center of the model is calculated using the values X_1 and X_2, as follows:

$$X_{center} = \frac{X_1 + X_2}{2} \tag{8.50}$$

Table 8.4 summarizes the test results and presents the values X_1, X_2 and the position of the model center. The amount of the axial displacement in each state is calculated by subtracting the position of the model center in the sound state from that position in that state. Finally, the percentage of test error is calculated by comparing the measured axial displacement with its actual value. The highest error is related to the third state, which equals 25%. Therefore, the SAR imaging method can help to estimate the amount of axial displacement.

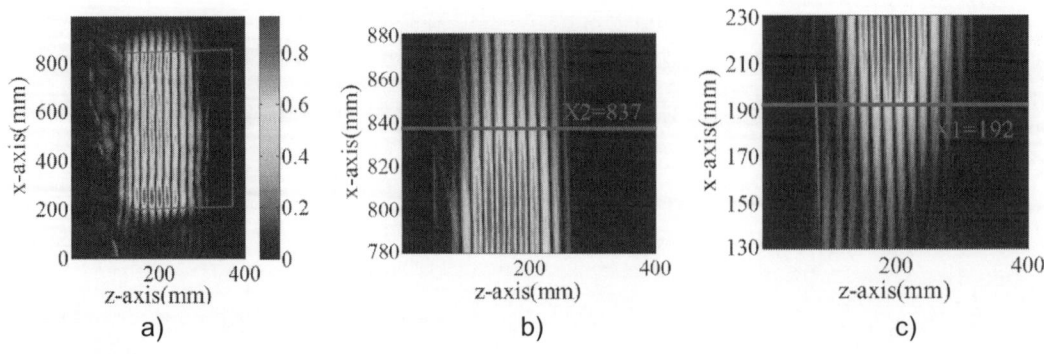

FIGURE 8.26

Results of 1 cm axial displacement, (A) main image, (B) upper part zoom, and (C) lower part zoom.

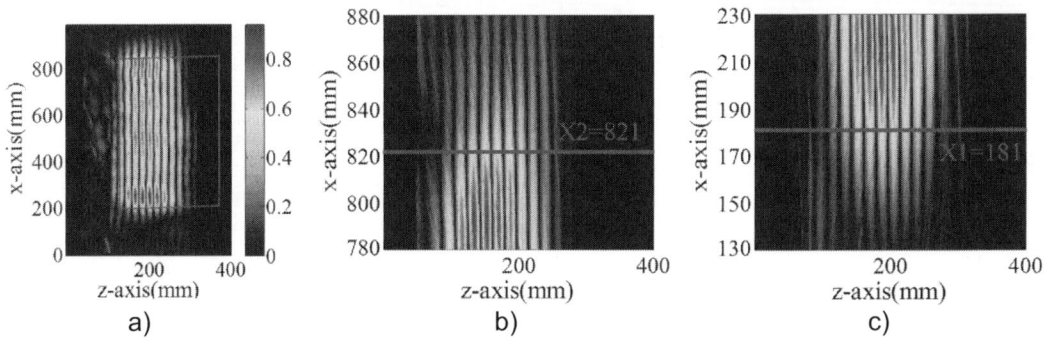

FIGURE 8.27

Results of 2 cm axial displacement, (A) main image, (B) upper part zoom, and (C) lower part zoom.

Table 8.4 Axial displacement approximation for different states.			
Axial displacement in mm	0	10	20
Location of the bottom of the cylinder in mm (X_1)	208	192	181
Top position of the cylinder in mm (X_2)	844	837	821
Location of the cylinder center in mm (X_{center})	526	514.5	501
Measured axial displacement in mm	0	11.5	25
Percentage error (%)	-	15	25

8.3.4 Detecting simultaneous radial and axial defects

The winding model, which is closer to the actual structure of transformers, is utilized in the aforementioned test since the actual winding is hard to be moved due to its weight and available facilities.

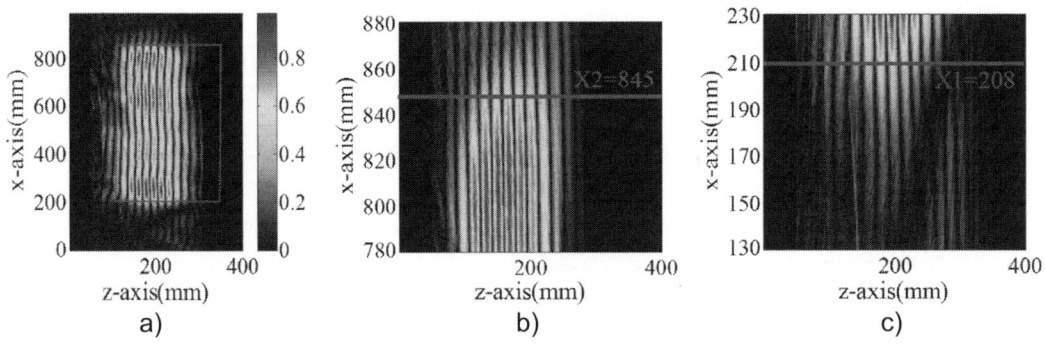

FIGURE 8.28

Images related to sound state for simultaneous detection study: (A) Whole image, (B) magnified upper part, and (C) magnified lower part.

In addition, the RD of the winding is modeled in its center, similar to Fig. 8.11. Three different cases are considered to evaluate the efficiency of the SAR imaging method in detecting radial and axial defects of the transformer winding simultaneously. In the first case, the winding model is in a sound state. In the second case, the model is moved by 20 mm, along the axis of the cylinder (X-axis), which is equivalent to 3% of the height of the model cylinder. In the third case, an RD with dimensions of $30 \times 40 \times 40$ mm^3 is created in the center of the cylinder, and the winding is moved by 20 mm in the axial direction.

Fig. 8.28 demonstrates the image of the winding model in the first state. In addition, the start and end parts of the model are displayed by a red rectangle. The upper and lower parts of the image, X_1 and X_2, are magnified to determine the exact position of the winding (Figs. 8.28B and C).

The images obtained from the second case are shown in Fig. 8.29, in which the model image is illustrated as (A) and the magnified ones are demonstrated as (B) and (C). In addition, the position of the model in each of the images is indicated by X_1 and X_2.

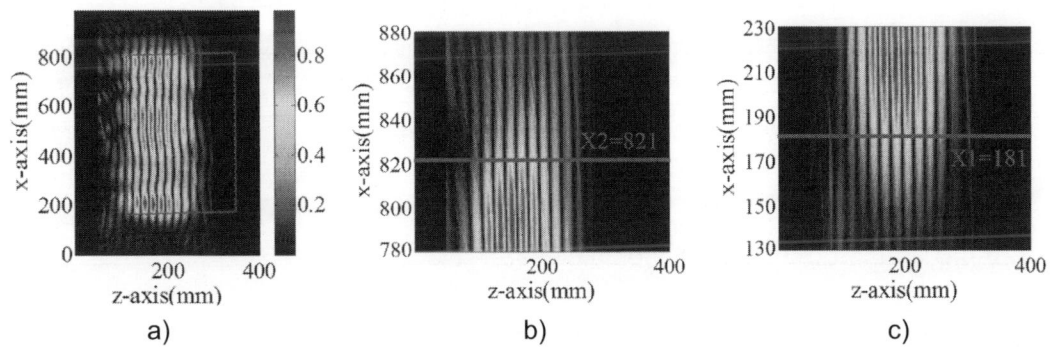

FIGURE 8.29

Images related to 20 mm axial displacement: (A) whole image, (B) magnified upper part, and (C) magnified lower part.

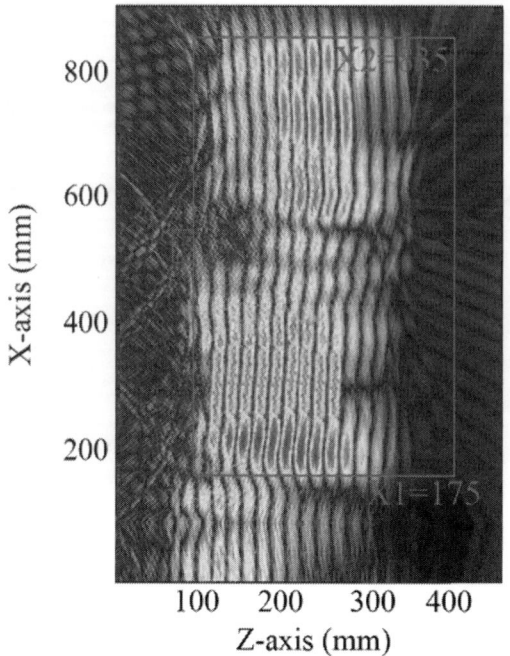

FIGURE 8.30

Image of third case with RD and axial displacement defects.

The image achieved in the third case is displayed in Fig. 8.30, in which the actual location of the winding is shown with a red rectangle. The RD is marked as a small rectangle. The model image appears as a few vertical lines, like in the previous cases. However, the appearance of the lines is regarded as chaotic compared to the previous cases, and a discontinuity is observed in the center. Note that the incidence of discontinuity indicates an RD in the model. In addition, the location of the discontinuity approximately equals that of the RD in the winding.

In the third case, the axial displacement of the model can be detected based on the location of the upper and lower parts of the model in addition to detecting radial deformation. Like the second case, the position of the model is marked with X_1 and X_2. Table 8.5 lists the test results. The errors in the second and third cases equal 25% and 5%, respectively. Therefore, the discussed method can simultaneously measure axial displacement and RD of the winding with an acceptable error.

8.3.5 RD detection on actual 3-phase repair transformer

8.3.5.1 Specifications of repair transformer

A 3-phase, damaged 30 MV A transformer is used to verify the method in actual dimensions. The diameter of each winding, shown in Fig. 8.31, is about 100 cm. The transformer specifications and test system configuration are listed in Table 8.6. To study the images related to sound and defective states,

Table 8.5 Results of simultaneous detection of radial and axial defects.

Parameter	Explanation	First case	Second case	Third case
Location of lower part of cylinder in mm (X_1)	Real	195	175	175
	Measured	208	181	175
Location of upper part of cylinder in mm (X_2)	Real	845	825	825
	Measured	844	821	835
Location of center of cylinder in mm	Real	520	500	500
	Measured	526	501	505
Amount of axial displacement in mm	Real	0	20	20
	Measured	0	25	21
Percentage of axial displacement measurement error		-	25	5

FIGURE 8.31

Transformer windings ready for repairing.

Table 8.6 Transformer and test system specifications.

Transformer cylinder diameter	100 cm
Distance of antennas from each other	80 cm
The height of the antennas from the ground at first	72 cm
Antenna height from the ground at the end	182 cm
The distance between two consecutive scans	2 cm
Number of measurement steps	55
Number of transformer phases	3 phases
The distance between antennas and transformer	50 cm
Number of scans per step	100
Time interval between transmitted pulses	50 ms
Antenna type	Vivaldi
Transformer power	30 MVA

FIGURE 8.32

Radial deformation on winding of 30MVA transformer.

the winding is considered as the sound state and a few millimeters concave RD is created on the winding, as shown in Fig. 8.32, after imaging the sound state. The sending and receiving process is performed at 55 points alongside the height of the winding as described. The distance between the antennas is about 80 cm. The distance between the antennas and the winding from the middle of the two antennas is about 50 cm, which equals the distance of the transformer tank to the winding surface.

8.3.5.2 Creating 2D image of transformer

The antennas are moved in steps of 2 cm to take images for each of the sound and defective states. The scanning procedure is repeated about 100 times in each step. After removing inappropriate data and passing the processing steps, including timing and selecting the appropriate window, averaging is performed for each step. The deformation is created at the top of the winding (Fig. 8.32). Figs. 8.33 and 8.34 demonstrate the sound and defective states of the understudy winding.

As displayed, the sound state has more warm points and indicates a concave RD in a part of the transformer. The number of points with the color number above 0.4 in the sound and defected states equals 165312 and 156593, respectively.

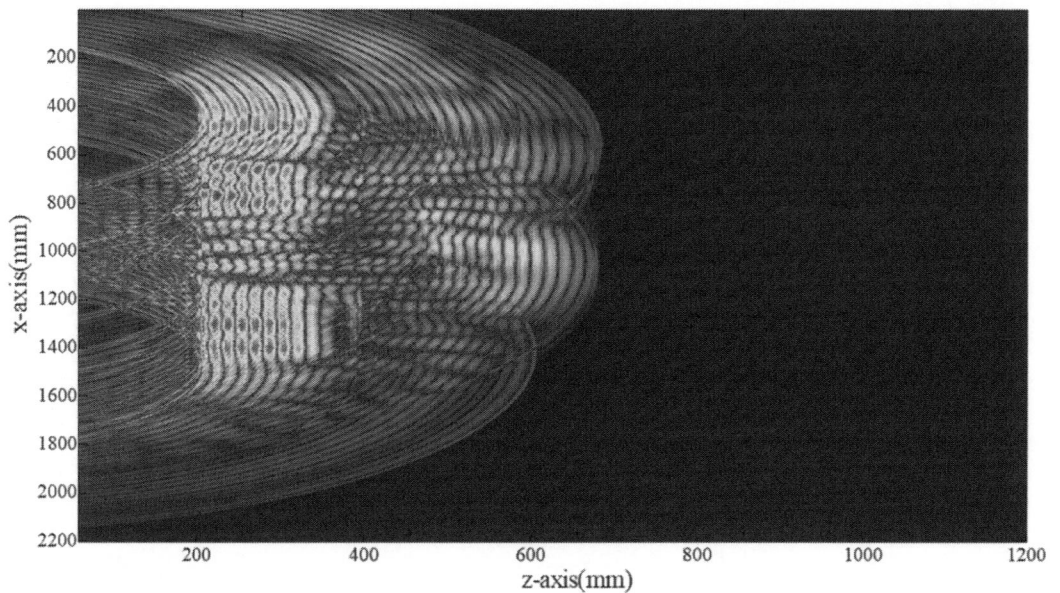

FIGURE 8.33

Generated image of actual 3-phase winding sound state.

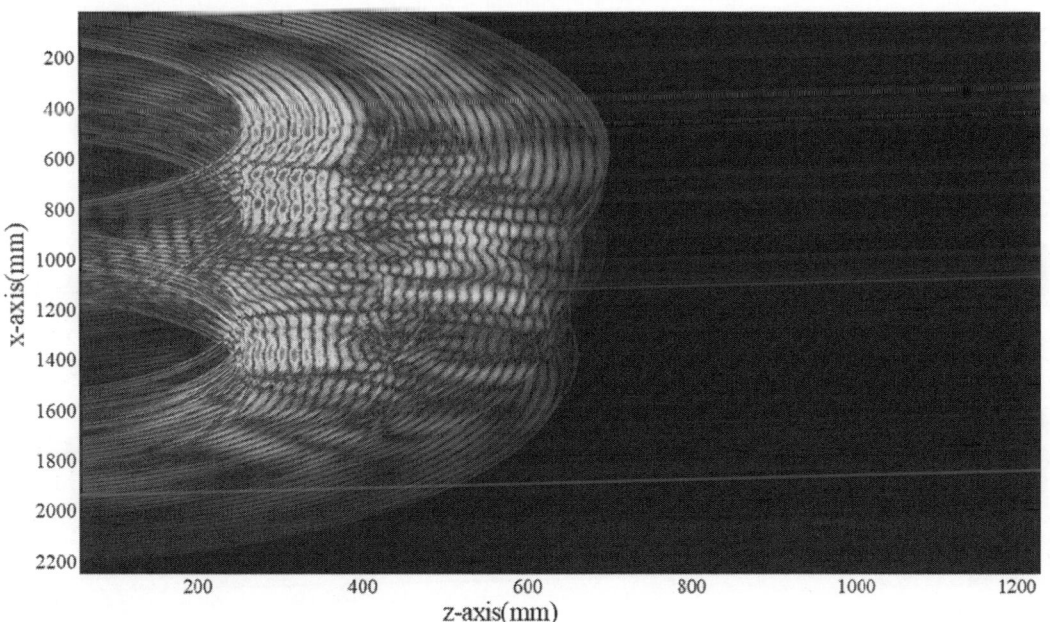

FIGURE 8.34

Generated image of actual 3-phase winding defected state.

8.3.6 Dielectric window

As discussed, mechanical defects including radial and axial ones can be detected by radar imaging of the winding. To this aim, the EMW should be transmitted toward the winding, and its reflection must be analyzed to monitor any deformation or displacement during the operation of the transformer. However, the transformer tank, as a conductor, does not allow waves to pass.

In fact, such a problem exists for detecting partial discharge (PD) defects. Different solutions have been proposed to monitor the PD defect in power transformers by EMWs and overcome this problem (Mahmoodi et al., 2020). In (Coenen, 2012; Judd et al., 2001, 2004), some examples of the afore-mentioned window are made to place an antenna for detecting the PD. In (Judd et al., 2004), a dielectric window (DW) is constructed to detect the PD in a single-phase 18 MVA transformer.

The DW allows UHF waves to exit from inside the transformer and be detected by an antenna installed in its behind, in addition to preventing oil leakage. Moreover, insulating the external environment of the DW and the installed antenna can ensure that the antenna does not measure the external noise. Further, the DWs help to monitor the transformer without a power outage. In (Judd et al., 2005b), a DW is made for PD detection in a 1000 MVA 275/400 kV transformer.

Power diagnostic service has recently developed a DW to monitor PD with UHF. The specifications of this window are listed in Table 8.7 (Judd et al., 2001). The DW has attracted a lot of attention, and some samples have been made to detect PD due to its easy implementation on the body of the transformer tank, no side effect on the performance of insulating components, easy installation and disassembly of the sensor during the operation of the transformer, easy maintenance, and most importantly its significant role in online monitoring of the PD and mechanical defects using electromagnetic antennas (Judd et al., 2005a).

To detect mechanical defects, a DW is needed to allow the antennas to move alongside the height of the transformer tank. However, the windows built so far are not considered appropriate for this purpose. The aim is to design a laboratory sample of such a window as a DW or valve. Designing such a window plays a significant role in terms of its impact on the useful life of the transformer tank by detecting PD and mechanical defects and can reduce the costs when it is included in the initial design of the tank (Karami et al., 2018). The window should have the following specifications:

- Impermeable to oil.
- Withstand the maximum temperature of the tank.
- Withstand the maximum pressure inside the tank due to oil, which is less than 1 bar for most transformers.
- Do not interfere with the integrity of the tank.
- Have sufficient mechanical strength against existing stresses.

Table 8.7 Specifications of DW made by power diagnostic service.	
Frequency range	200 MHz−1.2 GHz
Material of UHF enclosure	Stainless steel
Material of DW	PA 66 chopped fiberglass-reinforced nylon
Connector	N-type
Connection box	Waterproof box (IP 67) within surge absorber
Test lead	5D coaxial cable

8.3.6.1 Appropriate places for installing DW in 3-phase 125 MVA transformer

Based on the information received from the technical and design office of a power transformer manufacturer, the actual dimensions of a 125 MVA 66/230 kV transformer are displayed in Fig. 8.35. According to the information of the aforementioned transformer, a series of steel sheets are insulated and pressed together on the tank body, known as magnetic sheets, in front of each HV winding. Fig. 8.36 shows the insulation distances of the windings to these sheets by red curves.

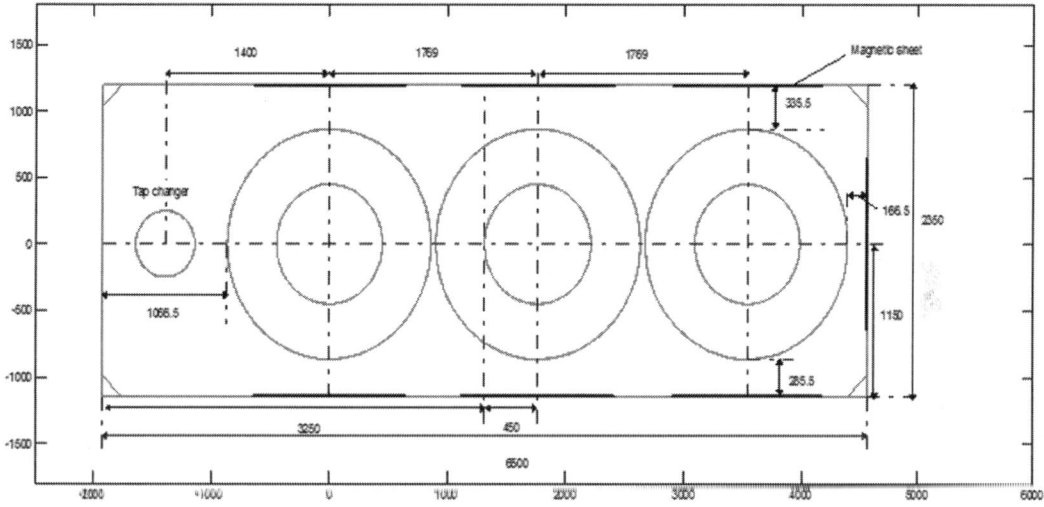

FIGURE 8.35

Actual dimensions of 125 MV 230/66 kV transformer.

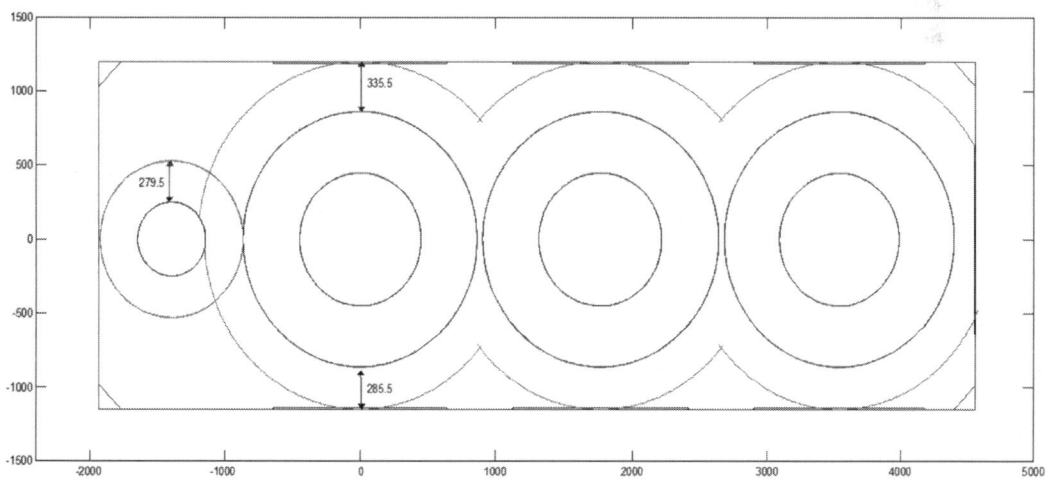

FIGURE 8.36

Insulation distances of MVA 125 66/230 kV transformer from steel sheets.

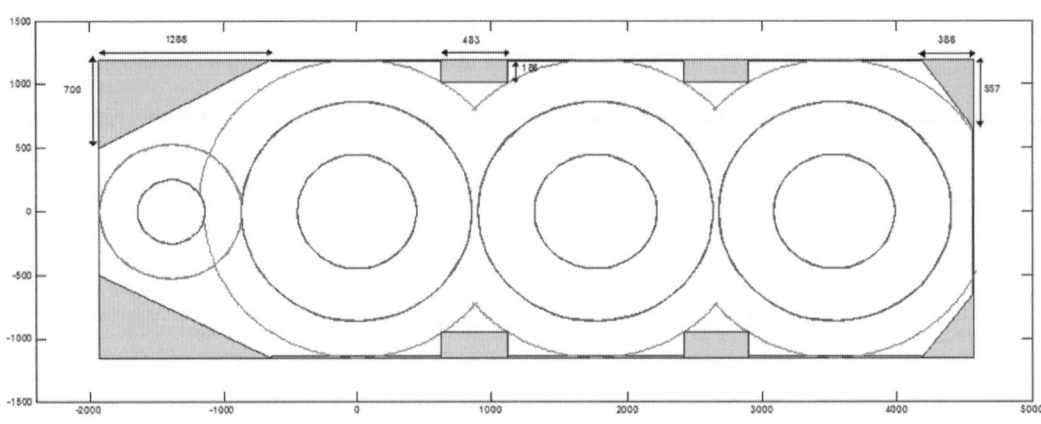

FIGURE 8.37

Candidate places for installing DW shown by green areas.

The candidate locations to install the DWs can be found by knowing the location of the magnetic sheets and the insulation distances of the windings. Fig. 8.37 illustrates the candidate places for installing DWs on the transformer tank.

8.3.6.2 Designing tank with DW for actual winding

Table 8.1 lists the characteristics of the winding. The minimum required tank dimension is 480 mm based on the information received from the manufacturer and is 63.5 mm for the dielectric distance. The corners of the tank, shown in Fig. 8.38, are considered appropriate locations for installing the DW.

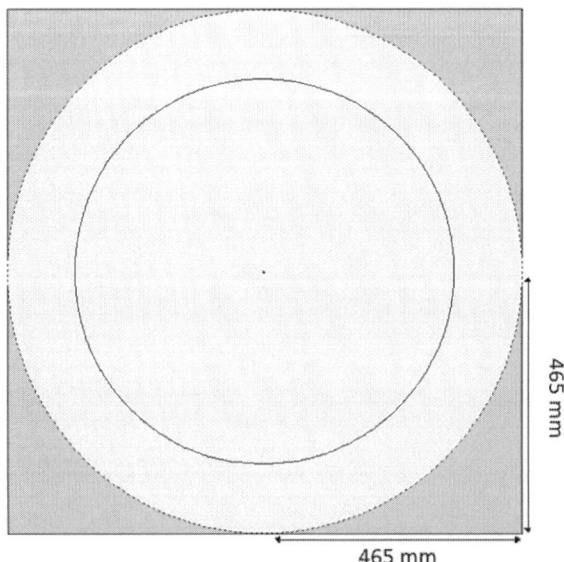

FIGURE 8.38

Appropriate locations of DW installation for 1-phase winding.

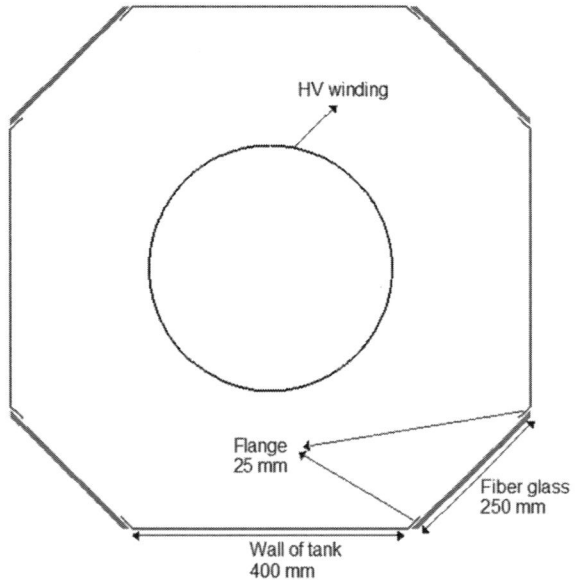

FIGURE 8.39

Final design of tank considering four DWs in its four corners.

The minimum distance between the antenna and winding should be 350 mm considering the field calculations around the Vivaldi antenna. Thus, the dimensions of the tank are those indicated in Fig. 8.39 regarding the four DWs in the four corners of the tank. Fig. 8.40 demonstrates the 3D view of the tank considering the DWs. The thickness of the tank is regarded to be 1 mm.

The mechanical analysis of the proposed design is performed by ABAQUS software. The simulations are performed at a temperature of 100°C and a pressure of 0.8 bar. As displayed in Fig. 8.41, only a quarter of the tank is modeled since the presented design for the tank is perfectly symmetrical. In addition, the models of the fiberglass sheet, flange, sealing washer, and screw are shown in Fig. 8.42.

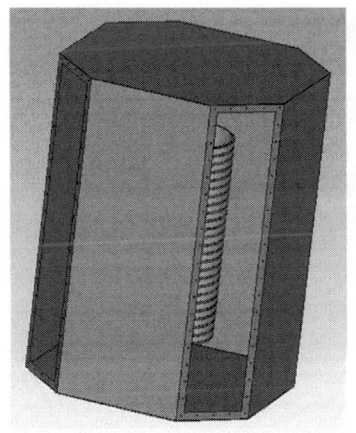 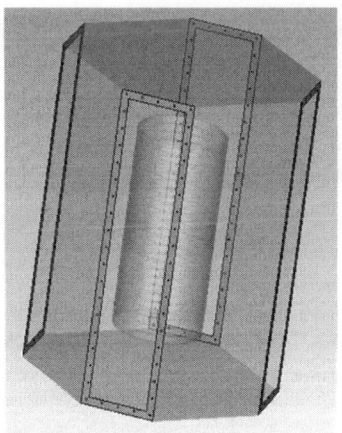 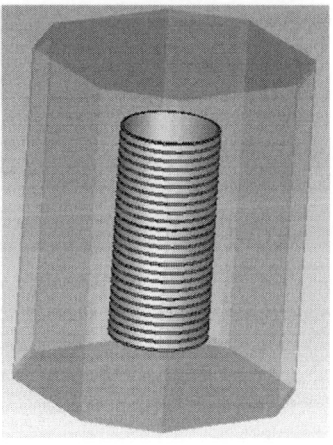

FIGURE 8.40

3D view of designed tank before placing fiberglass.

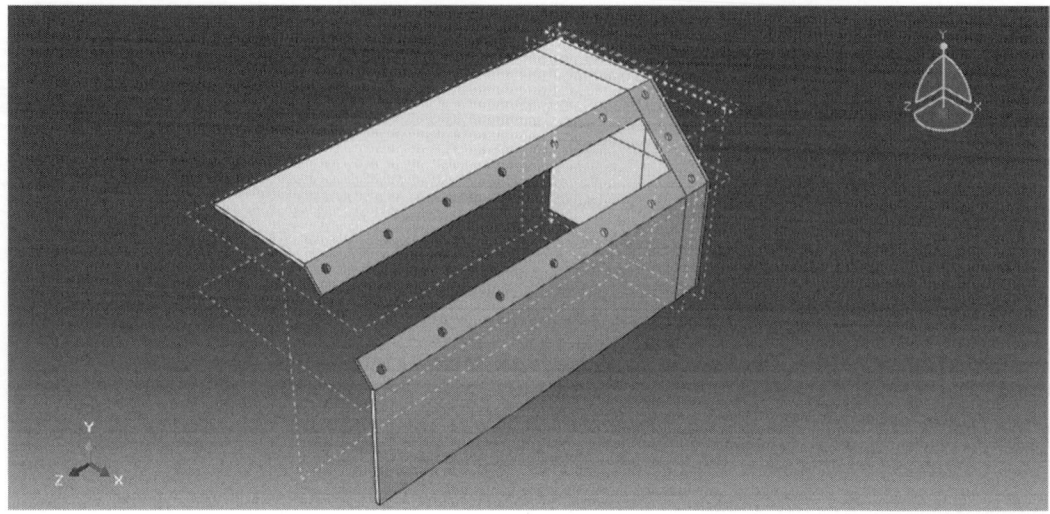

FIGURE 8.41

Model of quarter of the offered tank in ABAQUS software.

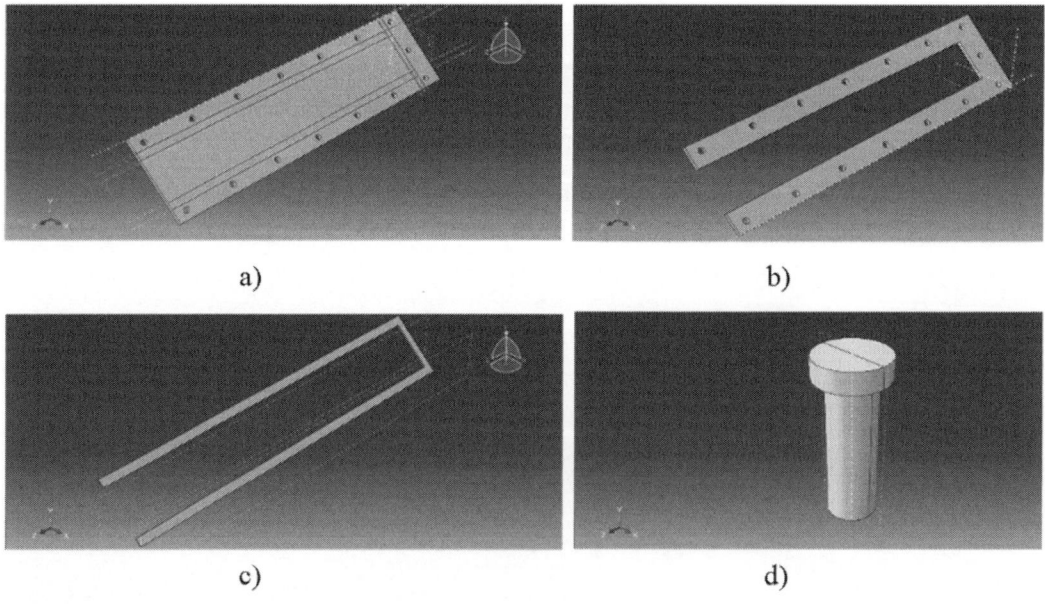

a)

b)

c)

d)

FIGURE 8.42

Models of different parts of proposed tank in ABAQUS software: (A) Fiberglass, (B) flange, (C) sealing washer, and (D) screw.

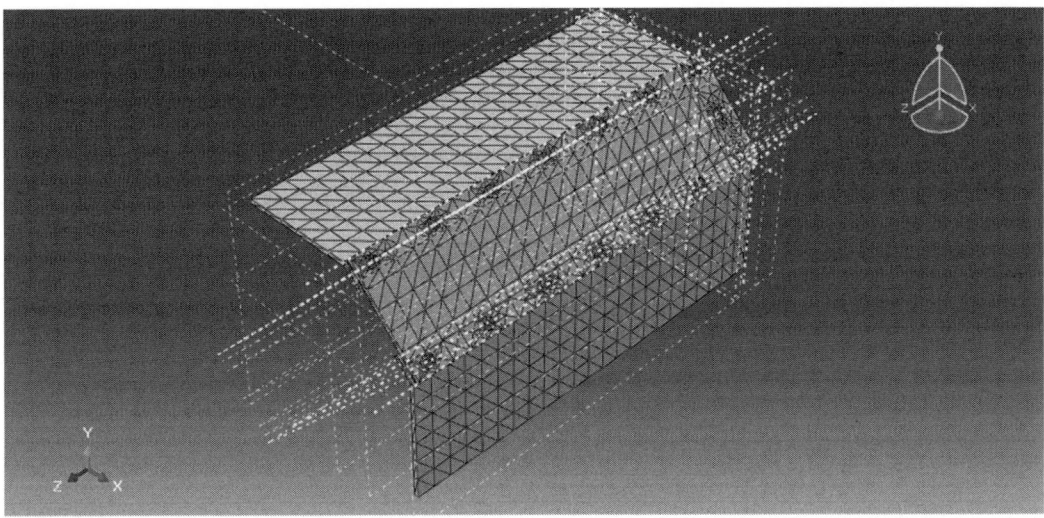

FIGURE 8.43

Meshed model.

Meshing the whole of model and its displacement rate at 100°C are illustrated in Fig. 8.43 and Fig. 8.44, respectively. Based on Fig. 8.44, the displacement rate of the whole set is extremely low and can be neglected. The stress contours for the chamber and the fiberglass sheet are demonstrated in Fig. 8.45 and Fig. 8.46, respectively.

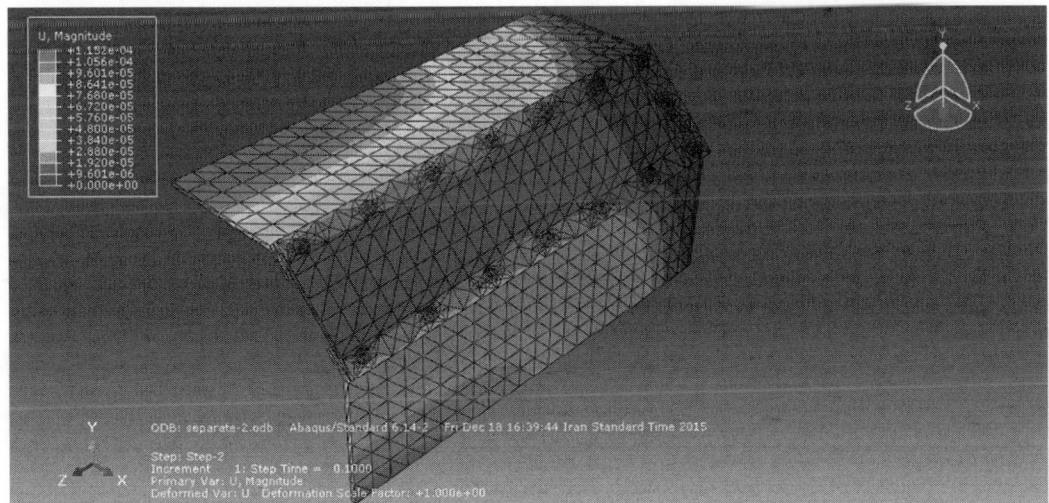

FIGURE 8.44

Displacement rate of model tank at 100°C.

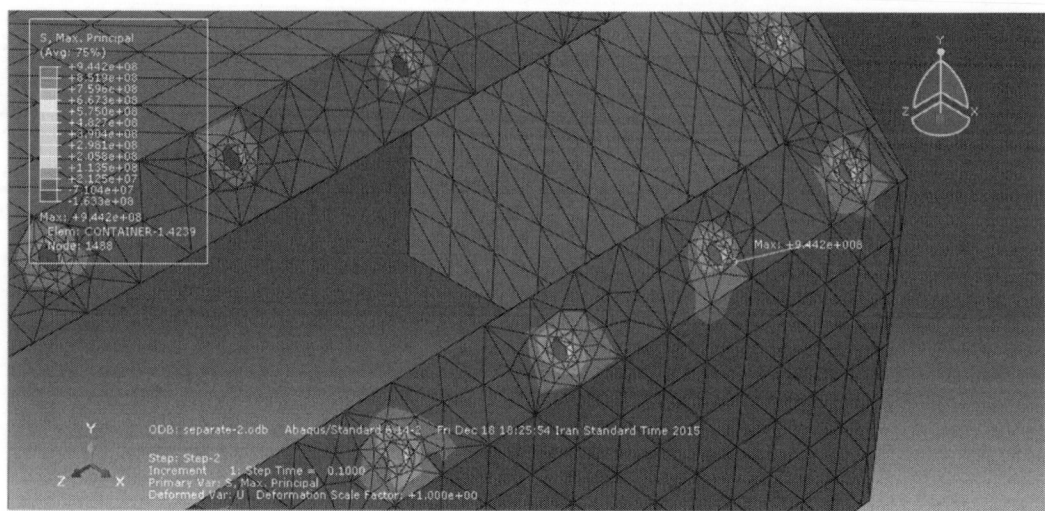

FIGURE 8.45

Stress contours for modeled tank chamber.

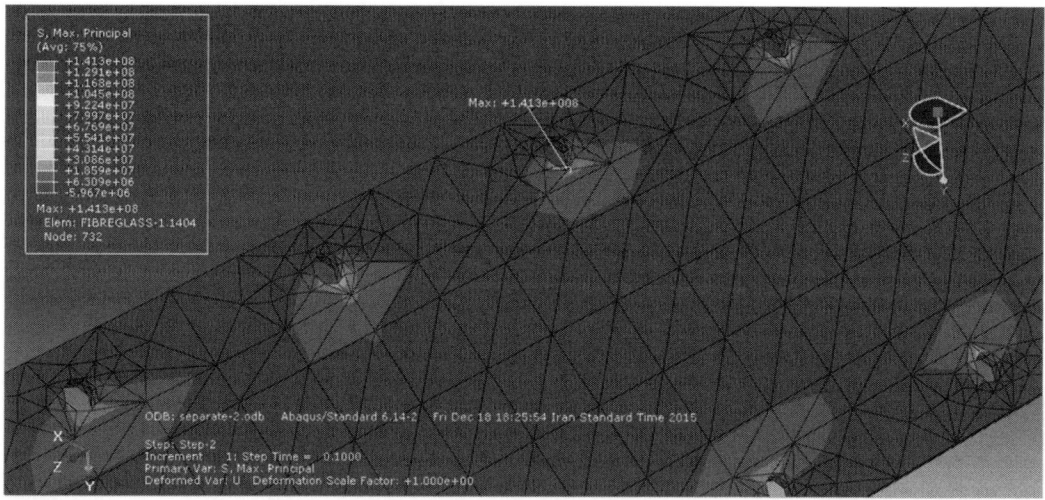

FIGURE 8.46

Stress contours for modeled fiberglass sheet.

The designed tank has a small displacement, and stress with sufficient strength based on the results. It is worth noting that the simulations are performed for 100°C and 0.8 bar pressure compared to outside air, while in actual conditions, the temperature and pressure are lower and so, the displacement and stress of the tank are extremely less.

8.3.6.3 Detecting and locating RD in the presence of DW

In this section, the effect of transformer tank presence on the results of radar imaging is studied. The negative effects of the metal tank on the received signal are predictable due to its metallic nature. The laboratory setup is illustrated in Fig. 8.47.

In the above setup, the antennas are located at a distance of 45 cm from each other each in front of the DW in the two corners of the tank. They are moved from a height of 25 cm above the ground to 77 cm. At 27 points with a step of 2 cm, the transmitter antenna sends the EMW into the transformer tank, and the receiver antenna records the reflected wave. In each position of the antennas, 100 scans are taken from the transformer winding, and 2D images are achieved in the sound and defected states based on the principles of the KMM. Fig. 8.48 displays the image of the sound state of the transformer in this setup, considering the effects of the tank and DW existence.

FIGURE 8.47

Laboratory setup for RD detection study with DW presence.

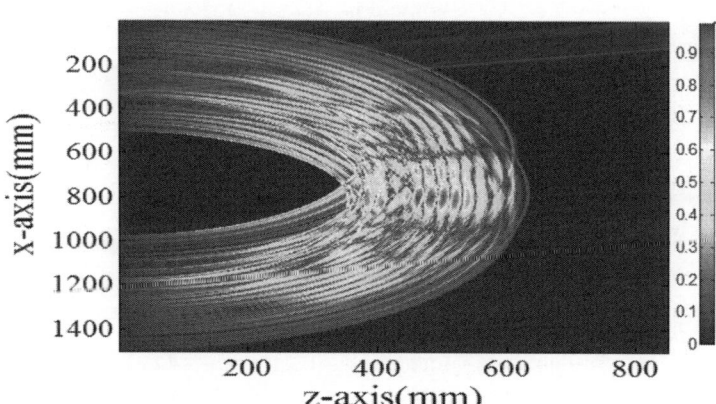

FIGURE 8.48

Generated image of sound state considering tank and DW presence.

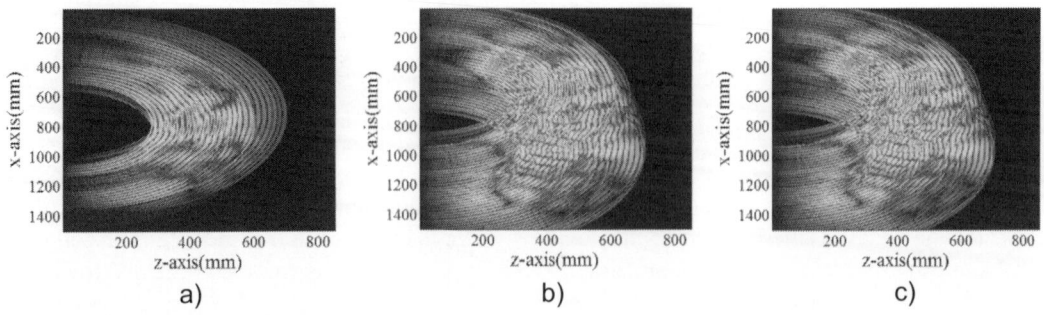

a)   b)   c)

FIGURE 8.49

Effect of tank and DW presence on the generated images for RD in (A) lower, (B) middle, and (C) upper parts.

Table 8.8 Number of points with colored number greater than 0.45 for different states of RD in presence of DW and tank.	
RD in lower part of transformer	Upper part of the image: 41,134
	Middle part of the image: 39,497
	Lower part of the image: 70,026
RD in middle of transformer	Upper part of the image: 38,026
	Middle part of the image: 43,566
	Lower part of the image: 40,185
RD in upper part of transformer	**Upper part of the image: 80,163**
	Middle part of the image: 35,077
	Lower part of the image: 44,279

The above image can be utilized as a reference for RD detection on the transformer considering the tank effect. In the next step, the imaging is performed in the same setup, but the RD is modeled in the upper, middle, and lower parts of the transformer. Fig. 8.49 shows the generated images of the defected states.

As observed, the presence of warm colors increases and more scattering is perceived in the colors compared to Fig. 8.48. Thus, the presence of an RD on the transformer winding is well detected, despite the inappropriate reflections sent to the antennas by the metal tank from the ambient. Table 8.8 lists the warm color number analysis according to the location of the defect in the presence of the tank and DW. Based on the performed tests, the radar imaging method can still detect the presence of an RD on the transformer with the actual metal tank.

8.4 PD effects on SAR imaging method

In the previous sections, the radar imaging method has been analyzed to detect and locate mechanical defects in HV winding, including RDs and axial displacement. The radar imaging method can fully be implemented on a transformer since all tests were performed on an actual transformer winding, even considering the tank.

The PD defect is another problem of a power transformer operation, which can be detected and located utilizing EMWs (Azadifar et al., 2020; Karami, Azadifar, Mostajabi, et al., 2020). To economically monitor a transformer, a solution is to simultaneously detect the mechanical and PD defects with one instrument. To this aim, changing the radar imaging frequency from UWB to UHF is investigated since the PD emits EMWs in the UHF range. Then, the effect of a PD defect is expressed after showing the possibility of detecting mechanical defects in the UHF frequency. Finally, the possibility of building an instrument to simultaneously detect both defects is studied by providing a solution to eliminate the obstacle.

8.4.1 Necessity of changing frequency of radar imaging from UWB to UHF

In the previous section, the mechanical defect of the transformer, including radial and axial, has been investigated by applying a transceiver in UWB waves. As shown, a Gaussian wave was sent in the range of 3.1−6.3 GHz and received for radar imaging of the transformer winding. Such process of sending and receiving was repeated in different steps along the height of the winding. The received waves presented a 2D image of the winding after the initial processing including synchronization of the signal start time, selecting the appropriate time window to remove inappropriate waves from the surroundings, and implementing KMM on the collected information.

It should be noted that based on the practical tests (Coenen et al., 2008; Convery & Judd, 2003; Judd et al., 2005a; Meijer et al., 2006), a PD defect in transformer oil emits an EMW in the frequency range of 300 MHz−3 GHz in the transformer oil. Therefore, a PD signal cannot be received using the UWB antennas in the abovementioned range. Two solutions are proposed to continue the work.

In the first solution, antennas with a bandwidth of 300 MHz−6.3 GHz should be utilized. Thus, the frequency ranges of 300 MHz−3 GHz and 3.1−6.3 GHz can be applied to detect PD and mechanical defects, respectively. The second solution focuses on changing the operating frequency in the radar imaging method from UWB to UHF. In other words, both the PD and mechanical defects can be detected using UHF antennas when the mechanical deformation defect can be performed utilizing the radar imaging method in a frequency range of 300 MHz−3 GHz.

The second solution appears preferable because an antenna with a frequency range between 300 MHz and 3 GHz is considered significantly smaller than one with a frequency range between 300 MHz and 6.3 GHz, which is regarded as a significant advantage due to the limitation of antenna entry into the transformer tank. Based on the radar systems, the resolution equals $c/2B$, where c indicates the wave velocity in the ambient and B represents the bandwidth. The bandwidth is equal to (6.3−3.1 GHz), i.e., 3.2 GHz, where the UWB range is used. When the transceiver is made in the UHF frequency range of 300 MHz−3 GHz, the bandwidth equals (3 GHz−300 MHz), i.e., 2.7 GHz. Therefore, the resolution changes significantly.

In order to demonstrate the possibility of applying radar imaging in the UHF frequency range, some simulations are performed using computer simulation technology (CST) software in the next section and, then the detection of RD and axial displacement are investigated.

8.4.2 Radar imaging in UHF range

8.4.2.1 RD detection of winding

A model is utilized in CST software to simulate this type of study as an alternative to implementing actual dimensions due to the large size of an actual transformer and the limitations of computer hardware, including high RAM and CPU occupation by CST software (Golsorkhi et al., 2012). Table 8.9 presents the simulated model dimensions shown in Figs. 8.50 and 8.51.

The bottom of the model is located at the height of −90 mm and its top is at 280 mm. The antennas are at a distance of 350 mm from the model and 346 mm from each other. In order to analyze the radial deformation of the winding, a bulgy RD is created at a position of (150.5, 0, 96.5) with dimensions of 60*40*25 mm^3. Based on the explanations, the antennas send and receive EMWs in 33 steps from −70 to 250 mm height. Fig. 8.52 shows the image created for the sound and defected states. Visually, the warm points in the defected state are more than that in the sound state. The number of points in the sound and defected states is equal to 43,192 and 47,632, respectively, when the number 0.4 is considered as a measure for comparing warm points. This difference indicates the existence of an RD and the ability to detect it from images.

Table 8.9 Parameters of simulated model.

Value	Parameter
300 mm	Disk diameter
14	Number of disks
20 mm	The height of each disk
6 mm	Height spacer made of pressboard
350 mm	Distance of antennas from disks

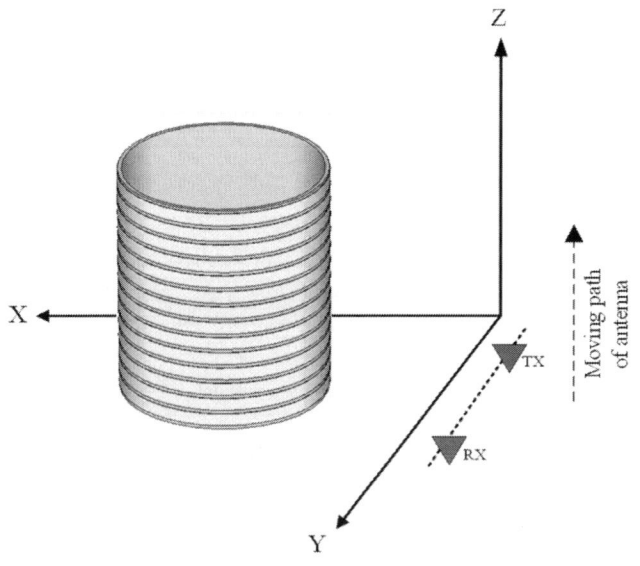

FIGURE 8.50

3D view of sound state in CST simulation for RD study.

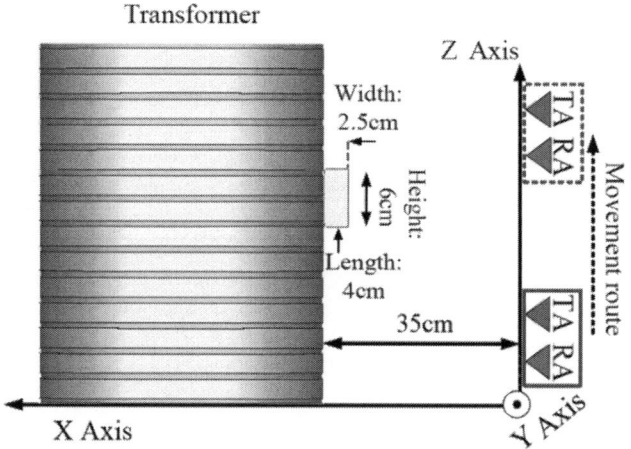

FIGURE 8.51

X-Z view of model with RD defect in CST software.

8.4.2.2 Detecting axial defect of winding

In order to assess the axial displacement of the winding, the model is moved downward by 20 mm and the antennas send and receive EMWs in 30 steps between the heights of −40 to 250 mm. Fig. 8.53 illustrates the image created for the sound and defected states. The lines corresponding to the upper limit (the first upper point of the figure with the colored number 0.4), lower limit (the last lower point of the figure with the colored number 0.4), and the average of the two lines are displayed as a criterion when the number 0.4 is used as a measure. As observed, axial displacement can be perceived by comparing the center lines. The simulations demonstrate a 16 mm downward displacement value.

8.4.3 Effect of PD during radar imaging

The present section demonstrates that the signals received in the antennas are disturbed and can no longer be utilized to detect mechanical defects when a PD occurs during radar imaging of the transformer winding because the PD signal leads to defect detection. The effects of PD on detecting each RD and axial displacement are evaluated in the next section.

8.4.3.1 PD effect on RD detection

As explained, the amount and number of warm points determine the RD in the transformer winding. However, the combination of the signal emitted from the PD and that related to the RD detection leads to a faulty detection in the RD. For instance, Fig. 8.54 shows that the shape obtained from the KMM is highly distorted when a PD defect occurs during some steps in the procedure of sending and receiving Gaussian signals. The number of warm points equals 25,223, which is extremely less than in Fig. 8.52A, indicating a concave RD, which is incorrect.

8.4.3.2 PD effect on axial displacement detection

As stated, the axial defect can be found in the transformer winding by finding the upper and lower limits of the winding. However, similar to RD, the combination of the signal emitted from the PD and that of the axial defect leads to a faulty detection of the axial defect. For example, the image shown in

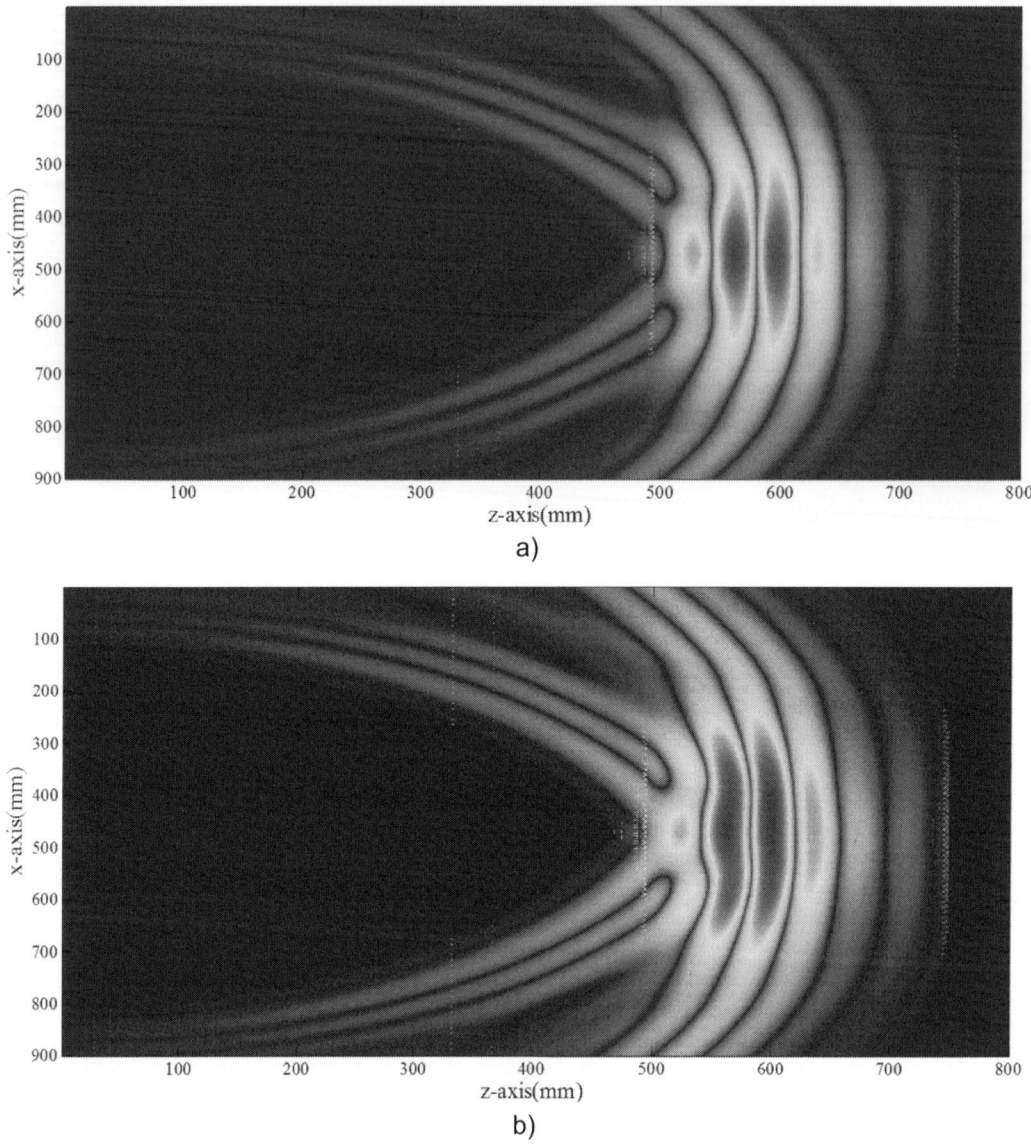

FIGURE 8.52

RD defect study in CST software: (A) Image of sound state, and (B) image of defected state.

Fig. 8.55 illustrates the defect incorrectly compared to Fig. 8.53A when a PD defect occurs during some steps in the procedure of sending and receiving the signals. Fig. 8.55 displays the defect as 8.5 mm upwards compared to Fig. 8.53A, while the actual defect is almost double and downwards.

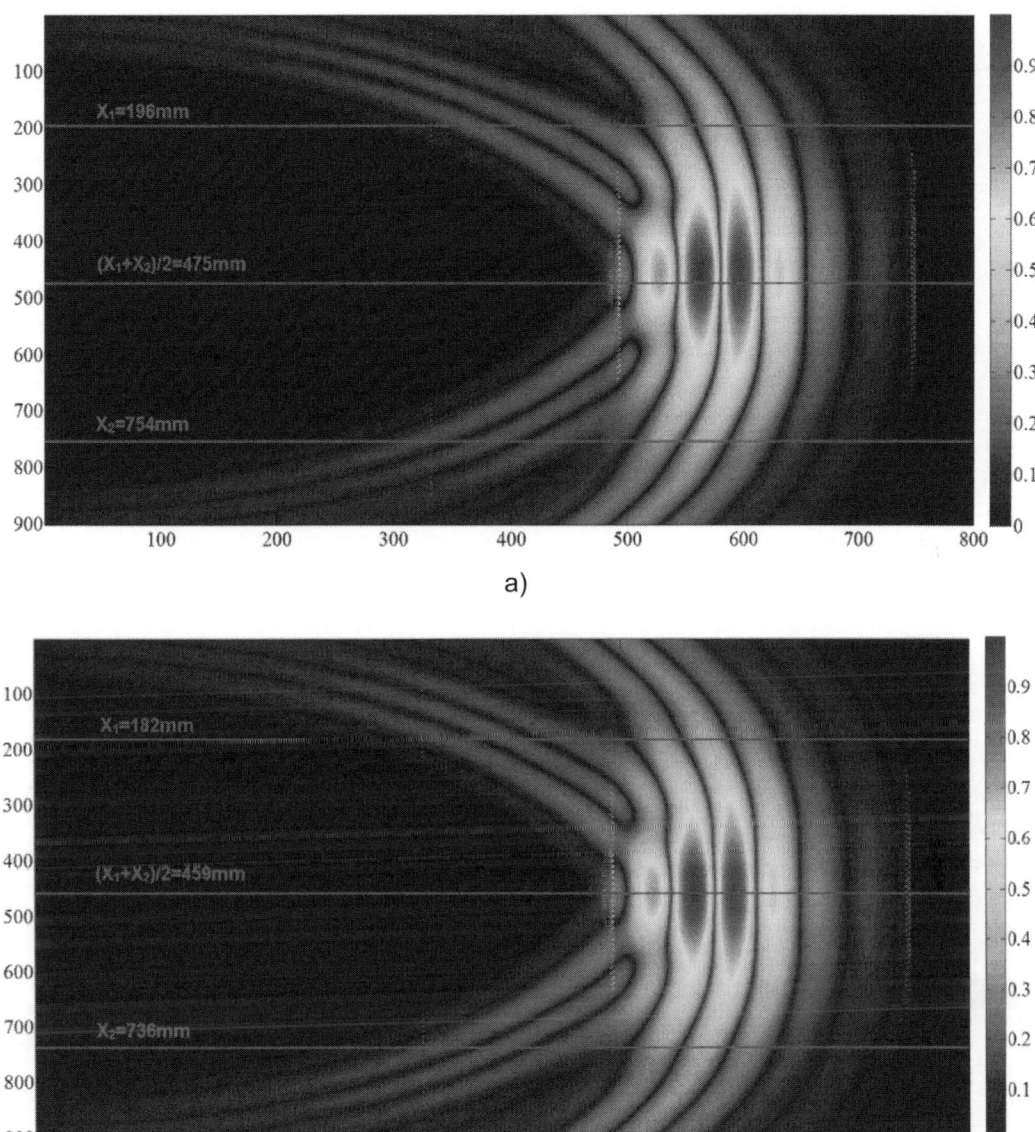

a)

b)

FIGURE 8.53

Axial displacement study in CST software: (A) Image of sound state, and (B) image with axial displacement.

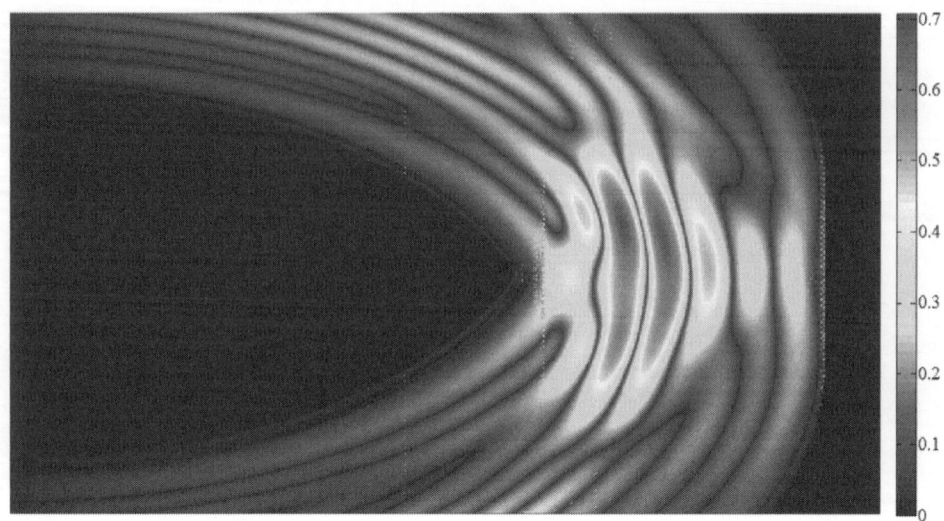

FIGURE 8.54

Effect of PD defect in sound state image.

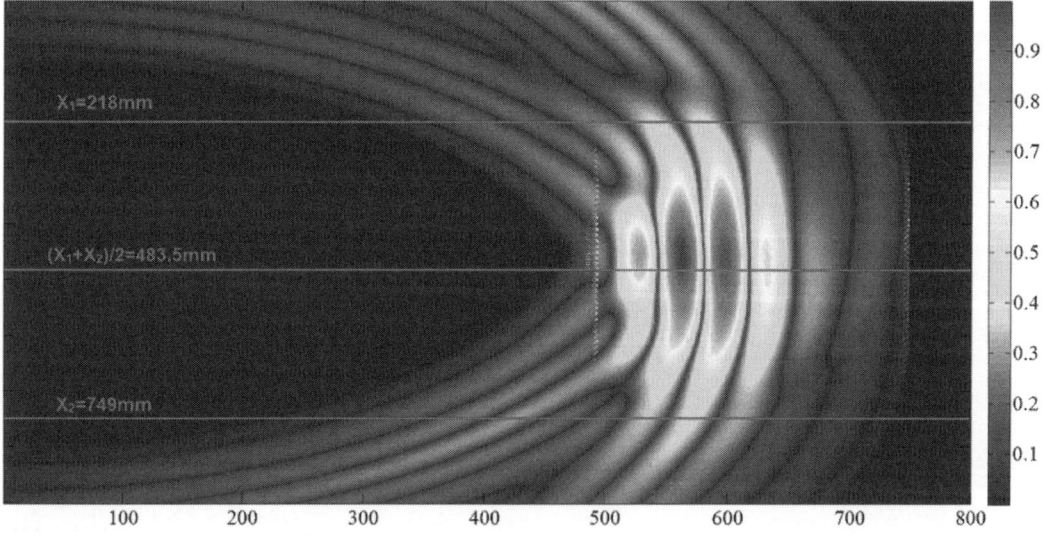

FIGURE 8.55

Effect of PD defect on axial displacement detection.

8.4.3.3 Stepped-frequency method in radar imaging

As discussed, the interference of a PD defect during detecting mechanical defects leads to the destruction of the resulting image and faulty detection. Thus, the information related to the PD should be separated from the data required to detect a mechanical defect. To this aim, it is proposed that corrections must be made in the method of detecting mechanical defects in order to prevent data interference. It is noteworthy that the problem is observed only during detecting the mechanical defects, not during detecting and locating the PD defect, since the PD signals are received and located as described when the transmitter and receiver are off to detect mechanical defects. As indicated, the method presented in radar imaging applies a Gaussian signal with all of the frequencies in the understudy interval, which stems from a set of sinusoidal signals with different frequencies in the interval when it is perceived in the frequency domain. In addition, the signal received after reflection from the object includes a set of sinusoidal signals with different frequencies, the result of which is shown as a Gaussian signal. In fact, the received signals for detecting mechanical defects are considered as the same as the transmitted sinusoidal signals, the size and phase of which change along the way. Therefore the same Gaussian signal is obtained when it is sent and received as several separate sinusoidal waves which are added together.

Based on the explanations, a set of several sinusoidal signals is proposed to be used for each step of the mechanical defect detection test instead of utilizing one Gaussian signal. Such a method is called stepped-frequency, in which the received signal is expected to be sinusoidal with frequency f, and only its amplitude and phase should be changed when a sinusoidal signal with frequency f is sent. The signal received in the antenna is not regarded as a sinusoidal one with a pure frequency when a PD defect occurs during sending and receiving one of the sinusoidal waves since the PD defect emits a wide range of UHF frequencies in the ambient (Karami et al., 2012, 2013), not a specific frequency. Thus, a PD occurs during the transmission and reception of the experiment, and the sending and receiving of a sinusoidal signal for the abovementioned frequency should be repeated. The repetition continues until a sinusoidal signal is sent and the received one is sinusoidal with the same frequency. In other words, the repetition continues until the received signal is detected without the presence of a PD (Karami, Gharehpetian, Hejazi, et al., 2020).

Suppose that the discharge defect occurs during sending and receiving in a step and the signal resulting from the PD is combined with the mechanical defect detection one. The previous section demonstrated that the final answer for mechanical defects is not considered as reliable when the signals are not separated. The combined sample signal is illustrated in Fig. 8.56. As observed, the signal received by the antenna is not regarded as pure sinusoidal at the aforementioned frequency, and the test should be repeated to detect mechanical defects correctly. The test is repeated until a pure sinusoidal signal is achieved. The feasibility of obtaining such a signal will be described.

Fig. 8.57 demonstrates changing the frequency to send and receive signals in the stepped-frequency method. Suppose that the antennas are in the first step. In the abovementioned method, a sinusoidal signal is sent in f_l frequency and its feedback is received from the winding. Then, the next frequency $(f_l + \Delta f)$ is emitted and the process is repeated. Similarly, the sending and receiving process continues by increasing the frequency with Δf steps until the upper-frequency limit (f_u) is achieved. Then, the antennas are moved and put in the second step, along with repeating the frequency increase process. Therefore, radar imaging is performed by the stepped-frequency method. Fig. 8.58 displays a schematic related to the effect of PD during stepped-frequency imaging. The received signal is expected to be received at the same frequency as transmitted with a different amplitude and phase in the absence of a PD defect.

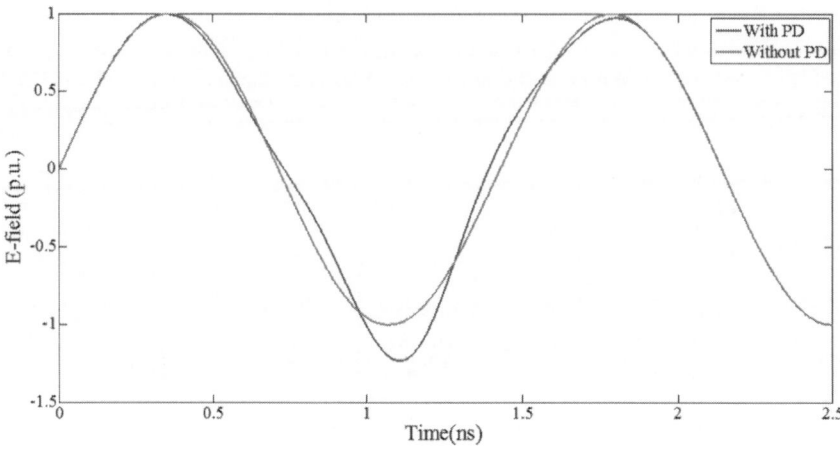

FIGURE 8.56

PD effect on received signal.

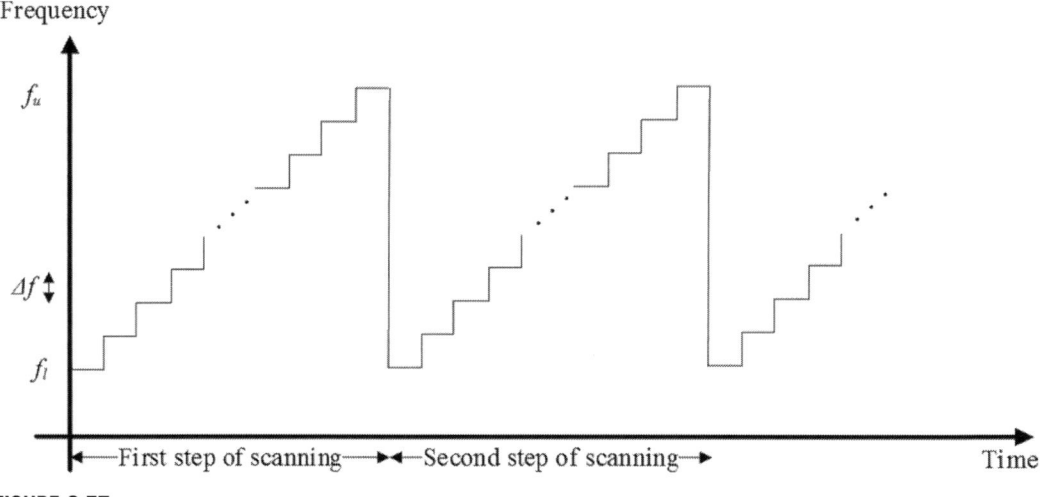

FIGURE 8.57

Schematic of frequency change in stepped-frequency method.

8.4.4 Generalized likelihood ratio test method for PD detection during radar imaging

The method of detecting the occurrence of PD during radar imaging was discussed in the previous section. However, no mathematical relations have been presented for the aforementioned method to prove its correctness and present a mathematical procedure for detecting the occurrence of a PD defect.

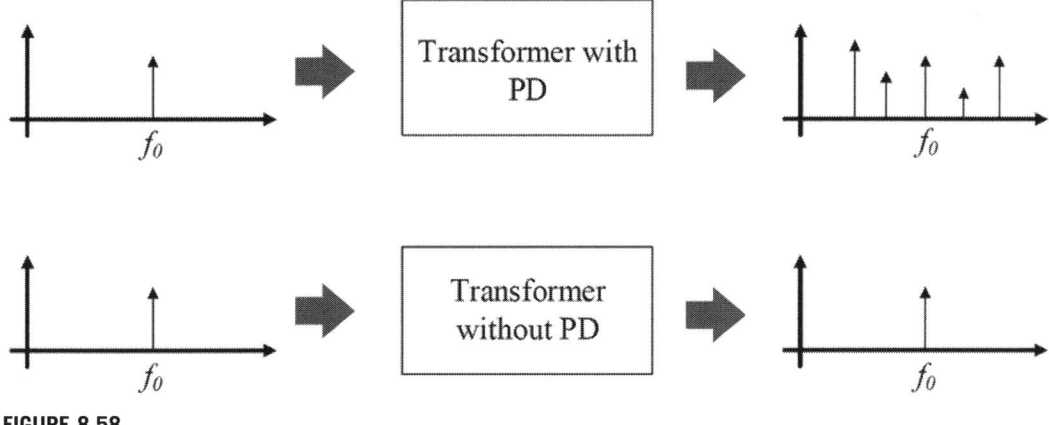

FIGURE 8.58

Schematic of PD effect in stepped-frequency method.

The present section proposes the generalized likelihood ratio test (GLRT) method to demonstrate the method of determining the frequency at which the sending and receiving process should be repeated.

The possibility of accessing a signal which does not interfere with the PD signal is among the significant issues in the abovementioned method. Such a question becomes even more significant when the PD can frequently occur over a period of 50 Hz. In other words, it should be clarified whether a sinusoidal signal can be sent without interference between the received signal and the PD defect one due to the repetition of the defect and the propagation of its signal in the ambient. No obstacle is observed when the time between the two PDs is so long that the sinusoidal signal sent from the antenna to detect an RD can be sent and received between two consecutive PDs. During RD detection, the signal travels the distance between the transmitter antenna to the winding surface and then to the receiver antenna at the speed of light in the ambient. Thus, the return of the signal lasts about a few nanoseconds, i.e., less than 10 ns. In addition, as indicated in (Li et al., 2009), about 50 PDs may occur in each cycle of the 50 Hz network frequency. In other words, a gap of more than a few hundred microseconds is observed between two consecutive PDs, the comparison of which demonstrates that a signal can be obtained without interfering with PD when a sinusoidal signal can be sent and received at least 10 times.

8.4.4.1 GLRT method

The GLRT method is utilized in telecommunications related to aircraft detection or any other object in the air (Derakhtian et al., 2007). Based on the GLRT method, the signal is hypothesized to be sent and received through the antenna to the K section, where each signal has N discrete data. The received signals (r_k) are received as noise when no object is observed in space (H_0). However, the received signal is observed as a combination of noise and the reflected signal when an object is perceived in space (H_1).

As explained, in order to simultaneously detect defects, the test should be repeated at the same frequency when the receiving antenna includes frequencies other than the transmitted one during radar imaging. Therefore, the frequency of the transmitted signal should be removed from the received one to analyze its rest. In the present study, the received signal in the frequency domain is assumed to be divided into the K regions instead of the K regions in space. The transmitted and received signals are displayed with $ae^{j\theta}$ and the received signal with r_n, respectively. Based on (Derakhtian et al., 2007; Papoulis & Pillai, 2002; Steven, 2003), the transmitted signal is observed in the k-th bin of the received signal frequency domain, and noise is perceived at other points when no PD is hypothesized to be in the received signal. Thus the probability density function (PDF) is calculated as follows.

$$f(r_k|H_0) = \frac{1}{\sigma_0\sqrt{2\pi}} \exp\left(-\frac{|r_k - ae^{j\theta}|^2}{2\sigma_0^2}\right) \tag{8.51}$$

$$f(r_n|H_0) = \frac{1}{\sigma_0\sqrt{2\pi}} \exp\left(-\frac{|r_n|^2}{2\sigma_0^2}\right), \quad n = 1, 2, ..., N, \; n \neq k \tag{8.52}$$

where σ_0 and σ_0^2 indicate the mean and variance of ambient noise, respectively. However, the PDF changes as follows when a PD is observed.

$$f(r_k|H_1) = \frac{1}{\sqrt{2\pi(\sigma_0^2 + \sigma_1^2)}} \exp\left(-\frac{|r_k - ae^{j\theta}|^2}{2(\sigma_0^2 + \sigma_1^2)}\right) \tag{8.53}$$

$$f(r_n|H_1) = \frac{1}{\sqrt{2\pi(\sigma_0^2 + \sigma_1^2)}} \exp\left(-\frac{|r_n|^2}{2(\sigma_0^2 + \sigma_1^2)}\right) \quad n = 1, 2, ..., N, \; n \neq k \tag{8.54}$$

where σ_1 and σ_1^2 represent the mean and variance of the PD signal, respectively. Based on Neyman-Pearson lemma, the similarity function should be calculated and compared with a threshold, as follows:

$$L(r_1, ..., r_N) = \prod_{n=1}^{N} \frac{f(r_n|H_1)}{f(r_n|H_0)} = \frac{f(r_n|H_1)}{f(r_n|H_0)} \times \prod_{n=1,n\neq k}^{N} \frac{f(r_n|H_1)}{f(r_n|H_0)} \tag{8.55}$$

$$\rightarrow L(r_1, ..., r_N) = \left(\frac{\sigma_0^2}{\sigma_0^2 + \sigma_1^2}\right)^{\frac{N}{2}} \times \exp\left(\frac{|r_k - ae^{j\theta}|^2}{2\sigma_0^2} - \frac{|r_k - ae^{j\theta}|^2}{2(\sigma_0^2 + \sigma_1^2)}\right)$$
$$\times \prod_{n=1,n\neq k}^{N} \exp\left(\frac{|r_n|^2}{2\sigma_0^2} - \frac{|r_n|^2}{2(\sigma_0^2 + \sigma_1^2)}\right) \tag{8.56}$$

$$\rightarrow L(r_1, ..., r_N) = \left(\frac{\sigma_0^2}{\sigma_0^2 + \sigma_1^2}\right)^{\frac{N}{2}} \times \exp\left(\frac{\sigma_1^2}{2\sigma_0^2(\sigma_0^2 + \sigma_1^2)}\left(|r_k - ae^{j\theta}|^2 + \sum_{n=1,n\neq k}^{N}|r_n|^2\right)\right) \tag{8.57}$$

Therefore the decision on the presence or absence of PD is made as follows:

$$L(r_1, ..., r_N) \underset{H_1}{\overset{H_0}{\gtrless}} \eta \tag{8.58}$$

By substituting Eq. (8.57) in Eq. (8.58) and after some calculations, we have:

$$\left(\left| r_k - ae^{j\theta} \right|^2 + \sum_{n=1,n\neq k}^{N} |r_n|^2 \right) \underset{H_1}{\overset{H_0}{\gtrless}} \frac{2\sigma_0^2 (\sigma_0^2 + \sigma_1^2)}{\sigma_1^2} \times \ln\left(\eta \left(\frac{\sigma_0^2 + \sigma_1^2}{\sigma_1^2} \right)^{\frac{N}{2}} \right) = \eta_1 \qquad (8.59)$$

Based on the abovementioned equation, a PD has occurred and the imaging process should be repeated at that frequency when the value on the left side of the equation exceeds a threshold (η_1). The probability of error in the aforementioned method is indicated by P_{fa} and the parameters such as a, θ, and σ_1 are considered as unknown. The following equation can be obtained based on the maximum likelihood (ML) method (Steven, 2003).

$$E\{ae^{j\theta}\} = r_k \qquad (8.60)$$

Thus, the following equations are achieved.

$$\left(|r_k - r_k|^2 + \sum_{n=1,n\neq k}^{N} |r_n|^2 \right) \underset{H_1}{\overset{H_0}{\gtrless}} \eta_1 \qquad (8.61)$$

$$\sum_{n=1,n\neq k}^{N} |r_n|^2 \underset{H_1}{\overset{H_0}{\gtrless}} \eta_1 \qquad (8.62)$$

The P_{fa} value should be acceptable for making the right decision. The P_{fa} value equals the probability of not occurrence of a PD, while a faulty detection occurs in this regard. An error may occur in every $3.3*10^3$ image when the P_{fa} equals 10^{-7} and 3000 frequencies are sent and received in each step since $10^{-7} = \frac{1}{(3000) \times (3.3 \times 10^3)}$. Therefore the value should be properly determined, which is selected as follows based on (Papoulis & Pillai, 2002).

$$P_{fa} = 1 - \frac{1}{(N-2)!} \gamma\left(N-1, \frac{\eta_1}{2\sigma_0^2} \right) \qquad (8.63)$$

where γ is determined, as follows:

$$\gamma(s,x) = \int_0^x t^{s-1} e^{-t} dt \qquad (8.64)$$

Thus, the following equation should numerically be solved to find the appropriate value of η_1 to have a given value of P_{fa}.

$$\int_0^{\frac{\eta_1}{2\sigma_0^2}} t^{N-2} e^{-t} dt = (N-2)!(1 - P_{fa}) \qquad (8.65)$$

8.4.4.2 Applying GLRT method

The model described in Section 8.4.2 is used to investigate the method in the simulation environment. The simulation is performed by 300 MHz−3 GHz sinusoidal signals instead of utilizing the Gaussian signal. In order to detect the RD in the winding, 2700 sinusoidal signals are examined at 2700 different frequencies in 33 steps with distances of 10 mm from each other to send and receive at each step. The received signal should differ from the transmitted one only in amplitude and phase for each sinusoidal signal. In order to investigate the effect of a PD defect during RD detection, a PD defect is hypothesized to occur during transmission and reception at 1 GHz frequency, and the PD signal is combined with the RD detection signal in one of the steps. The abovementioned process is only studied for RD detection, and it is the same in axial displacement due to its similarity between the radial and axial defect detection.

To apply GLRT method, the value of $N = 300$ is considered, which means that the received signals are divided into 300 different frequency ranges in the frequency domain and the GLRT value is calculated. The signal-to-noise ratio (SNR) value for PD equals about 6 dB (Moore et al., 2005). The value of η_1 becomes 1105 by selecting P_{fa} equal to 10^{-7}. The value of $\sum_{n=1,n\neq k}^{N} |r_n|^2$ is 1209 after ignoring the frequency of the transmitted signal, which warns of a PD occurrence.

The sending and receiving process should be repeated for the aforementioned frequency and step when a PD is detected during the process. The test is repeated until a pure sinusoidal signal is achieved; the feasibility of which is described. The images obtained after separating the signals are similar to those achieved in 8.4.2 by applying the abovementioned method for all of the steps and frequencies. It means that the GLRT method can correctly distinguish and detect the mechanical defect despite the existence of a PD defect.

8.4.4.3 Parameters change effect on GLRT method

Fig. 8.59 shows the changes in P_{fa} in terms of the SNR of PD. The P_f parameter is optionally selected based on the conditions of each test and system sensitivity. Extremely low value selection of P_f leads to

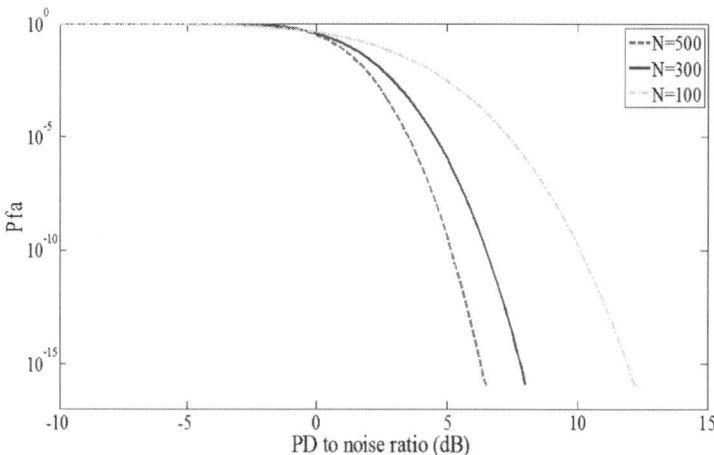

FIGURE 8.59

Changes in error probability versus PD SNR for different values of N.

an unnecessary warning, while its high value selection creates a non warning when a PD occurs during sending and receiving procedure.

As illustrated, the error probability decreases as the SNR increases in the ambient. In other words, the high possibility of distinguishing the PD signal from the ambient noise increases the possibility of detecting the PD signal during radar imaging.

Fig. 8.59 demonstrates the effects of selecting the value of N, as well. The accuracy of the method can be increased by raising the value of N. In other words, the probability of fault detection reduces when the frequency domain is divided into more ranges to detect the occurrence of a PD because a PD with a lower frequency range and a signal with worse noise can more accurately be detected.

8.5 Designing and implementing monitoring system

The previous section demonstrated that the installed antennas could simultaneously be used to detect mechanical defects and PD when the radar imaging method is implemented in the UHF frequency range. In addition, a method was presented to prevent the interference of the signals created by the abovementioned defects. The present section focuses on designing a system that can implement the method presented previously. In other words, the mechanical defects and PD can be detected in the transformer winding by utilizing a system and a set of antennas called the AUTPDMD system.[1]

The AUTPDMD system includes a transmitter-receiver set and a sampling unit to sample the signals in the 100—3000 MHz frequency band. The set transmits the frequency of the received signals to the base band, and is sampled by the sampling unit to transmit the information to a computer for storage. The components of the AUTPDMD and the tasks of its important parts are reviewed in the next section. Finally, the application results to detect PD and mechanical defects are presented.

8.5.1 General description of system

The AUTPDMD system has two main parts, including a transmitter-receiver and a sampling unit. The set has two modes, active and inactive, which will be described here. Fig. 8.60 displays the functional block diagram of the set. Based on the definition of system performance, the transmitter-receiver has three parts, transmitter, receivers, and local oscillator.

8.5.1.1 Functional modes of system

The AUTPDMD system has two modes, which are called active and inactive, respectively, since the transmitter sends a signal in the first mode, while it receives the signal only from the ambient in the second one. In addition, the operating frequency in the first mode is in the range between 800 and 3000 MHz, while it is in the range between 100 and 3000 MHz in the second one. Figs. 8.61 and 8.62 show an overview for the performance of the two modes.

In the active mode, all of the components, including the transmitter, receiver, and local oscillator, are considered active and the system sends signals from the frequency of 800 MHz—3 GHz. In addition, the maximum transmission power equals +10 dBm, which can be reduced and adjusted up to 0 dBm by the user interface. In passive mode, only the receivers and local oscillator section are regarded as active.

---

[1]Amirkabir University of Technology Partial Discharge Mechanical Defect monitoring system.

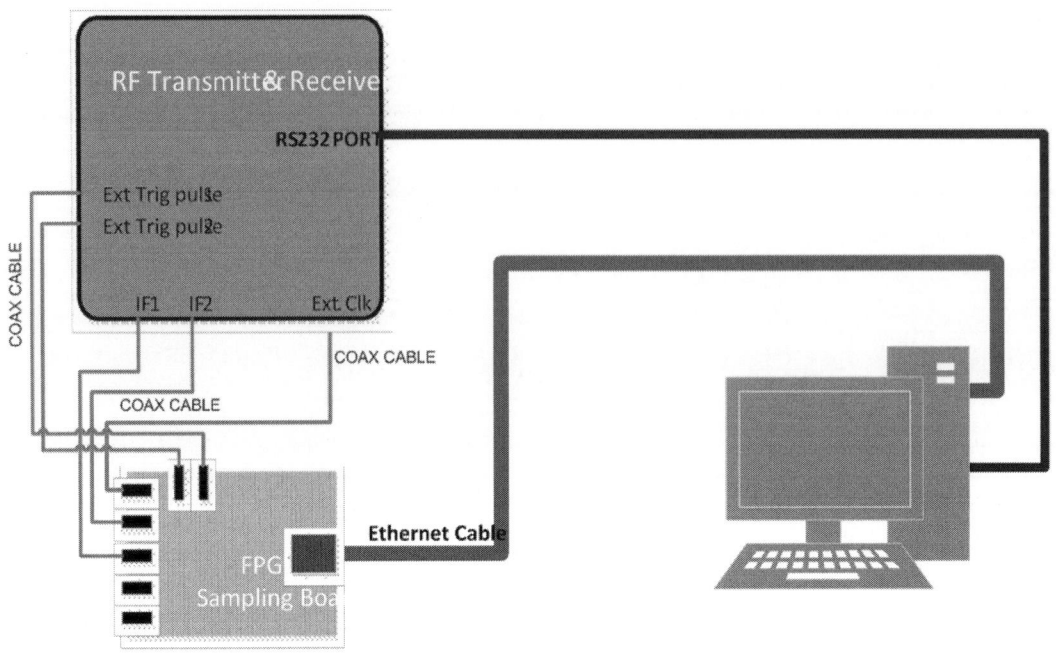

FIGURE 8.60

Different parts of AUTPDMD system.

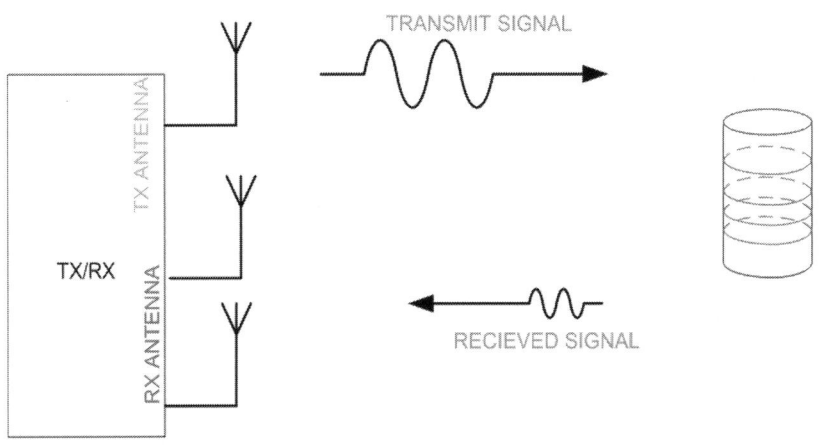

FIGURE 8.61

View of active mode operation in AUTPDMD system.

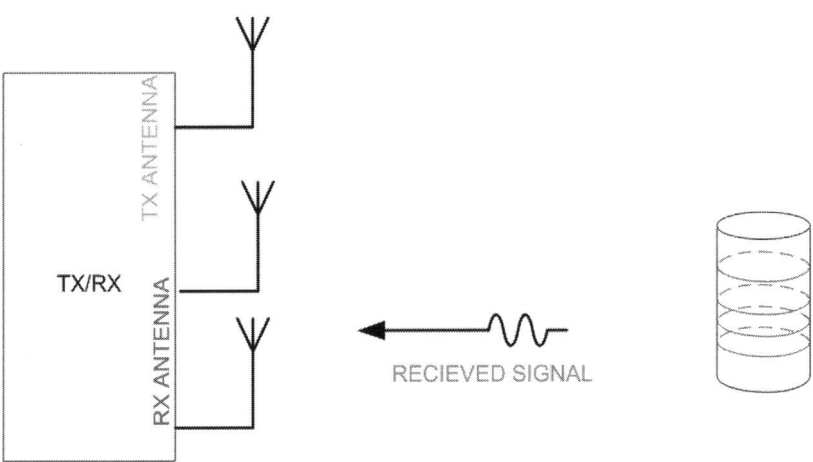

FIGURE 8.62

View of inactive mode operation in AUTPDMD system.

8.5.1.2 Transmitter section

The transmitter basically includes two phase-locked loop (PLL) blocks that create a sinusoidal signal. The PLLs are located in the main output path and operate based on the frequency or range selected through the graphical output interface.

8.5.1.3 Local oscillator and receiver

The local oscillator in the AUTPDMD system includes three separate PLLs, which create the signal related to the local oscillator of the receiver mixer. The receiver in the system uses a down converter to transmit the high-frequency signal to the middle one for sampling.

There are two paths after the input signal. The above path is related to weak signals and is called "high gain" mode, while the down path supplied by a coupler is related to strong signals and is called "low gain."

8.5.1.4 Software and hardware tasks of processing section

The processing section in the AUTPDMD system samples the IF output signals from the radio transmitter and receiver sections and sends different sampled values to the computer (Fig. 8.60).

The hardware of the processing section in the AUTPDMD system samples the signals received from the two IF outputs. The signals are sampled at each change in sending and receiving frequency and sent to a computer by a local area network interface to be stored on the PC. The software in the PC sends the appropriate commands to control the RF section according to the user settings in the software. The FPGA board starts sampling the IF input values utilizing "ext trigger pulse 1" and "ext trigger pulse 2" after sending the user commands to the RF section and sending the samples to the software to be stored on the PC for analysis.

8.5.2 Results of applying AUTPDMD system

Necessary measurements were performed to analyze the system performance in detecting partial and mechanical discharge defects. The tests were performed in three scenarios: RD, axial defect, and PD detection. The fourth scenario focused on PD detection without the presence of radar imaging signals. The winding model of the transformer was used for tests. The model was selected due to the ease of modeling axial displacement.

8.5.2.1 Radial defect detection

The sending and receiving process is performed in 16 steps with 3 cm intervals. Sending and receiving processes start at 800 MHz and continue up to 3 GHz. One second is a rest interval to ensure that the signal is received, and the previously sent signal is damped to diminish interference with the next frequency signal. The distance between the antennas is 34 cm, and the distance between the middle antenna and the transformer is 70 cm. One antenna acts as the transmitter and two can act as receivers. Storing the signal from one antenna suffices in radar imaging tests. Fig. 8.63 displays the test setup.

A bulgy defect is applied to the winding as in the previous sections to model the RD. Figs. 8.64 and 8.65 display the imaging results for the sound and defected states, respectively. The number of warm points with a colored number greater than 0.4 in sound and defected states equals 294 and 936, respectively. In addition, the highest color number has a smaller value than in previous studies with simulation. To improve the criterion of 0.4 in the images, the criterion for selecting a warm color number can be proposed to be 40% of the highest value in the colored number, despite its responsiveness. In other words, if the maximum color number equals 0.62, the color number criterion for detecting the number of warm points equals 0.25. In this case, the number of warm points in sound and defected states equals 6387 and 8635, respectively. However, it is emphasized again that the previous

FIGURE 8.63

Test layout utilizing AUTPDMD system to detect RD.

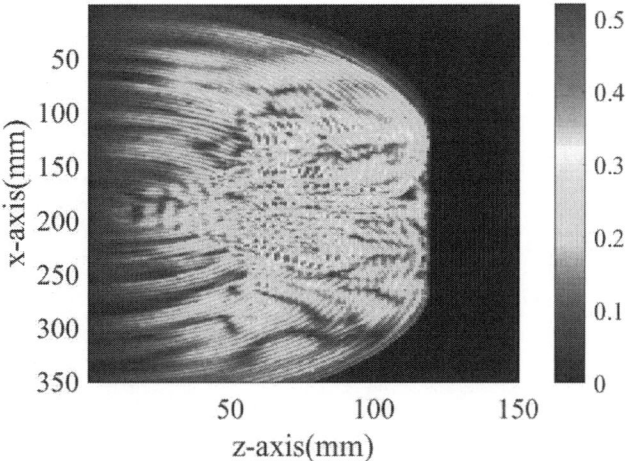

FIGURE 8.64

Imaging results in sound state using AUTPDMD.

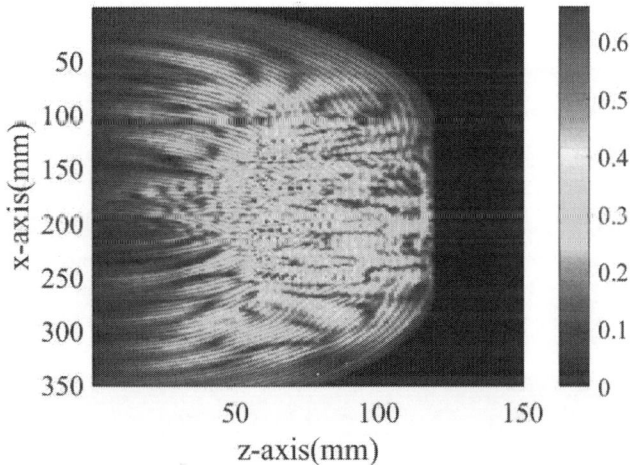

FIGURE 8.65

Imaging results in RD state using AUTPDMD.

criterion is still responsive and there is no need to change the criterion of warm-colored numbers. The abovementioned proposition can be considered an improvement in the defect detection process since sometimes colored points may not exceed a certain number such as 0.4 in certain cases.

It is worth noting that the numbers are assigned to each color in the images based on their maximum values. For instance, the maximum color numbers available in Figs. 8.64 and 8.65 equal approximately 0.5 and 0.65, respectively, corresponding to the red color.

8.5.2.2 Axial displacement detection

To model the axial defect, the winding is moved 3 cm, and imaging is performed in 15 steps. Figs. 8.66 and 8.67 show the results of radar imaging using the AUTPDMD system for sound and displaced

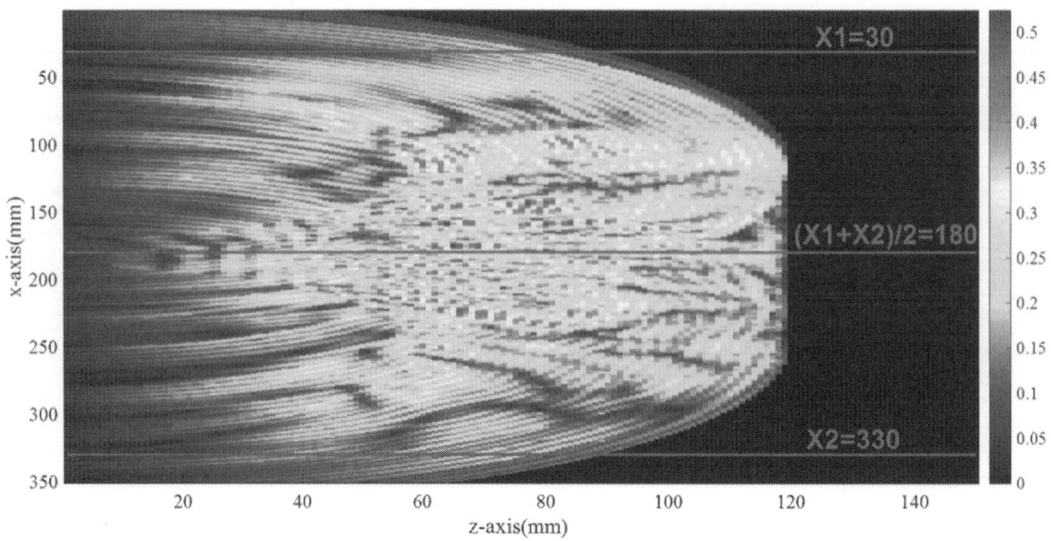

FIGURE 8.66

Imaging results using AUTPDMD without axial displacement.

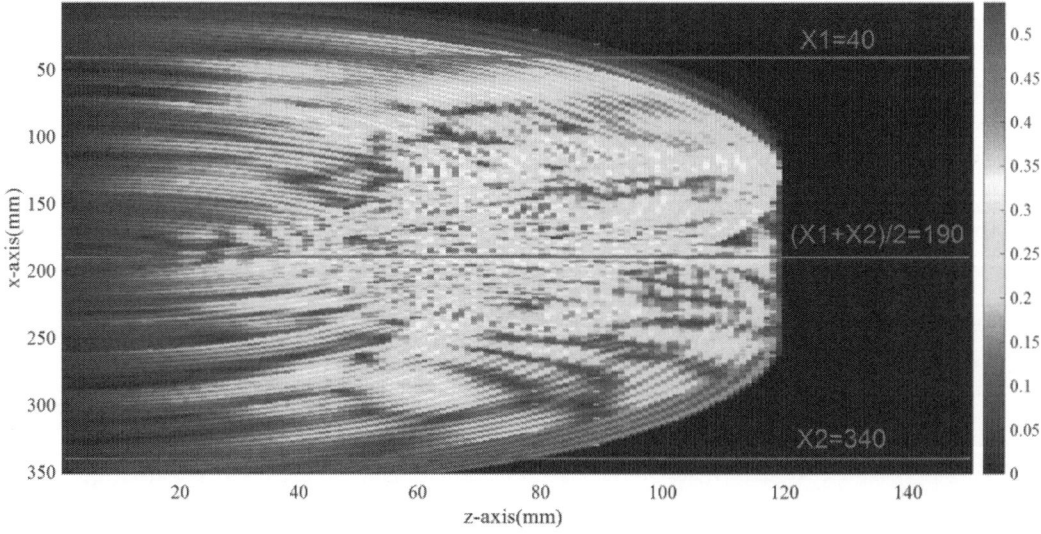

FIGURE 8.67

Imaging results using AUTPDMD with axial displacement.

states, respectively. As observed, the number of warm points almost equals each other, meaning that no RD occurs. Based on the numbers, the position of the winding center in the defected state differs from the sound state, meaning axial displacement. In other words, the axial displacement defects can be detected by the AUTPDMD system.

8.5.2.3 PD detection during radar imaging

It is assumed that a PD occurs in one of the imaging steps while sending and receiving a frequency. The PD is created by two needle-shaped electrodes which are about 3 mm apart. Fig. 8.68 illustrates the area of the received signal where the PD signal is combined with the radar imaging one. As observed, the received signal is out of sinusoidal mode.

Figs. 8.69 and 8.70 displays the frequency range of the received signals before and after interfering with the PD, respectively. The reason is that the transmitted frequency amplitude is extremely large

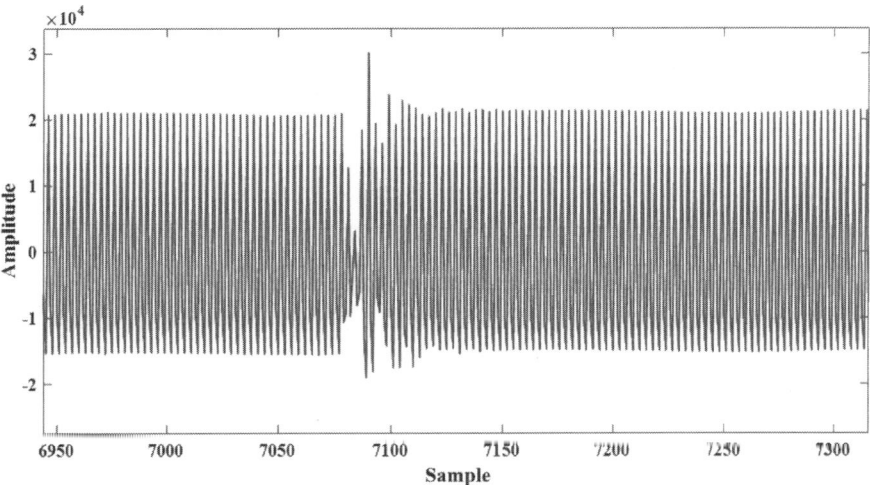

FIGURE 8.68

Effect of PD on signal received by AUTPDMD.

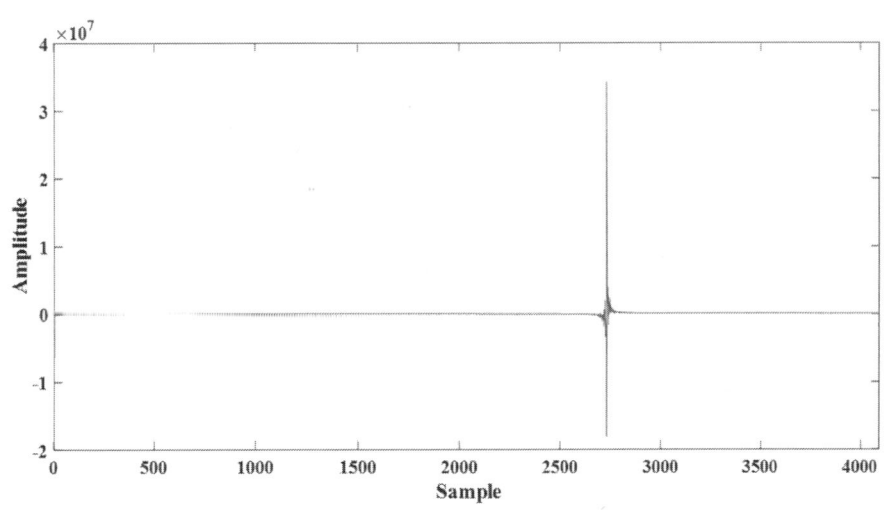

FIGURE 8.69

Signal received by AUTPDMD in frequency domain before interfering with PD.

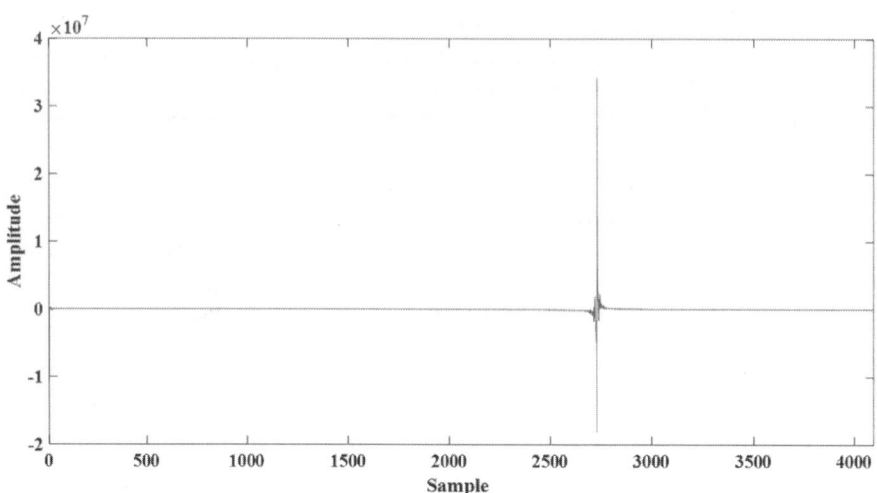

FIGURE 8.70

Signal received by AUTPDMD in frequency domain after interfering with PD.

compared to the PD amplitude, and what is added to the signal in the frequency domain can be observed in Fig. 8.71, when the transmitted frequency is eliminated. It is noteworthy that the appropriate value of η_l in Eq. (8.62) or Eq. (8.65) should be selected in order to detect the PD correctly. The method of obtaining the right value was explained in detailin previous sections.

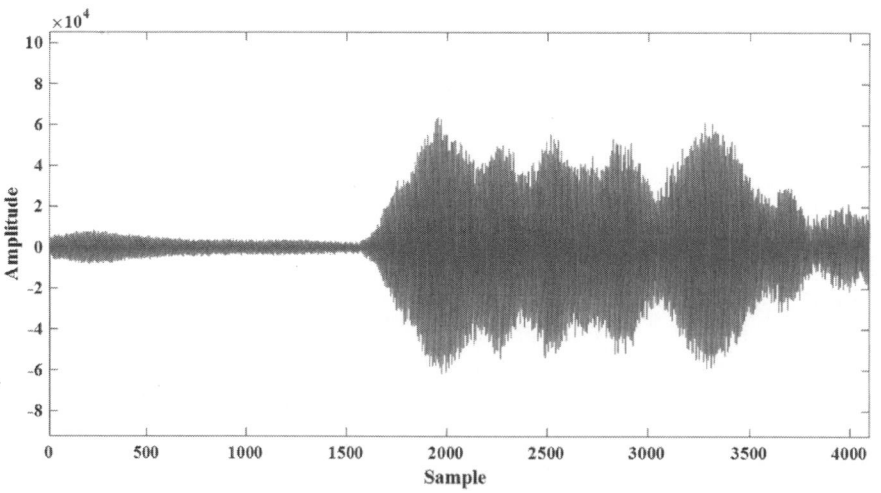

FIGURE 8.71

Added signal due to PD in frequency domain.

8.5.2.4 *PD detection without presence of imaging signals*

As indicated, the AUTPDMD system receives the signal at the frequencies of 100 MHz–3 GHz in the off mode, which is not regarded as a wide bandwidth reception. In other words, the reception changes from 100 MHz to 3 GHz with 10 MHz steps. The system can capture signal information with a bandwidth of 20 MHz at each frequency and perform sampling operations by reducing its frequency to adapt the sampling board.

The received signal is observed in Fig. 8.72 when a PD occurs in the system. Two needle-shaped electrodes create the PD near the transformer winding. Thus, the amplitude of the signal is extremely significant, so that a PD defect can be easily warned.

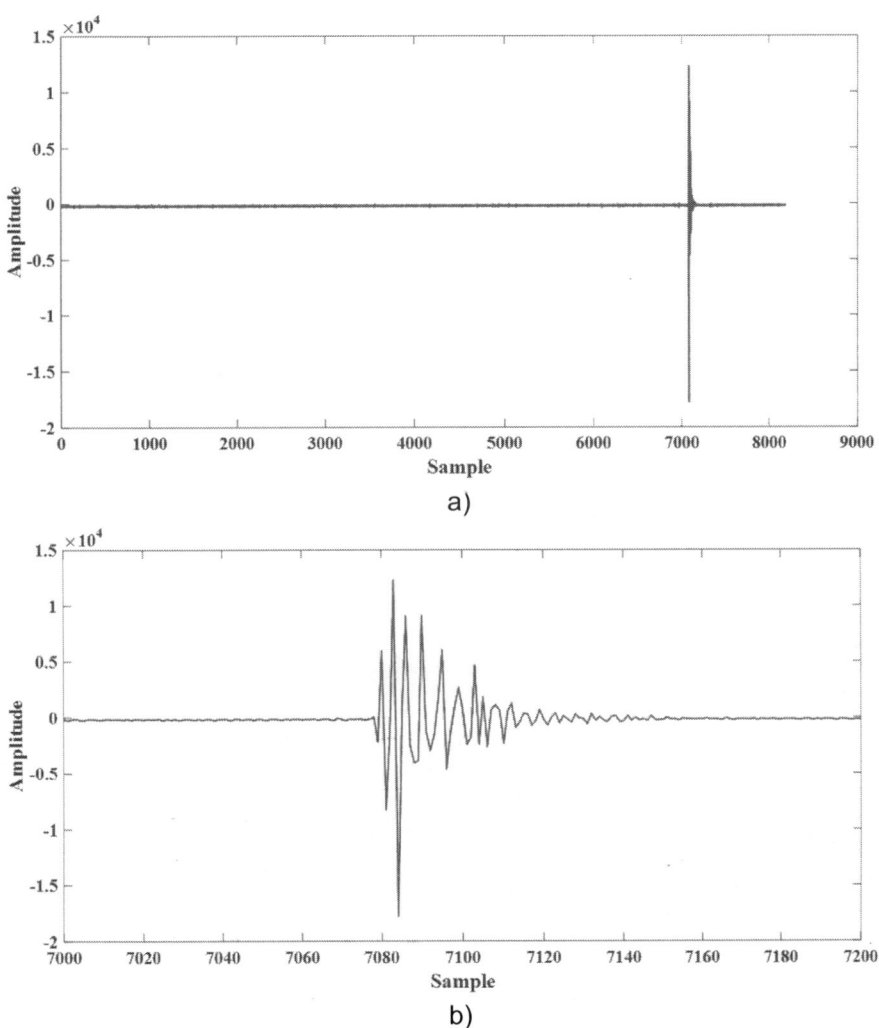

FIGURE 8.72

Received PD signal by AUTPDMD system: (A) whole signal, and (B) magnified view of whole signal.

Based on the results, the AUTPDMD system can detect PD defects in addition to the radial and axial defects of the transformer winding. In addition, the system can separate the signals utilizing the method presented in Chapter 5 in case of simultaneous occurrence of both defects. As a significant advantage, the mechanical defects including radial and axial defects can be detected and a PD defect can be warned in the case of occurrence using a set of antennas and applying the AUTPDMD system, resulting in a reduction of transformer defect monitoring costs.

References

Aftanas, M. (2009). *Through wall imaging with UWB radar system.* Department of Electronics and Multimedia Communications, Technical University of Kosice.

Azadifar, M., Karami, H., Wang, Z., Rubinstein, M., Rachidi, F., Karami, H., Ghasemi, A., & Gharehpetian, G. B. (2020). Partial discharge localization using electromagnetic time reversal: a performance analysis. *IEEE Access, 8*, 147507—147515.

Coenen, S. (2012). *Measurement of partial discharges in power transformers using electromagnetic signals.* BoD—Books on Demand.

Coenen, S., Tenbohlen, S., Markalous, S. M., & Strehl, T. (2008). Sensitivity of UHF PD measurements in power transformers. *IEEE Transactions on Dielectrics and Electrical Insulation, 15*(6), 1553—1558.

Convery, A. R., & Judd, M. D. (2003). Measurement of propagation characteristics for UHF signals in transformer insulation materials. In *2003 13th international symposium on high voltage engineering (ISH).*

Derakhtian, M., Tadaion, A. A., Gazor, S., & Nayebi, M. M. (2007). Invariant tests for rapid-fluctuating radar signal detection with unknown arrival time. *Signal Processing, 87*(3), 441—452.

Golsorkhi, M. S., Hejazi, M. S. A., Gharehpetian, G. B., & Dehmollaian, M. (2012). A feasibility study on the application of radar imaging for the detection of transformer winding radial deformation. *IEEE Transactions on Power Delivery, 27*(4), 2113—2121.

Golsorkhi, M. S., Mosayebi, R., Hejazi, M. A., Gharehpetian, G. B., & Sheikhzadeh, H. (2021). *Detection of transformer winding axial displacement by Kirchhoff and delay and sum radar imaging algorithms. ArXiv Preprint ArXiv:2102.10519.*

Goodman, J. W. (1996). *Introduction to fourier optics.* secondMcGraw-Hill.

Herman, G. T. (1995). Image reconstruction from projections. *Real-Time Imaging, 1*(1), 3—18.

Ji, K., Zhao, P., Zhuo, C., Jin, H., Chen, M., Chen, J., Ye, S., & Fu, J. (2022). Efficient phase shift migration for ultrasonic full-matrix imaging of multilayer composite structures. *Mechanical Systems and Signal Processing, 174*, 109114.

Judd, M. D., Farish, O., Pearson, J. S., & Hampton, B. F. (2001). Dielectric windows for UHF partial discharge detection. *IEEE Transactions on Dielectrics and Electrical Insulation, 8*(6), 953—958.

Judd, M. D., Yang, L., Bennoch, C. J., & Hunter, I. (2004). UHF diagnostic monitoring techniques for power transformers. In *EPRI substation equipment diagnostics conference XII.*

Judd, M. D., Yang, L., & Hunter, I. B. (2005a). Partial Discharge Monitoring for Power Transformers Using UHF Sensors Part 2: Field Experience-The results of PD tests on power transformers provide sufficient evidence to justify making provision. *IEEE Electrical Insulation Magazine, 21*(3), 5—13.

Judd, M. D., Yang, L., & Hunter, I. B. (2005b). Partial discharge monitoring of power transformers using UHF sensors. Part I: Sensors and signal interpretation. *IEEE Electrical Insulation Magazine, 21*(2), 5—14.

Karami, H., Azadifar, M., Mostajabi, A., Rubinstein, M., Karami, H., Gharehpetian, G. B., & Rachidi, F. (2020). Partial discharge localization using time reversal: Application to power transformers. *Sensors, 20*(5), 1419.

Karami, H., Gharehpetian, G. B., Hejazi, M. A. A., & Norouzi, Y. (2020). *Detection of radial deformations of transformers.* Google Patents.

Karami, H., Gharehpetian, G. B., Norouzi, Y., & Akhavan-Hejazi, M. (2020). Simultaneous radial deformation and partial discharge detection of high-voltage winding of power transformer. *IET Electric Power Applications, 14*(3), 383–390.

Karami, H., Gharehpetian, G. B., Norouzi, Y., & Hejazi, M. A. (2016). GLRT-based mitigation of partial discharge effect on detection of radial deformation of transformer HV winding using SAR imaging method. *IEEE Sensors Journal, 16*(19), 7234–7241.

Karami, H., Gharehpetian, G. B., Norouzi, Y., & Hejazi, M. A. (2018). Experimental study on elimination of partial discharge effect on detection of radial deformation of high voltage transformer winding using electromagnetic waves. In *2018 IEEE international conference on environment and electrical engineering and 2018 IEEE industrial and commercial power systems europe (EEEIC/I&CPS europe)* (pp. 1–5).

Karami, H., Hejazi, M. S. A., & Gharehpetian, G. B. (2013). Simulation of transformer oil effect on PD source allocation. In *4th conference on partial discharge in electrical apparatus (PDC'13)* (pp. 26–27).

Karami, H., Hejazi, M. S. A., Naderi, M. S., Gharehpetian, G. B., & Mortazavian, S. (2012). Three-dimensional simulation of PD source allocation through TDOA method. In *The 4th conference on thermal power plants* (pp. 1–4).

Karami, H., Tabarsa, H., Gharehpetian, G. B., Norouzi, Y., & Hejazi, M. A. (2019). Feasibility study on simultaneous detection of partial discharge and axial displacement of HV transformer winding using electromagnetic waves. *IEEE Transactions on Industrial Informatics, 16*(1), 67–76.

Li, J., Si, W., Yao, X., & Li, Y. (2009). Partial discharge characteristics over differently aged oil/pressboard interfaces. *IEEE Transactions on Dielectrics and Electrical Insulation, 16*(6), 1640–1647.

Mahmoodi, M., Abadi, S. M. N., Karami, H., Hejazi, M. A., & Gharehpetian, G. B. (2020). Design and implementation of dielectric windows for detection of radial deformation of HV transformer winding using radar imaging. *IET Science, Measurement & Technology, 14*(4), 478–485.

Margrave, G. F., & Lamoureux, M. P. (2019). *Numerical methods of exploration seismology: With algorithms in MATLAB®.* Cambridge University Press.

Meijer, S., Agoris, P. D., Smit, J. J., Judd, M. D., & Yang, L. (2006). Application of UHF diagnostics to detect PD during power transformer acceptance tests. In *Conference record of the 2006 IEEE international symposium on electrical insulation* (pp. 416–419).

Moore, P. J., Portugues, I. E., & Glover, I. A. (2005). Radiometric location of partial discharge sources on energized high-voltage plant. *IEEE Transactions on Power Delivery, 20*(3), 2264–2272.

Mortazavian, S., Gharehpetian, G. B., Hejazi, M. A., Golsorkhi, M. S., & Karami, H. (2012). A simultaneous method for detection of radial deformation and axial displacement in transformer winding using UWB SAR imaging. In *The 4th conference on thermal power plants* (pp. 1–6).

Mortazavian, S., Shabestary, M. M., Mohamed, Y. A.-R. I., & Gharehpetian, G. B. (2015). Experimental studies on monitoring and metering of radial deformations on transformer HV winding using image processing and UWB transceivers. *IEEE Transactions on Industrial Informatics, 11*(6), 1334–1345.

Mosayebi, R., Sheikhzadeh, H., Golsorkhi, M. S., Hejazi, M. S. A., & Gharehpetian, G. B. (2014). Detection of winding radial deformation in power transformers by confocal microwave imaging. *Electric Power Components and Systems, 42*(6), 605–611.

Papoulis, A., & Pillai, S. U. (2002). *Probability, random variables, and stochastic processes.* Tata McGraw-Hill Education.

Rahbarimagham, H., Esmaeili, S., & Gharehpetian, G. B. (2017). Localization of radial deformation and its extent in power transformer HV winding using stationary UWB antennas. *IEEE Sensors Journal, 17*(10), 3184–3192.

Rahbarimagham, H., Karami, H., Esmaeili, S., & Gharehpetian, G. B. (2019). Determination of transformer HV winding axial displacement extent using hyperbolic method—a feasibility study. *IET Electric Power Applications, 13*(7), 1004–1013.

Rahbarimagham, H., Porzani, H. K., Hejazi, M. S. A., Naderi, M. S., & Gharehpetian, G. B. (2015). Determination of transformer winding radial deformation using UWB system and hyperboloid method. *IEEE Sensors Journal, 15*(8), 4194–4202.

Scales, J. A. (1995). *Theory of seismic imaging. 2*. Springer-Verlag Berlin.

Schneider, W. A. (1978). Integral formulation for migration in two and three dimensions. *Geophysics, 43*(1), 49–76.

Steven, M. K. (2003). *Fundamentals of statistical signal processing, volume I: Estimation theory/volume II: Detection theory*. Prentice Hall PTR.

Ulander, L. M., Hellsten, H., & Stenstrom, G. (2003). Synthetic-aperture radar processing using fast factorized back-projection. *IEEE Transactions on Aerospace and Electronic Systems, 39*(3), 760–776.

Hyperbolic method

Gevork B. Gharehpetian[1], Hossein Karami[2] and Seyed-Alireza Ahmadi[3]

[1]*Electrical Engineering Department, Amirkabir University of Technology (AUT), Tehran, Iran;* [2]*High Voltage Studies Research Department, Niroo Research Institute (NRI), Tehran, Iran;* [3]*School of Electrical and Computer Engineering, College of Engineering, University of Tehran, Tehran, Iran*

9.1 Description of method

A summary of the hyperbolic method in the ultra wideband (UWB) frequency band, to detect the radial deformation (RD) and axial displacement (AD) of the transformer winding is presented in this section. Then, the feasibility of simultaneous detection of two defects is investigated.

9.1.1 Radial deformation detection

9.1.1.1 Signal

The transmitted waves hit the target and are reflected uniformly in all directions, similar to the radar imaging method (Fig. 9.1). The receiving antenna records the signal reflected from the target. In fact,

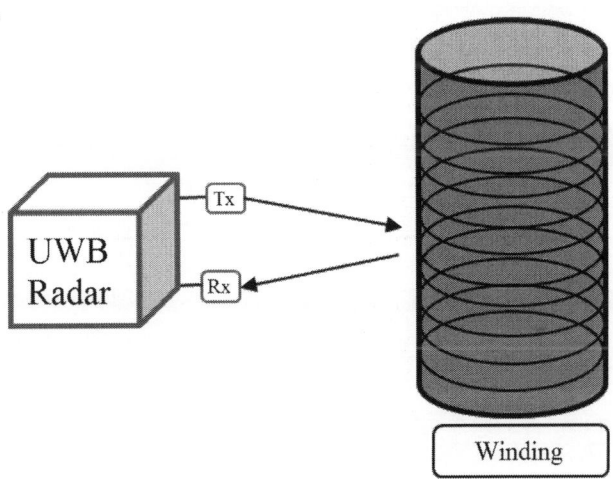

FIGURE 9.1

Simple layout for detecting defect occurrence.

Power Transformer Online Monitoring Using Electromagnetic Waves. https://doi.org/10.1016/B978-0-12-822801-2.00005-7

the UWB pulse sent by the transmitter hits various objects, and its reflection approaches the receiver. Therefore, the transmitted wave travels several different paths from the transmitter to the receiver, called multipath, as described previously (Golsorkhi et al., 2012). The different parts of the received signal were reviewed in detail in the previous chapter.

9.1.1.2 Detecting defect occurrence

In most methods, a reference is needed to detect defect occurrence and compare the obtained data with that reference, called a fingerprint (Karami, Gharehpetian, Hejazi, et al., 2020). In the hyperbolic method, the UWB signal is first sent and received in a sound condition, and the data are recorded and stored as a reference. The received and reference signals are compared while monitoring the transformer state, and the occurrence of a defect in the target is indicated when there is a difference between the two signals. Fig. 9.2 shows a sample received signal in case of no deformation in the transformer winding. The received signal differs from the sound condition when a radial defect occurs in the transformer winding, as shown in Fig. 9.3.

In Fig. 9.4, the received signals in the sound and defected conditions are subtracted to determine the full effect of the RD defect. As observed, T_C is considered as the starting point for the changes in the subtracted signal, which displays the effect of the defect, resulting in detection of the RD occurrence in the transformer winding.

9.1.1.3 Hyperbolic algorithm for defect location

The subtracted signal is achieved with a delay in proportion to the distance from the defect. Thus, when a defect occurs in the transformer winding, two receiving antennas in different locations observe the defect-induced changes in the recorded signal with different time delays (intervals), e.g., t_1 and t_2. Such a feature can be utilized to identify the defect location (Rahbarimagham et al., 2015, 2019). The time difference (TD) of the received signals can be converted into a distance difference by having the propagation speed of signals in the ambient. Assume that the defect is located at a distance of x_1 and x_2 from the perspective of the first and second receiving antennas, respectively. Fig. 9.5 illustrates a simple view for understanding this concept better.

The two receiving antennas are located at the top and bottom, and the transmitting antenna is placed in the middle. The upper receiving antenna detects sooner the defect that occurred in the upper half of the winding. Such a process occurs for the lower receiving antenna if the defect is in the lower

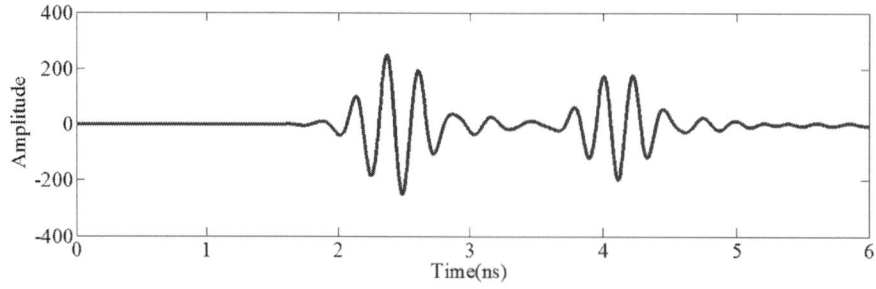

FIGURE 9.2

Signal received in sound condition of transformer.

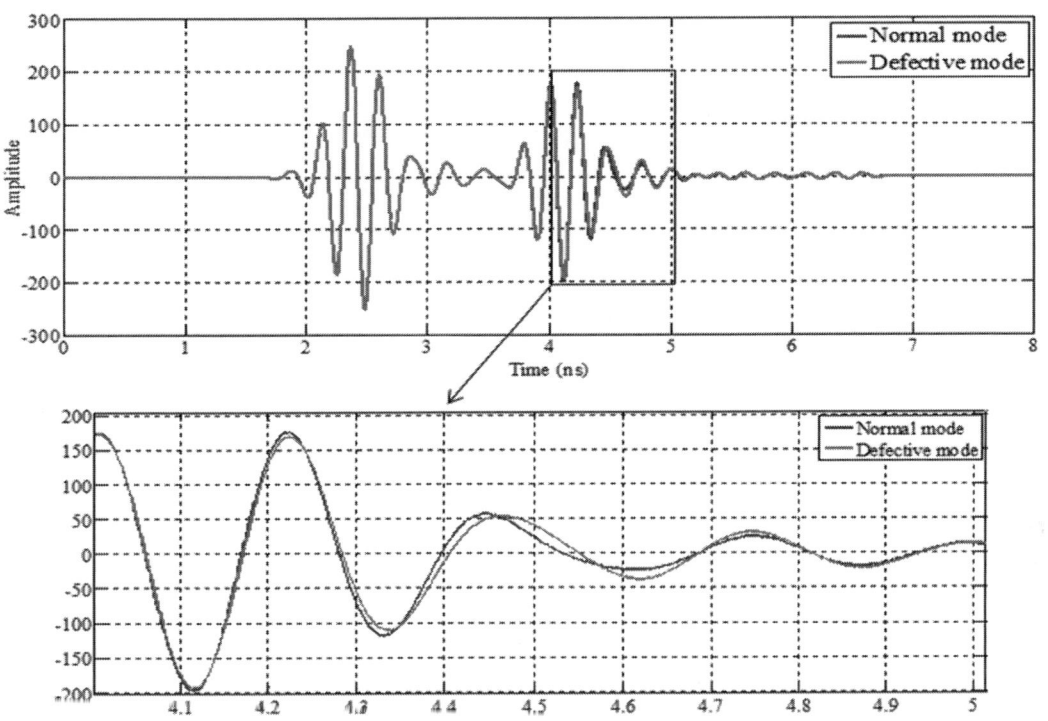

FIGURE 9.3

Difference between received signals in sound and defected conditions for sample RD.

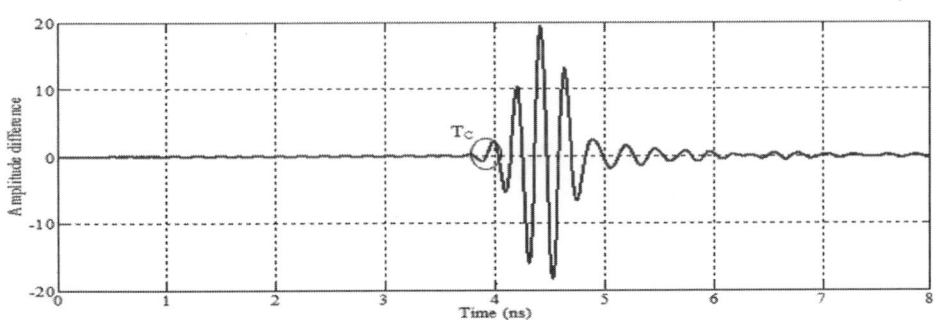

FIGURE 9.4

Variations in subtracted signal for sample RD defect.

half of the winding. The *TD* for detecting the defect by the two receiving antennas can be converted to the distance difference by Eq. (9.1).

$$x_1 - x_2 = c \times (t_1 - t_2) \tag{9.1}$$

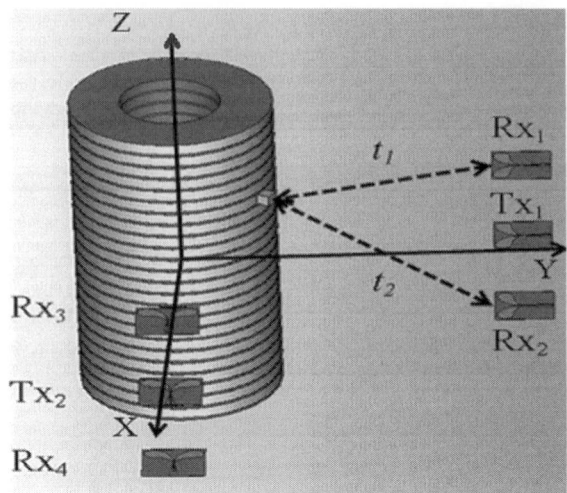

FIGURE 9.5

TD in signal propagation until reaching receiving antennas relative to defect.

where, c indicates the speed of light, which equals 3×10^8 m/s. The distance difference between the defect to the two receiving antennas can be obtained by having $(t_1 - t_2)$. Therefore, the distance difference between the two fixed points, i.e., receiving antennas, is achieved for the defect in a certain place. The geometric location of points whose distance difference from the two fixed points is a constant value in space is a hyperbola, whose foci are regarded as the location of the receiving antennas (Tabarsa et al., 2019). Thus, a hyperbola can be considered in proportion to an RD defect, in which the defect in question is located at one point.

The present chapter aims to locate the RD defect on the entire surface of the power transformer winding. As shown in Fig. 9.5, two sets of antennas can be considered to cover the desired defect. Based on the time difference of arrival (TDOA) theory (Karami et al., 2012), a hyperbola can be considered in three-dimensional (3D) space for each antenna set relative to the defect. In addition, a cylinder is regarded as the geometric location of the transformer winding, which is hypothesized to have the defect of RD on the high-voltage (HV) winding as the external layer. Fig. 9.6 illustrates different views of the main layout and problem parameters. The following comprehensive algorithm is used step by step to locate the RD defect.

Step 1: The signals are transmitted and received for four sets of antennas. Received signals are stored and processed.

Step 2: During this step, it is first determined whether there is a defect in the transformer or not. Then, the quadrant in which the deformation occurred is identified. To this aim, the signals for all of the receiving antennas are compared with the sound condition of the transformer. Two sets of antennas show changes in the received signal compared to the sound condition when a defect occurs. Such changes appear evident by subtracting the sound condition signal from the defective one. The quarter in which the defect occurred is identified after detecting the two sets of antennas. For example, the signal changes by the sets r_1 and r_2 in Fig. 9.6 indicate that the defect occurred in the first quarter of the coordinates.

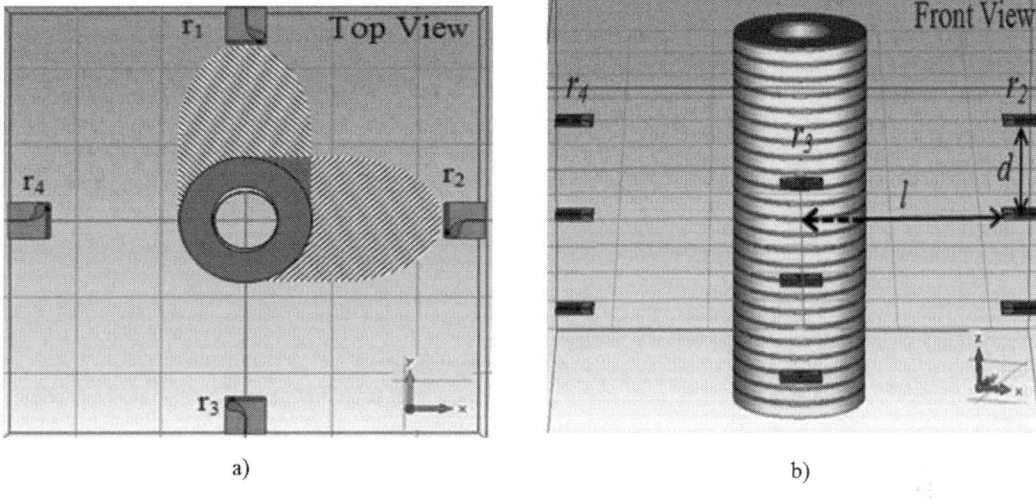

a)   b)

FIGURE 9.6

Different views of 3D layout: (A) Top view, and (B) Front view.

Step 3: The *TD* of the receiving antennas relative to the defect in each set is obtained by analyzing the received signals of the two sets after determining the set of antennas that detect the defect simultaneously.

Step 4: The *TD* between the two receiving antennas and the defect is converted into the distance difference by Eq. (9.1).

Step 5: A hyperbolic equation in space is achieved, corresponding to each set of antennas detecting the defect, as shown in Fig. 9.7.

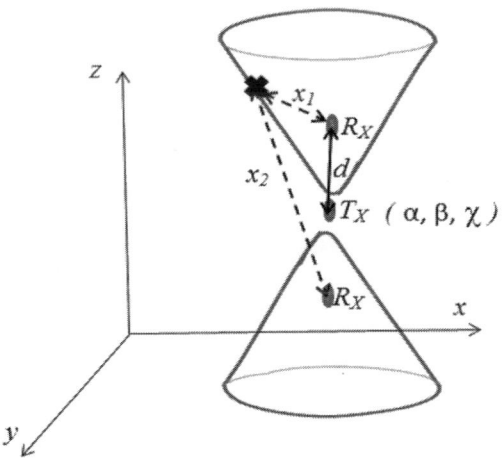

FIGURE 9.7

Hyperbolic corresponding to location of antennas with nonorigin center of symmetry.

$$\frac{(z-\alpha)^2}{a^2} - \frac{(x-\beta)^2}{b^2} - \frac{(y-\chi)^2}{c^2} = 1 \tag{9.2}$$

where, α, β, and χ indicate the centers of hyperbolic symmetry in the Cartesian coordinates. In addition, a, b, and c are calculated from Eqs. (9.3) and (9.4), respectively.

$$2a = x_1 - x_2 \tag{9.3}$$

$$b^2 = c^2 = d^2 - a^2 \tag{9.4}$$

A hyperbola is obtained as a possible geometric location of the defect proportionate to each of the abovementioned sets of antennas when the sets of antennas r_1 and r_2 display the signal changing. The following hyperbolic equations are expressed based on the location of the antennas and the test layout shown in Fig. 9.6:

$$\frac{z^2}{a_0^2} - \frac{(x-l)^2}{b_0^2} - \frac{y^2}{c_0^2} = 1 \tag{9.5}$$

$$\frac{z^2}{a_1^2} - \frac{x^2}{b_1^2} - \frac{(y-l)^2}{c_1^2} = 1 \tag{9.6}$$

where, l represents the distance between the transformer center and the sets of antennas r_1 and r_2. In addition, a_0 and a_1 are calculated by the fourth step, and the parameters b_0, c_0, b_1, and c_1 are calculated from the following equations:

$$b_0^2 = c_0^2 = d^2 - a_0^2 \tag{9.7}$$

$$b_1^2 = c_1^2 = d^2 - a_1^2 \tag{9.8}$$

where, d is considered as the distance between the receiving and transmitting antennas. Accordingly, two hyperbolic equations are achieved during the fifth step, corresponding to the two sets of antennas.

Step 6: During this step, the geometric location of the transformer winding at the HV level is considered a cylinder. The corresponding equation, according to its dimensions and 3D layout shown in Fig. 9.6, is as follows:

$$x^2 + y^2 = r^2, -h/2 < z < +h/2 \tag{9.9}$$

where, r indicates the radius of the cylinder of the transformer winding and h represents its height.

Step 7: The possible geometric location of the defect from sets of antennas, and geometric location related to the transformer winding were obtained during the fifth and sixth steps, respectively. Therefore, four points are calculated by intersecting these three geometric locations.

Step 8: The acceptable location of the defect among four possible points is determined in this step. Two points are regarded as acceptable on the side of the set of antennas. It means that the two points located in the first quarter (x, y > 0) are considered acceptable when the antennas r_1 and r_2 detect the defect. One of the receivers detects the defect earlier in each antenna set. Analyzing the received signals indicates that which one of the receiving antennas detects the start of changes in the subtracted signal in the sound and defective conditions sooner. Thus, one point is regarded as acceptable among the remaining points.

Step 9: The closest receiving antenna to RD is determined after detecting the defect location. The amount of defect can be determined by analyzing the received signals in the closer antenna. The returned time-domain signal related to the winding is defined. The amplitude euclidean distance (AED) is utilized to compare and measure the changes in the signal target section. The amplitude of the target signal during the i-th sampling with a defect length of x mm, i.e., $Tg_x(i)$, is compared with the reference signal, i.e., $Tg_0(i)$, by applying Eq. (9.10).

$$AED = \sqrt{\frac{\sum\limits_{i=\left(T_1/T_s\right)}^{T_2/T_s} (|Tg_x(i)| - |Tg_0(i)|)^2}{N}} \qquad (9.10)$$

where, T_1 and T_2 are the start and end times of the target signal, respectively. In addition, N is the number of samples, defined as follows.

$$N = \frac{T_2 - T_1}{T_s} \qquad (9.11)$$

where, T_S indicates the sampling time.

As observed, the AED is a comparative index that compares an online measured signal with a reference one. The AED index and defect length relationship are obtained using simulation and laboratory results in different RD sizes. Based on the results, the relationship between the AED index and defect length is considered as proportional, meaning that increasing the defect length makes the AED index larger. Therefore, such a feature is utilized in the next section. The MATLAB software is applied by a database obtained with different measurements, and the curves are achieved by fitting the data. The defect extent is estimated using the aforementioned curves, and so is the amount of the AED index obtained online.

9.1.2 Detecting AD defect

9.1.2.1 Comparative method

The AD is among the common defects of the windings of power transformers (Karami et al., 2019). The UWB system and time-domain analysis of received signals are applied to detect and determine the AD extent. The layout used here is the same as the layout of the previous one, includes one transmitting and two receiving antennas. Utilizing the AED index, which is based on the difference in the amplitude of the received signal in the sound and defective conditions for each receiving antenna, is regarded as the criterion for comparing the sound and defective conditions of the winding. The AED differs for the two receiving antennas since they are placed in two different locations relative to the winding. The ratio of AED in the two receiving antennas has a relationship in the simulation and laboratory modes. A method will be presented here, one of its main advantages is that it does not need a database and large training data, which must be obtained from the measurement of the actual winding.

Fig. 9.8 displays the sound and displaced position of the winding. As shown in Fig. 9.8, the transmitted pulse, i.e., $P(t)$, is emitted in the ambient, and a part of the reflected energy from the winding is received by the top and bottom antennas, which are shown by RA_U and RA_L, respectively. The signals received in RA_U for the sound and defective conditions of the winding are presented with

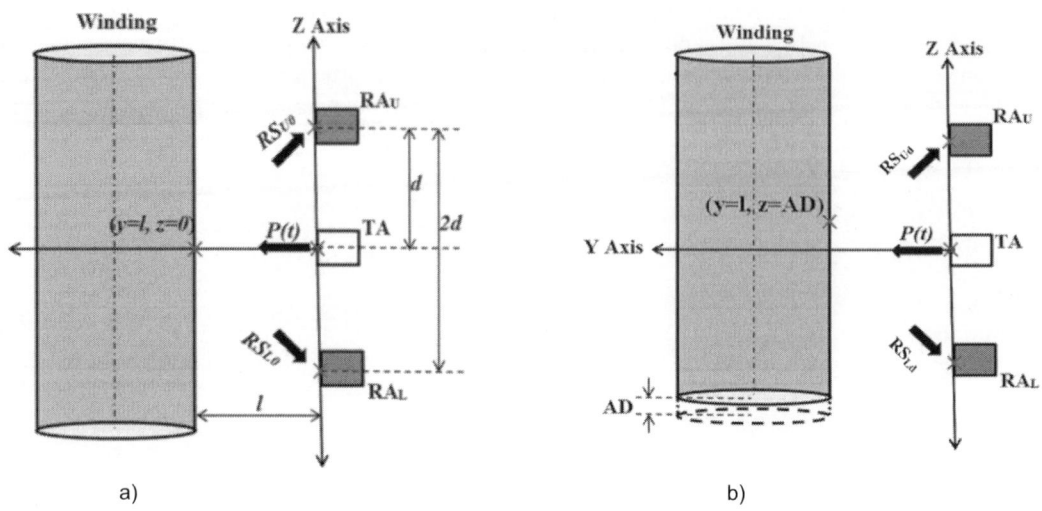

FIGURE 9.8

Schematic diagram for AD monitoring: (A) Sound condition of winding, and (B) Displaced condition of winding.

RS_{U0} and RS_{Ud}, respectively, and those received in RA_L for the above-mentioned modes are represented by RS_{L0} and RS_{Ld}, respectively.

The following equations show the subtracted signals in the sound and defective conditions.

$$Sb_{Us} = RS_{Ud} - RS_{U0} \tag{9.12}$$

$$Sb_{Ls} = RS_{Ld} - RS_{L0} \tag{9.13}$$

where, Sb_{Us} and Sb_{Ls} are the subtracted signals for RA_U and RA_L receiving antennas, respectively.

Complete AD monitoring includes defect detection, AD direction, and AD extent. The following equations can be used to detect the occurrence of displacement:

$$AED_U > AED_U^{Thresh} \tag{9.14}$$

$$AED_L > AED_L^{Thresh} \tag{9.15}$$

where, AED_U and AED_L indicate the calculated AED values for RA_U and RA_L receiving antennas, respectively. In addition, AED_U^{Thresh} and AED_L^{Thresh} represent reasonable thresholds for RA_U and RA_L, respectively. The threshold value is equal to zero for simulations under ideal conditions, and a small value for experimental measurements should be considered.

The direction of the displacement should be determined after the detection step. For this aim, the calculated AED value in RA_U and RA_L receiving antennas can be compared. Based on the results, AED_U is greater than AED_L in the positive direction of the Z-axis, i.e., upward displacement, and vice versa for downward displacement. Therefore, comparing the aforementioned index in RA_U and RA_L receiving antennas can be an appropriate and reliable criterion for detecting displacement direction.

For determining the displacement extent, the AED_U and AED_L indices are combined. It is noteworthy that the two receiving antennas are located in two different and symmetrical positions relative to the winding in the proposed model. Therefore, the results and database achieved from the simulation can be used to estimate the AD in the actual winding by combining the results obtained in RA_U and RA_L antennas. As mentioned before, the UWB waves do not penetrate the HV winding. Thus, only the external dimensions of the winding and the distances of the test elements should correctly be simulated for the real winding in the simulation software. There is no need for accurate simulation inside the winding. At first, the sum of AED in RA_U and RA_L antennas is calculated according to the following equation.

$$AED_S^{Sim}(x) = AED_U^{Sim}(x) + AED_L^{Sim}(x) \tag{9.16}$$

where, $AED_S^{Sim}(x)$ indicates the sum of the AED in the two receiving antennas for the x mm displacement condition in the simulation environment. Accordingly, a complete database of different displacement conditions in the simulated winding is achieved to detect the displacement extent in the actual winding with an acceptable error. The sum of the AED in the two receiving antennas in the case of x mm displacement in the laboratory is achieved by the following equation:

$$AED_S^{Exp}(x) = AED_U^{Exp}(x) + AED_L^{Exp}(x) \tag{9.17}$$

Accordingly, the value of $AED_S^{Exp}(x)$ is compared with $AED_S^{Sim}(x)$ and the AD extent is estimated.

9.1.2.2 TDOA analytical method

Fig. 9.9 displays the layout of UWB antennas for detecting the displacement defect of the transformer winding using the TDOA method. Two scenarios are defined in this method. In the first scenario, i.e., S_l, the lower antenna A_1 acts as a transmitter, and the other two antennas perform as receivers. In the

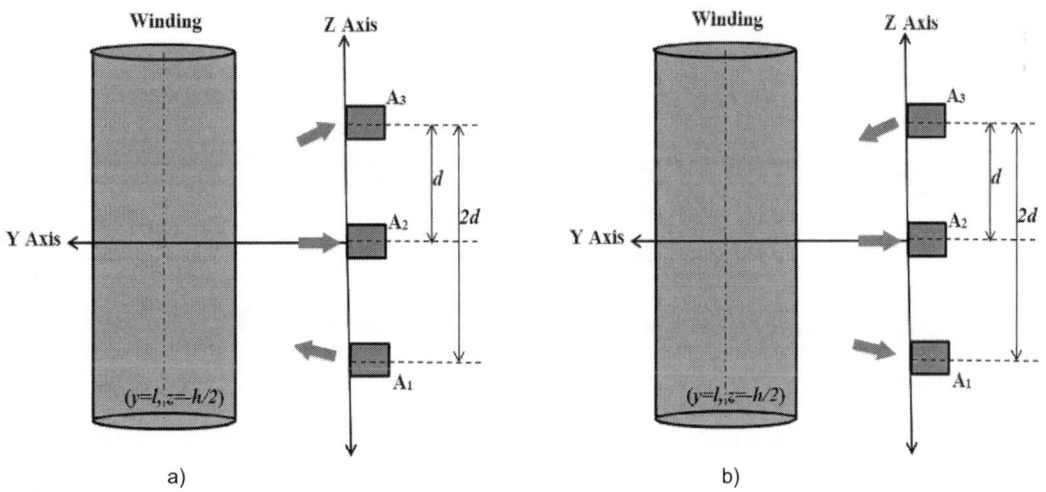

a)   b)

FIGURE 9.9

AD monitoring layout utilizing TDOA: (A) S_l scenario, and (B) S_U scenario.

second scenario, i.e., S_u, the antenna A_3 acts as a transmitter, and the other antennas, i.e., A_1 and A_2, perform as receivers. In the first scenario, the received signals in A2 and A3 are shown with $R_L^{S_l}$ and $R_U^{S_l}$, respectively. Accordingly, the received signals in the second scenario in A_1 and A_2 are defined as $R_L^{S_u}$ and $R_U^{S_u}$, respectively. In both S_l and S_u scenarios, the received signals for sound and displaced conditions are known using subscribes 0 and d, respectively.

To detect the occurrence of a defect, the received signals should be compared with the reference received signals. The following equations demonstrate the subtracted signal for the possible conditions.

$$Sb_{Ls}^{S_l} = R_{Ld}^{S_l} - R_{L0}^{S_l} \tag{9.18}$$

$$Sb_{Ls}^{S_u} = R_{Ld}^{S_u} - R_{L0}^{S_u} \tag{9.19}$$

$$Sb_{Us}^{S_l} = R_{Ud}^{S_l} - R_{U0}^{S_l} \tag{9.20}$$

$$Sb_{Us}^{S_u} = R_{Ud}^{S_u} - R_{U0}^{S_u} \tag{9.21}$$

where, $Sb_{Ls}^{S_l}$ and $Sb_{Us}^{S_l}$ indicate the subtracted signals of antennas A_2 and A_3 in the S_l scenario, respectively. $Sb_{Ls}^{S_u}$ and $Sb_{Us}^{S_u}$ represent the subtracted signals of antennas A_1 and A_2 in the S_u scenario, respectively. The above-mentioned equations detect the occurrence of displacement in the winding.

The displacement direction should be determined once the defect is detected. Comparing the peak time related to the subtracted signals obtained from Eqs. (9.18) and (9.21), determines the displacement direction. Fig. 9.10 displays the subtracted signals for a sample AD in the upward direction.

The peak time of the subtracted signal for $Sb_{Us}^{S_u}$ is ahead of $Sb_{Ls}^{S_l}$. It is regarded as an appropriate criterion for determining the displacement direction. The subtracted signal is affected by the amount of displaced (upper and lower) parts of the winding. The lower part of the winding can reflect fewer signals to the middle antenna by upward displacement. Accordingly, the upward displacement effect is

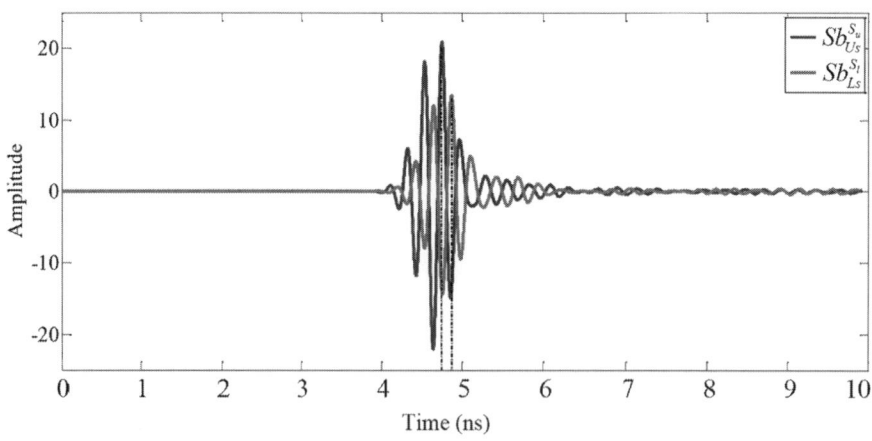

FIGURE 9.10

Subtracted signals achieved for S_l and S_u scenarios in antenna A_2 for displaced condition.

detected earlier in the subtracted signal in the S_u scenario relative to the S_l one. In the downward case, the abovementioned effect is reversed, and the subtracted signal of the center antenna approaches the peak earlier in the S_l mode than in the S_u mode.

The TDOA method can be used to determine the AD extent. The signals in the S_u mode are used for analysis when the displacement is upward, while the signals in the S_l mode are utilized in the TDOA method when the displacement is downward.

The received subtracted signals at the two receiving antennas have a *TD* when a sample upward AD occurs.

Then, the obtained *TD* is converted into a spatial difference by applying Eq. (9.1). Based on the TDOA method, one hyperbola can be considered for two receiving antennas in 2D space. As illustrated in Fig. 9.9, the lower part of the winding is demonstrated with coordinates ($y = l$, $z = -h/2$). Based on the position of the antenna in Fig. 9.9, the hyperbolic equation is achieved as follows:

$$b^2 \left(z - \frac{d}{2} \right)^2 - a^2 y^2 = a^2 b^2 \tag{9.22}$$

However, Eq. (9.22) changes considering the S_u scenario and when the movement is downward, as follows:

$$b^2 \left(z + \frac{d}{2} \right)^2 - a^2 y^2 = a^2 b^2 \tag{9.23}$$

where, the parameters a and b were defined previously. Thus, the AD extent in the winding for upward and downward movement is estimated using Eqs. (9.22) and (9.23), respectively, after analyzing the received signals and obtaining the *TD*.

$$Z_i = \sqrt{\frac{a^2 b^2 + a^2 y^2}{b^2}} + \left(\frac{d}{2} \right) \Rightarrow AD = \Delta Z_i = \frac{h}{2} + Z_i \tag{9.24}$$

$$Z_i = \sqrt{\frac{a^2 b^2 + a^2 y^2}{b^2}} - \left(\frac{d}{2} \right) \Rightarrow AD = \Delta Z_i = -\frac{h}{2} + Z_i \tag{9.25}$$

9.1.3 Distinguishing simultaneous RD and AD defects

During monitoring, the type of defect should be identified after detecting its occurrence, and then its location and extent should be determined. Defects in RD and AD may occur simultaneously (Karami et al., 2016; Karami, Gharehpetian, Norouzi, et al., 2020; Mortazavian et al., 2012). The present section investigates a comparative method for distinguishing between the RD and AD of a transformer winding utilizing electromagnetic waves (EMWs). The *AED* criterion is applied to compare the received signals in the sound and defective condition of the winding. The equations related to the AED criterion are not repeated here because the criterion was fully described previously.

As described in Section 9.1.2.1, the summation of *AED* for two receiving antennas is calculated. Similarly, the following equations can be utilized to detect the occurrence of each RD and AD defect.

$$AED_S(RD) > AED_S^{Thresh}(RD) \tag{9.26}$$

$$AED_S(AD) > AED_S^{Thresh}(AD) \tag{9.27}$$

The AED_S of a simulated RD with a length of 2 mm, approximately 5.4% of the winding diameter, is calculated and assumed in this section as the threshold for the RD. For the AD, the AED_S of a simulated AD with a length of 5 mm, which is approximately 5.4% of the winding height, is assumed as the threshold. The aforementioned values can be increased or decreased in the simulation environment, depending on the predetermined accuracy.

Based on the different simulation studies, the calculated value for AED_S during the AD defect is significantly higher than that of RD. In fact, the whole or a large part of the winding is displaced during the AD occurrence, while only a part of the winding is deformed during the RD occurrence. Therefore, the AED_S comparison criterion is considered an appropriate criterion for distinguishing defects. In other words, the AED_S threshold value for AD defects is higher than those calculated for RD defects with different lengths.

9.2 Simulation results

9.2.1 Detecting and locating RD

The simulation setup was described in Section 9.1.1.3. The specification of the winding was mentioned in Chapter 8. The RD defect with dimensions of 2×2 cm^2 is simulated at two different angles and heights to assess the algorithm.

9.2.1.1 Radial deformation defect at 45° angle

The defect is simulated as a bulgy RD at the angle of 45°, as shown in Fig. 9.11, at the height of 11.2 cm from the transformer center. The coordinates of the defect are ($x = 10.6$, $y = 10.6$, $z = 11.2$).

First, the signal is sent and received for the whole set of antennas. Then, the defective quarter should be determined. Fig. 9.12 displays the subtracted signals. The signal changes for the set of antennas in r_1 and r_2 indicate that the defect occurred in the first quarter, resulting in evaluating the received signals related to these antennas.

The TD of the receiving antennas is calculated in each set for the defect. The same behavior is perceived for the defect at an angle of 45° due to the symmetry of the defect to both sets of antennas. In addition, the upper antenna detects the changes in the subtracted signal earlier; because the defect is at a closer height. The distance difference is calculated for the two sets r_1 and r_2 by applying Eq. (9.1), as follows:

$$x_1 - x_2 = y_1 - y_2 = 7.5 cm \tag{9.28}$$

where, x_1-x_2 indicates the distance difference between the receiving antennas r_1, and y_1-y_2 represents that between the receiving antennas r_2.

The necessary parameters for writing hyperbolic equations are obtained using Eqs. (9.5) and (9.6), as follows:

$$\frac{z^2}{3.75^2} - \frac{(x-55)^2}{16.58^2} - \frac{y^2}{16.58^2} = 1 \tag{9.29}$$

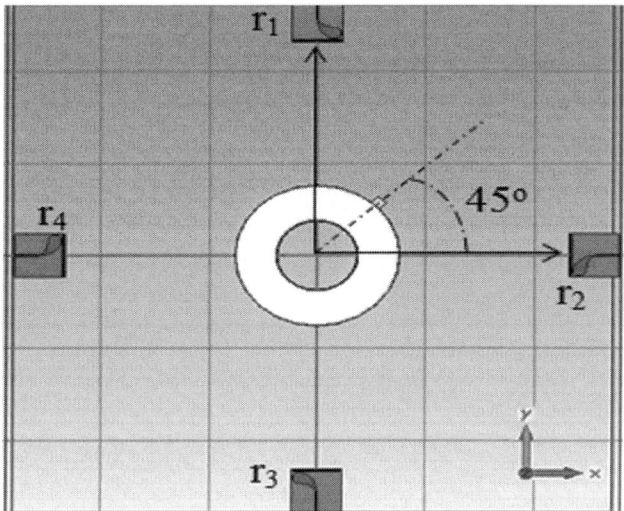

FIGURE 9.11

Top view of simulated RD at angle of 45°.

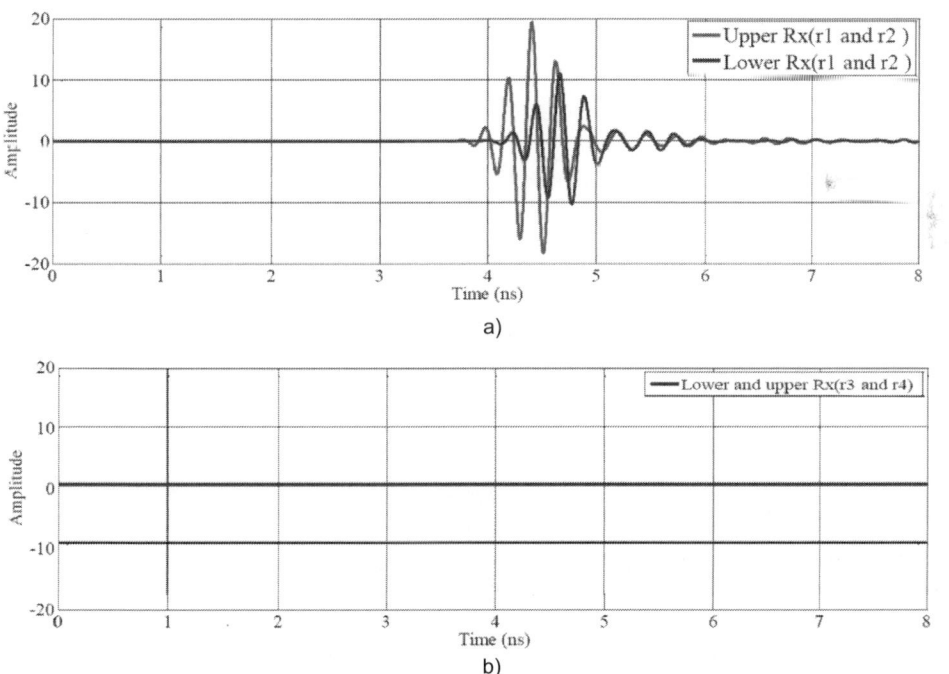

FIGURE 9.12

Subtracted signals from transformer model with defect at angle of 45°: (A) Set of antennas in r_1 and r_2, and (B) Set of antennas in r_3 and r_4.

$$\frac{z^2}{3.75^2} - \frac{x^2}{16.58^2} - \frac{(y-55)^2}{16.58^2} = 1 \qquad (9.30)$$

The geometric location of the transformer winding model, which was regarded as a cylinder, is achieved as follows:

$$x^2 + y^2 = 15^2, \; -30 < z < +30 \qquad (9.31)$$

Thus, all of the equations which should solve the problem in 3D space are obtained up to this point. So, four points, listed in Table 9.1, are obtained as possible locations for the RD defect after the intersection of three geometric locations. Point P_4 is considered as the only acceptable one because the quadrant of occurrence is previously determined in the first quadrant. Fig. 9.13 shows a schematic diagram related to the intersection of the aforementioned geometric locations.

9.2.1.2 Radial deformation defect at 30° angle

In this section, the defect is simulated as a bulgy at an angle of 30° to the plane $y = 0$ and a height of 14.6 cm from the transformer center, with coordination of $x = 12.99$, $y = 7.5$, and $z = 14.6$, demonstrated from the top view in Fig. 9.14.

Going through the same procedure in Section 9.2.1.1 shows that the defect occurs in the first quarter. As observed in Fig. 9.15, the symmetry of the previous condition is lost in this part, and the

Table 9.1 RD location results in 3D space for simulated winding defect with 45° angle.

Estimated location (l_I)	Points calculated from intersection of geometric locations				Defect location (l_0)
	p_1	p_2	p_3	p_4	
(10.6; 10.6; 10.98)	(−10.6; −10.6; −15.49)	(−10.6; −10.6; 15.49)	(10.6; 10.6; −10.98)	(10.6; 10.6; 10.98)	(10.6; 10.6; 11.2)

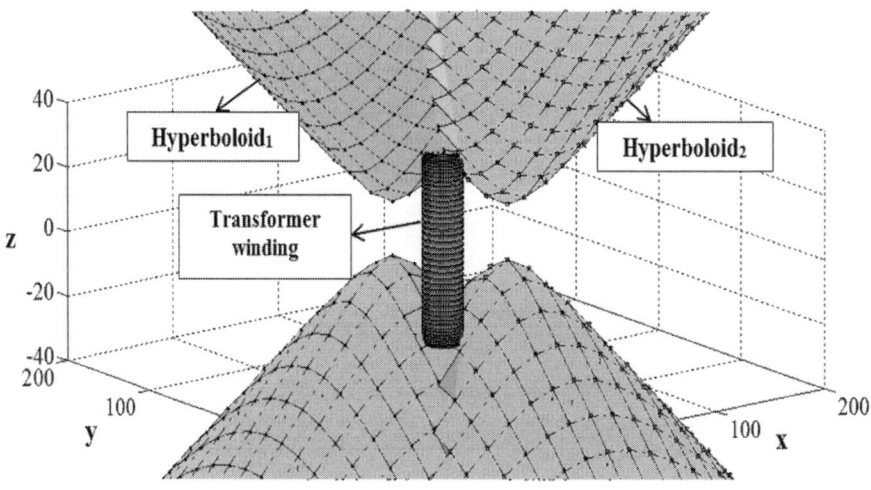

FIGURE 9.13

Intersection of geometric locations for simulated winding.

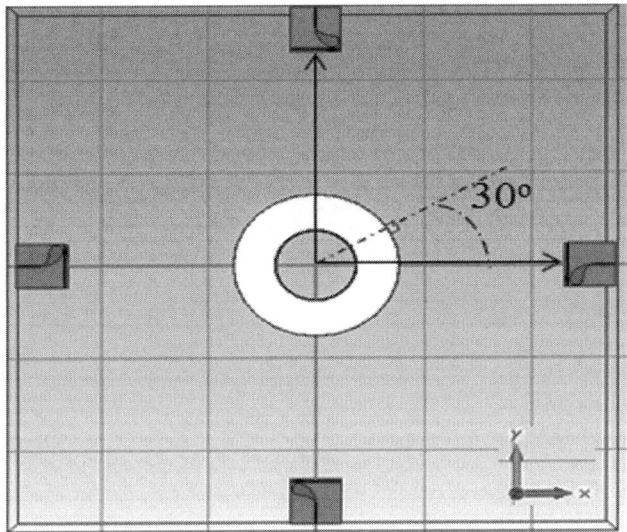

FIGURE 9.14

Top view of simulated defect at angle of 30°.

behavior of the received signals related to the antennas r_1 and r_2 differs. The *TD* for the receiving antennas r_1 and r_2 equal 0.3 and 0.27 ns, respectively. In addition, the *TD* for the receiving antennas in r_1 and r_2 compared to the defect differs from the previous case due to the asymmetry of the defect location in the model of this section.

Similar to Section 9.2.1.1, the geometric locations of the hyperboles are obtained as follows:

$$\frac{z^2}{4.5^2} - \frac{(x-55)^2}{14.31^2} - \frac{y^2}{14.31^2} = 1 \tag{9.32}$$

$$\frac{z^2}{4.05^2} - \frac{x^2}{14.44^2} - \frac{(y-55)^2}{14.44^2} = 1 \tag{9.33}$$

The intersection of the geometries in this condition results in four points, among which only one in the first quarter of the coordinates is regarded as acceptable. Table 9.2 lists all the obtained and acceptable points.

9.2.2 Detecting AD extent

This section examines the detection and determination of the displacement value in computer simulation technology (CST) simulation software. A database is created by simulation of layout and applying AD in different values to determine the displacement extent. Then, the database is used to estimate the AD extent according to the prepared algorithm.

To study the AD defect, the results related to the simulation of two ADs are thoroughly studied with the values of +3 cm and +5 cm, where the positive sign indicates the upward and positive displacement along the Z-axis. The simulation is performed for each displacement in sound (without

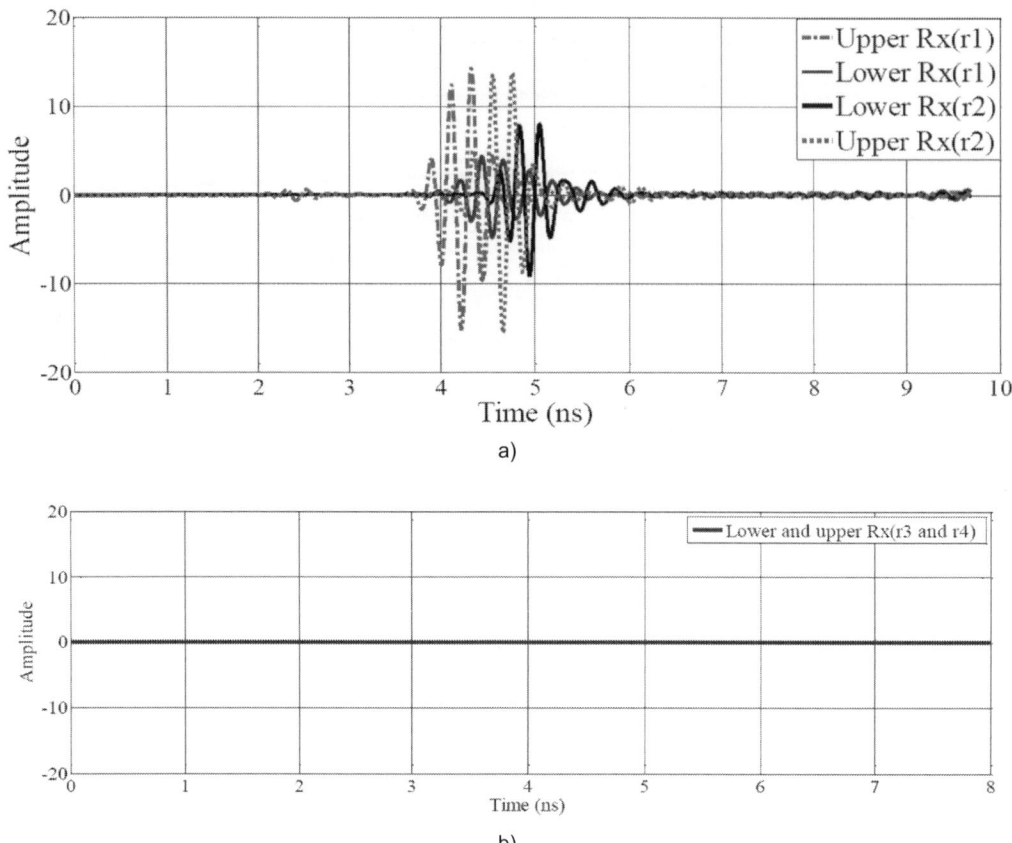

FIGURE 9.15

Subtracted signals for defect at angle of 30°: (A) r_1 and r_2 antennas, and (B) r_3 and r_4 antennas.

Table 9.2 RD location results in 3D space for simulated winding defect with 30° angle.

Estimated location (l_1)	Points calculated from intersection of geometric locations				Defect location (l_0)
	p_1	p_2	p_3	p_4	
(12.72; 7.94; 14.25)	(−4.87; −14.18; 19.8)	(−4.87; −14.18;−19.8)	(12.72; 7.94; −14.25)	(12.72; 7.94; 14.25)	(12.99; 7.5; 14.6)

displacement) and defective condition (with displacement). Fig. 9.16 illustrates the signals received at the receiving antennas for the displacement of +3 cm. As observed, the AD defect leads to some changes in the target section of the signal relative to the sound condition.

Similarly, Figs. 9.17 shows the received signals in RA$_U$ and RA$_L$ for the displacement of +5 cm. To create a complete database, AD modeling is performed on the simulated winding at different positions (from 2 to 58 mm). Fig. 9.18 illustrates the *AED* value for different AD extents. By applying the methods described in Section 9.1.2, the error rate in determining the defect extent for the simulation results is equal to zero since the conditions are ideal in simulations.

9.2.3 Distinguishing between radial and axial defects

Similar to the previous sections, the RD defect with different sizes and the AD with different values in the positive direction of the Z-axis are modeled. Fig. 9.19 displays the *AED* values calculated for the abovementioned defects. As observed, the *AED* value calculated in all samples exceeds the defined

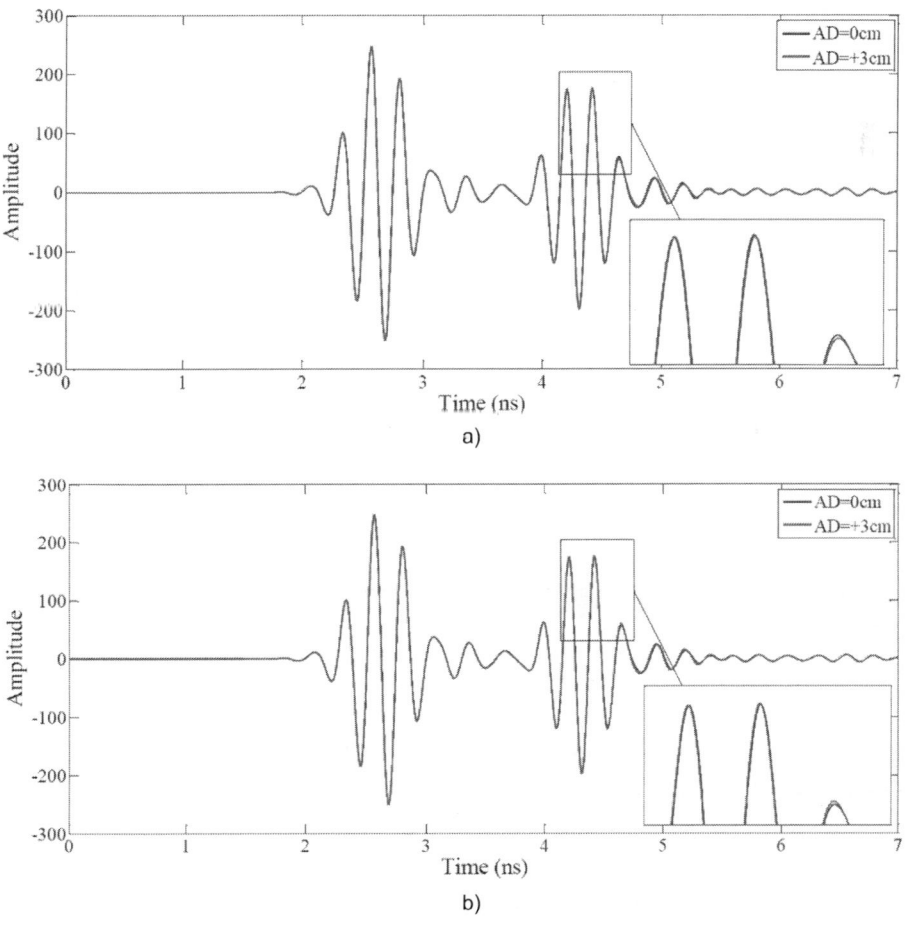

FIGURE 9.16

Received signals for simulated AD = +3 cm, (A) RA$_U$ antenna, and (B) RA$_L$ antenna.

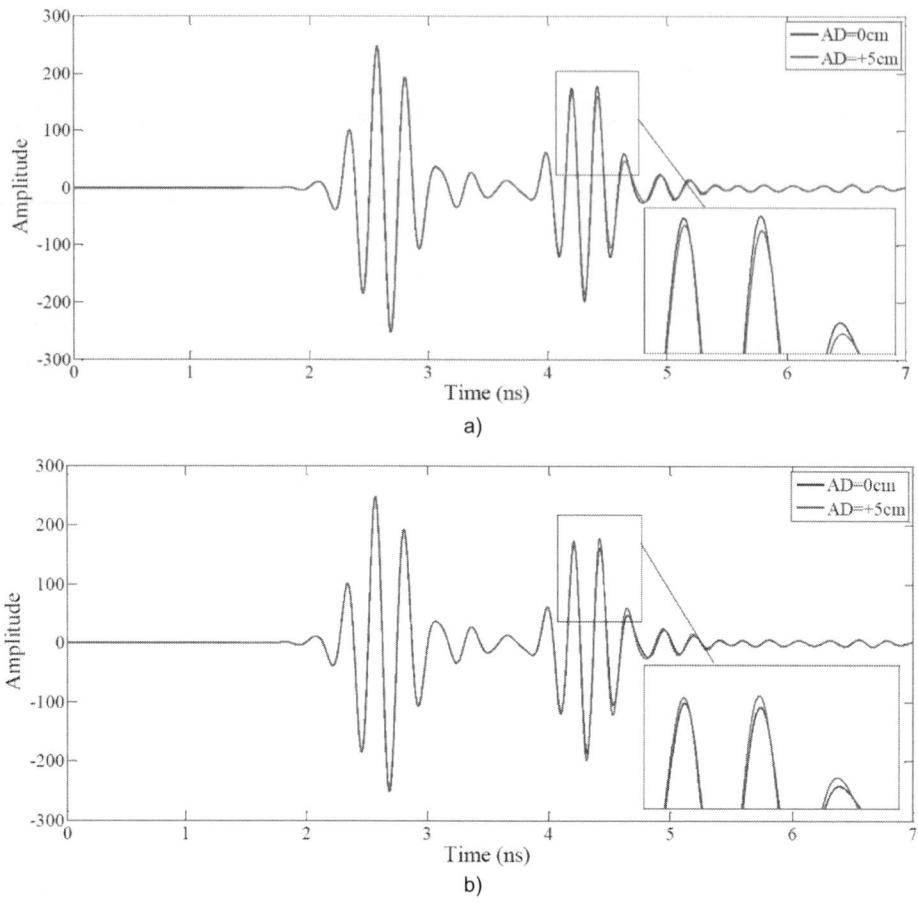

FIGURE 9.17

Received signals for simulated AD = +5 cm: (A) RA_U antenna, and (B) RA_L antenna.

threshold for both RD and AD defects. Therefore, defects in all the conditions are correctly detected. In addition, the extent of *AEDs* calculated for all the AD samples is greater than RD, indicating that the algorithm can distinguish the *AEDs* in addition to detecting the occurrence of defects.

9.3 Assessing laboratory results

The specifications of the antenna and transceiver were described, along with those of the understudy winding in the previous chapter. Fig. 9.20 shows the laboratory setup used in this section.

As illustrated in Fig. 9.20, (1) and (2) are the PulsON 220 device, (3) and (5) act as receiving antennas and (4) act as transmitting antennas; (6) is used to hold the antennas, and finally, (7) is the understudy winding.

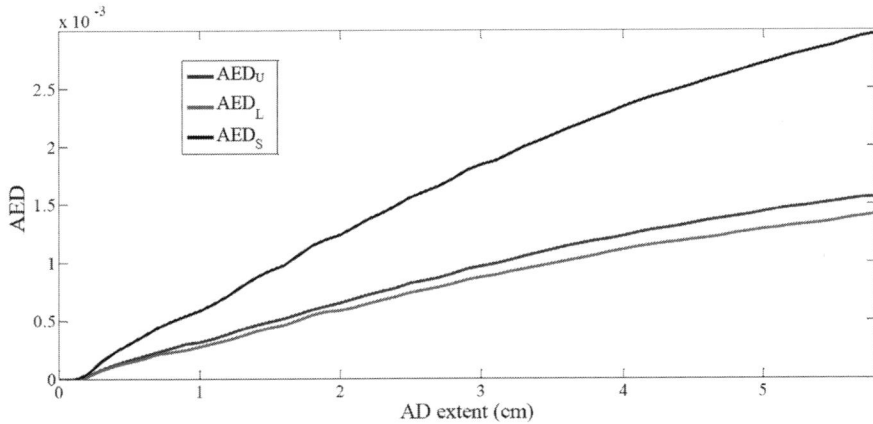

FIGURE 9.18

AED variations versus different AD values (simulation results).

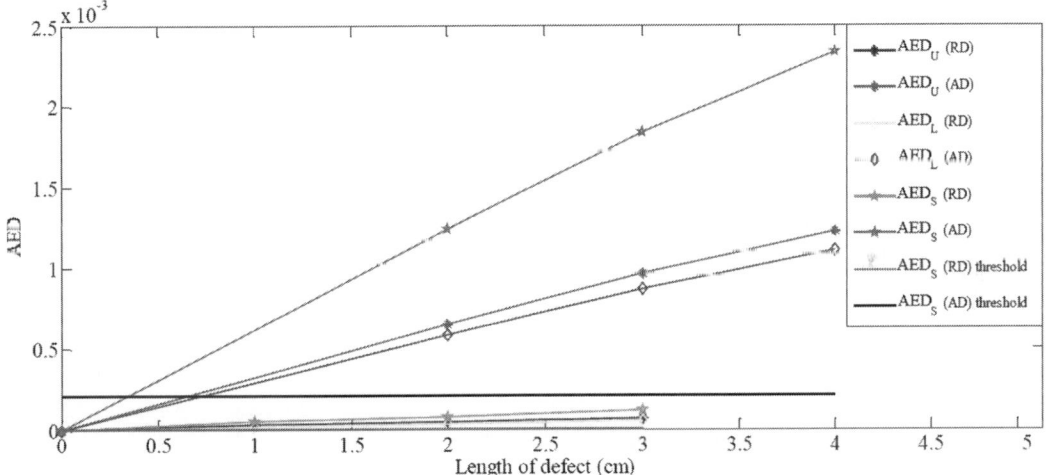

FIGURE 9.19

Comparison of simulated *AED* variations for AD and RD defects.

9.3.1 Radial defect

As shown in Fig. 8.13 in the previous chapter, a bulgy defect is modeled in the upper and lower parts of the winding, at $z = +20$ cm and $z = -17$ cm, respectively, with different extents of 3, 5, and 7 cm. Fig. 9.21 illustrates the received signals of the RA_1 and RA_2 antennas in the sound and defective conditions of the winding for the defect modeled in the upper part. The target section of the signals received by the RA_1 antenna undergoes minor changes by creating various defects in the upper part. In

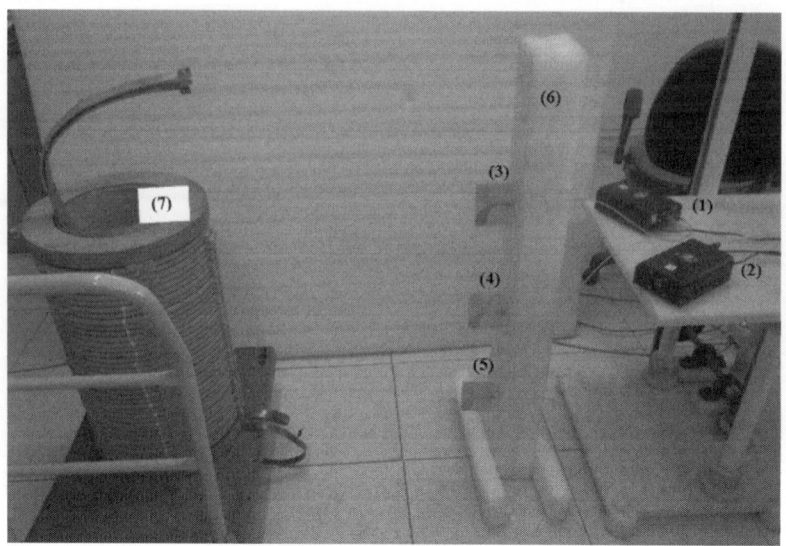

FIGURE 9.20

Laboratory setup.

addition, the changes are visible from the beginning of the target due to the proximity of the defect to the RA_1 antenna. The changes in the target section of the received signal start with a slight delay since the RA_2 antenna is farther away from the defect.

Fig. 9.22 shows the subtracted signal for the RD defect with 3 cm extent in the upper part. As observed, the effects of the defect are more clear on the subtracted signal for both receiving antennas.

Fig. 9.23 displays the received signals of the RA_1 and RA_2 antennas in the sound and defective conditions of the winding for the defect modeled in the lower part. As observed, the defect creates some changes in the target section of the signal, starting faster in RA2 than in RA1 due to the proximity. Fig. 9.24 demonstrates the subtracted signal for a defect with the extent of 3 cm in the lower part.

The RD location should be determined once its occurrence is detected. For a sample, the RD with a length of 3 cm at $z = +20$ cm and $z = -17$ cm is analyzed. To locate the RD, the TD should first be calculated. The TD_1 and TD_2 equal 0.5661 and 0.46,073 ns for the defect in the upper and lower parts, respectively. Thus, the spatial difference is calculated for RD at $z = +20$ cm based on Eqs. (9.1) and (9.3), as follows:

$$x_1 - x_2 = c \times TD_1 = 16.098 cm = 2a \tag{9.34}$$

The hyperbolic equation representing the possible geometric location of the defect, is obtained as follows:

$$\frac{z^2}{64.79} - \frac{x^2}{335.21} - \frac{y^2}{335.21} = 1 \tag{9.35}$$

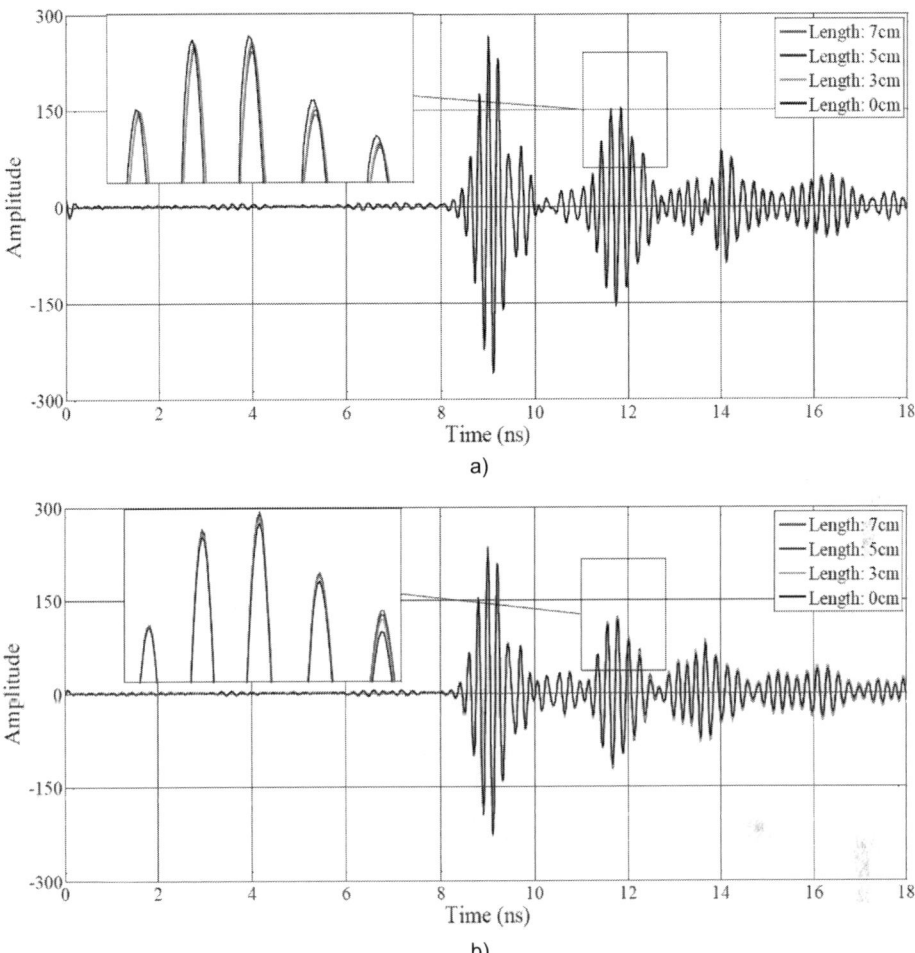

FIGURE 9.21

Received signals for laboratory setup RD at $z = +20$ cm: (A) RA_1 antenna and (B) RA_2 antenna.

As indicated, to calculate the RD location, the hyperbolic equations intersect with that of the HV winding in two dimensions considering that the UWB waves do not pass through the winding surface and the defect should be located on the surface.

$$\begin{cases} x = 0 \\ l \leq y \leq l + r \\ -h/2 \leq z \leq +h/2 \end{cases} \qquad (9.36)$$

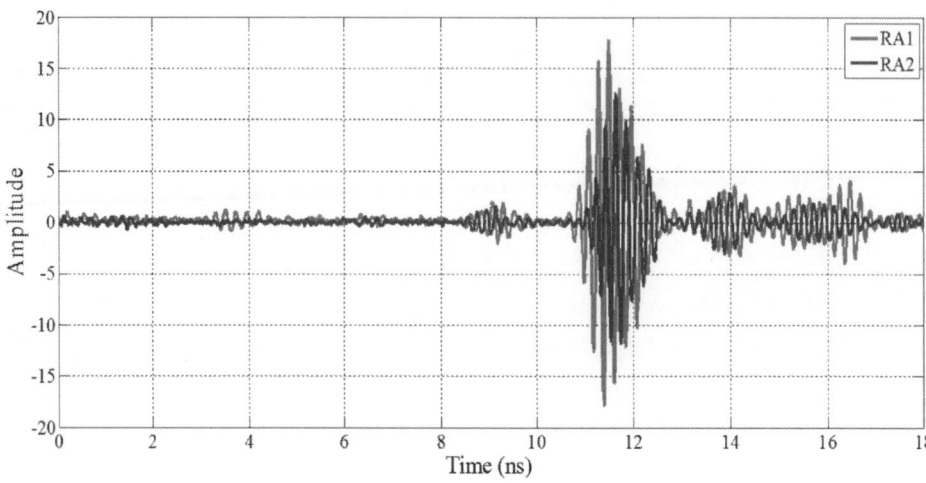

FIGURE 9.22

Subtracted signals for RA_1 and RA_2 antennas due to laboratory setup RD at $z = +20$ cm.

Similarly, the possible geometric location of the defect is achieved for RD at $z = -17$ cm, as follows:

$$\frac{z^2}{47.76} - \frac{x^2}{352.24} - \frac{y^2}{352.24} = 1 \tag{9.37}$$

Similarly, by intersecting the equations, the defect location is estimated and the results are listed in Table 9.3.

The error percentage is calculated based on the ratio of the absolute error to the height of the winding. Thus, the RD defect can be detected in the UWB frequency band and located with high accuracy, considering the presented simulations.

9.3.2 Axial displacement defect

The AD modeling is performed in 3 and 5 cm upwards. For easier modeling of the AD on the actual winding in the laboratory, the antennas are displaced downwards by 3 and 5 cm instead of moving the entire winding upwards. The tests are performed in two modes. In the first mode, the antennas are placed in their original location, related to the sound condition without any displacement. In the second mode, the antennas are shifted down by 3 cm, indicating a displaced condition of the winding. The same method is utilized to model the displacement of 5 cm, and the antenna position relative to the winding changes instead of its displacement. Fig. 9.25 shows the signals recorded in the receiving antenna for the displacement of $+3$ cm.

Similarly, Fig. 9.26 displays the signals received in RA_U and RA_L for the displacement of $+5$ cm.

The changes resulting from the AD of the winding in the target part of the received signals are seen in the experimental results. Based on the described algorithm, the AED_U and AED_L are calculated for

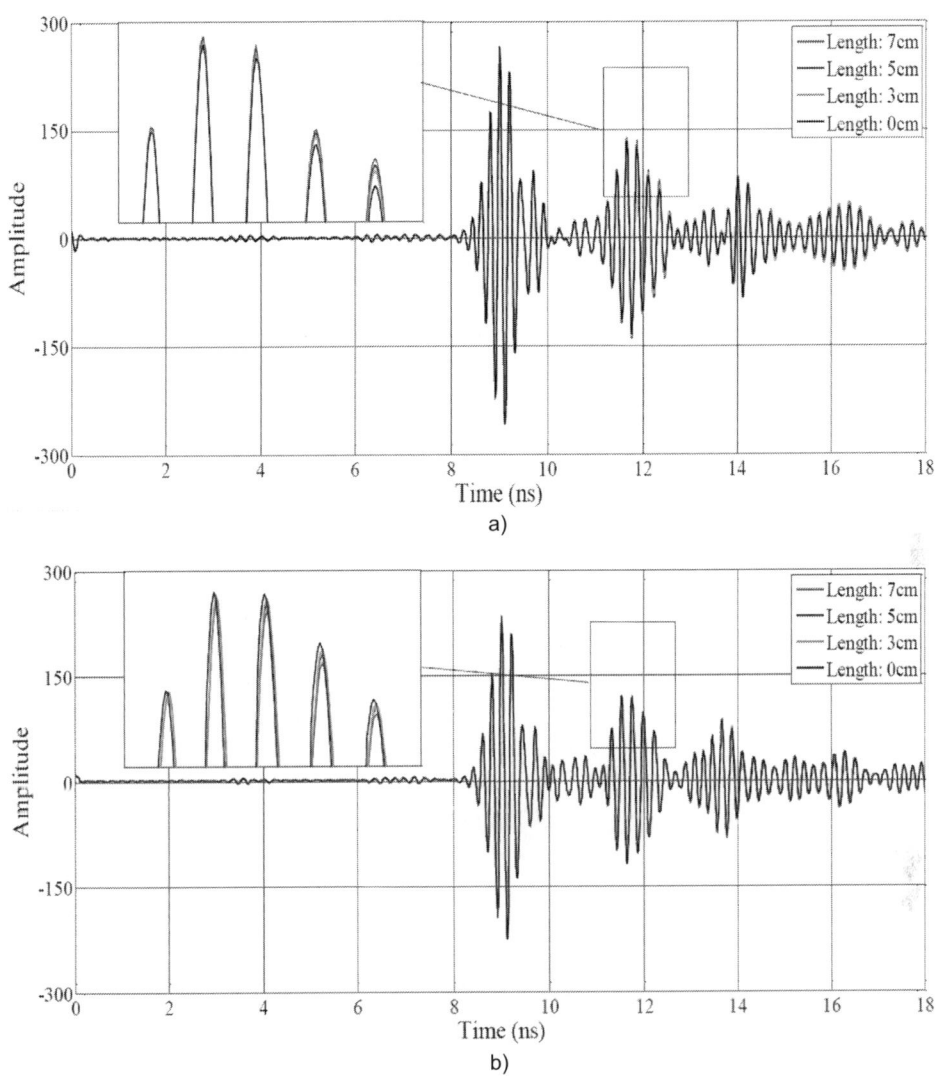

FIGURE 9.23

Received signals for laboratory setup RD at $z = -17$ cm: (A) RA_1 antenna, and (B) RA_2 antenna.

each test, resulting in determining the AD occurrence and direction. Then, the value of AED_s^{Exp} is compared with AED_s^{Sim} obtained from simulation results, and the AD extent is estimated. Table 9.4 lists all of the results related to the test and the error percentage in estimating the AD extent of the winding. The relative error percentage measured relative to the height of the winding is small and acceptable.

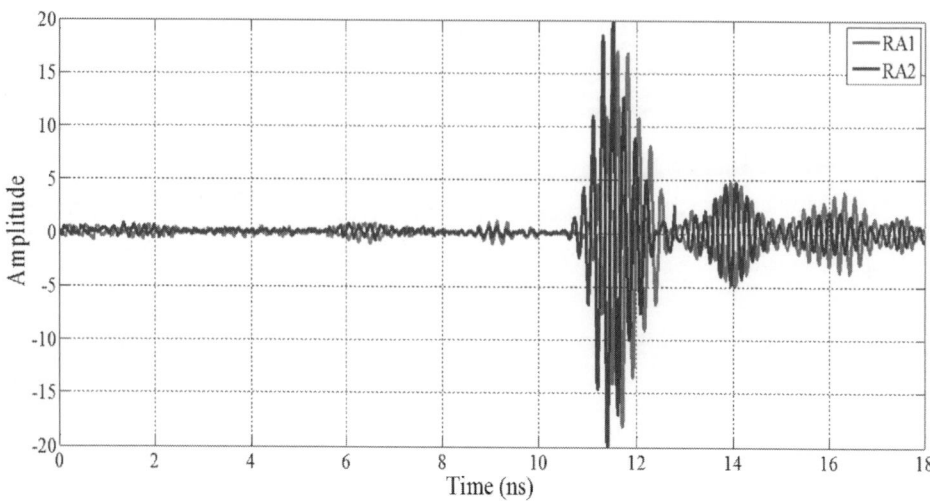

FIGURE 9.24

Subtracted signals for RA$_1$ and RA$_2$ antennas due to laboratory setup RD at $z = -17$ cm.

Table 9.3 Calculated defect location and error for RD laboratory setup.

Actual location	Calculated location	Absolute error (cm)	Error (%)
+20 cm	+19.33993 cm	0.66007	1.213
−17 cm	−16.26992 cm	0.73008	1.342

9.3.3 Distinguishing between AD and RD defects

Fig. 9.27 shows all the *AED* variations related to the tests. Based on the results, the described algorithm can detect and distinguish between the RD and AD defects in the power transformer winding due to the difference in *AED* values of AD and RD defects.

9.4 Effects of PD on hyperbolic method

The hyperbolic method was assessed for detection and extent determination of the RD and AD defects of the transformer HV winding. As indicated, the PD is regarded as one of the other defects in the power transformer, which can be detected and located by EMWs (Azadifar et al., 2020; Karami, Azadifar, Mostajabi, et al., 2020; Karami et al., 2021). This section proposes a solution to simultaneously detect the RD and PD defects. The need to change the UWB to ultra high frequency (UHF) band was discussed in Section 8.4.1. Many topics are not repeated here due to their similarity to those presented in Section 8.4.1. Then, the synchronism of the RD and PD defects is evaluated, while using the hyperbolic method. The issues related to simultaneously detecting AD and PD are not examined

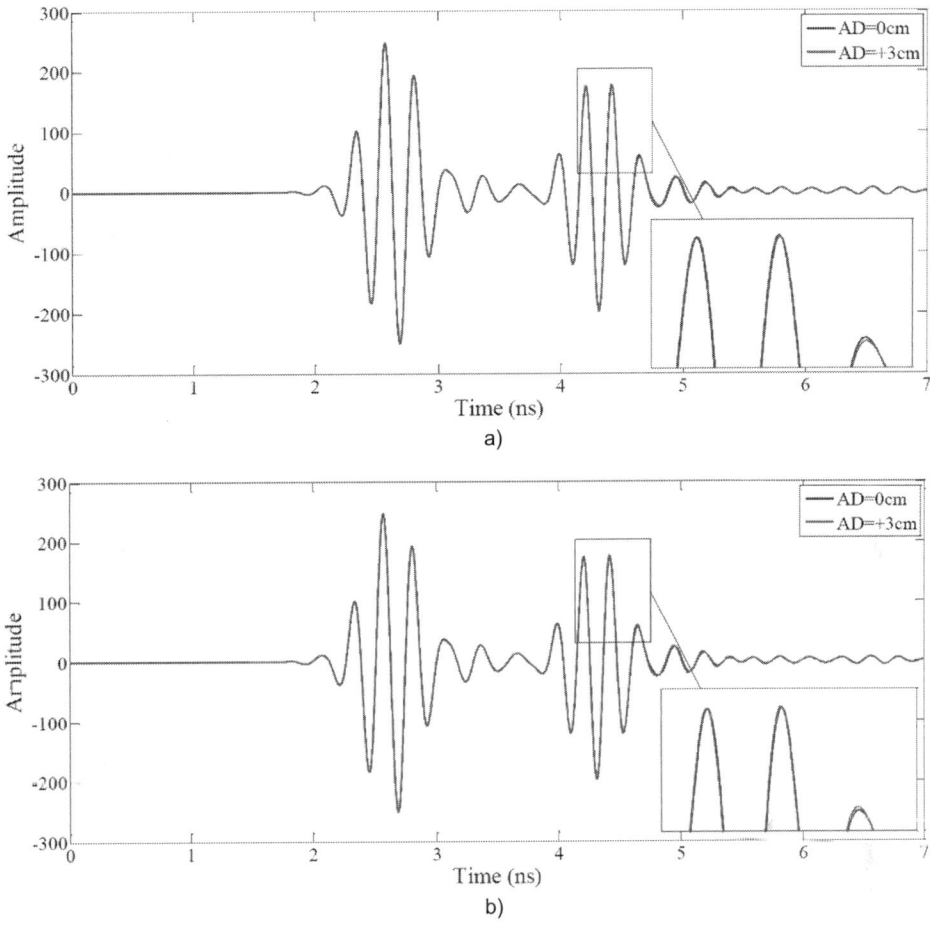

FIGURE 9.25

Received signals at AD = +3 cm: (A) RA$_U$ antenna, and (B) RA$_L$ antenna.

here due to their similarity to those of the previous chapter and the ability to generalize the topics related to synchronism of the RD and PD defects.

In the next step, the effect of the PD is investigated when the frequency is the same after showing the possibility of detecting the RD defect in the UHF band using the hyperbolic method. Finally, the possibility of simultaneous detection of both defects is studied by providing a solution to eliminate the obstacles.

9.4.1 Detecting RD by hyperbolic method in UHF band

Fig. 9.28 illustrates the 3D model of the winding in the sound condition simulated in CST software [37]. A winding model with an overall height of 60 cm is created by 25 disks with a copper-made layer

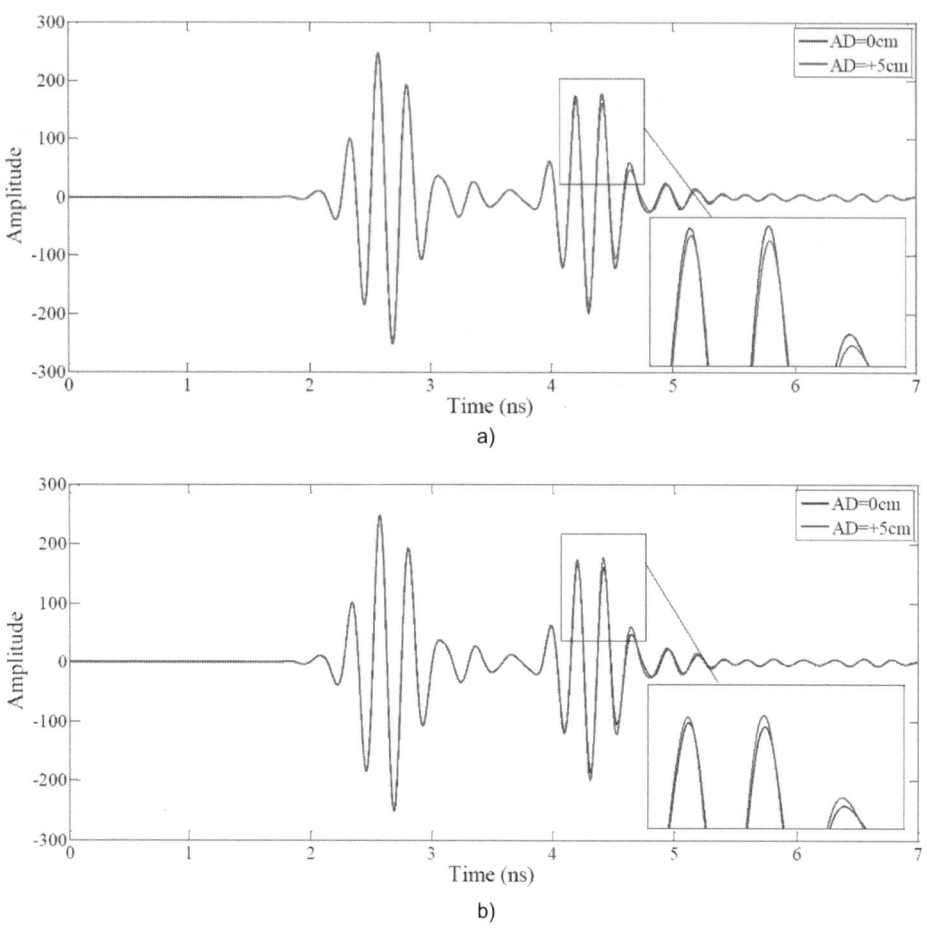

FIGURE 9.26

Received signals at AD = +5 cm: (A) RA_U antenna, and (B) RA_L antenna.

Actual AD extent	AED_U	AED_L	AED_S	AD detection	AD direction	Estimated AD extent	Absolute error	Error (%)
Table 9.4 Results of laboratory AD study.								
+3 cm	7.22e −4	6.46e −4	1.368e −3	Properly	Upward	+2.2 *cm*	5.8 cm	1.47
+5 cm	1.311e −3	1.18e −3	2.49e −3	Properly	Upward	+4.4 *cm*	5.6 cm	1.1

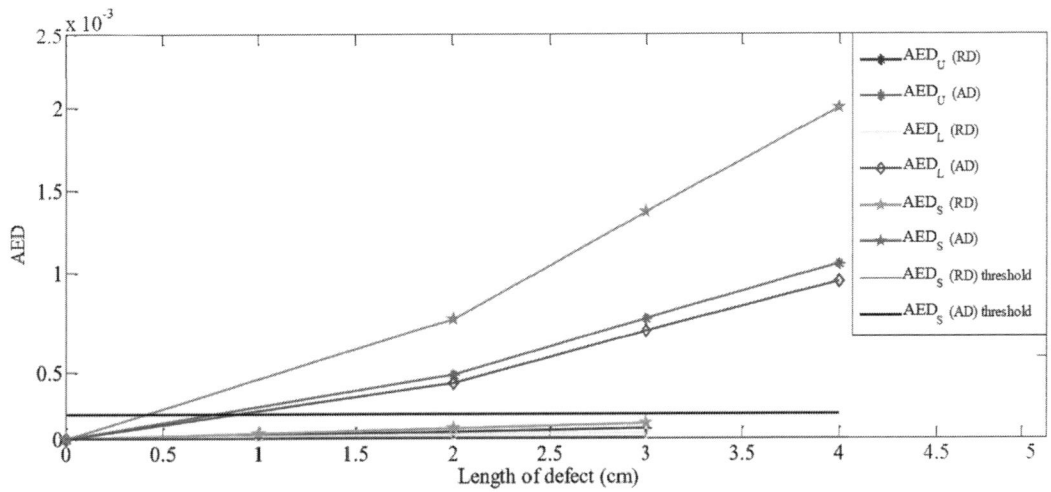

FIGURE 9.27

Comparison of laboratory *AED* variations for AD and RD defects.

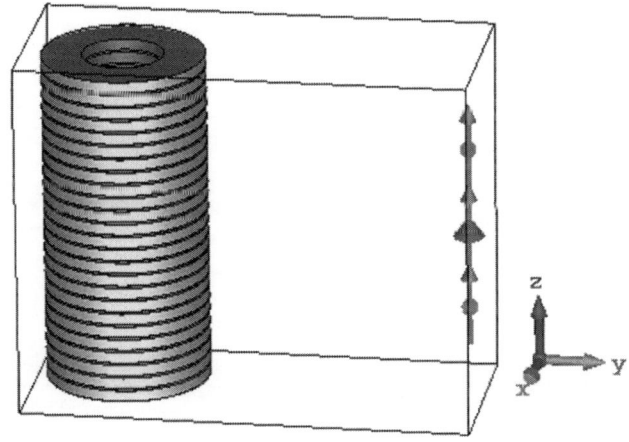

FIGURE 9.28

3D view of a sound condition of winding model in simulation environment.

with an external radius of 30 cm and a thickness of 2 cm and 24 spacers with the same radius and thickness of 0.5 cm. The transmitting and two receiving antennas are located at (0, 825 mm, 300 mm), (0, 825 mm, 150 mm), and (0, 825 mm, 450 mm), respectively. The transformer surface is located 485 mm from the antennas.

The transmitted signal should be changed from UWB to UHF to simulate the hyperbolic method in the UHF band. Therefore, a Gaussian signal is generated in the band of 300 MHz−3 GHz, sent by the transmitter. Fig. 9.29 shows the signal recorded by the receiving antennas in the sound condition.

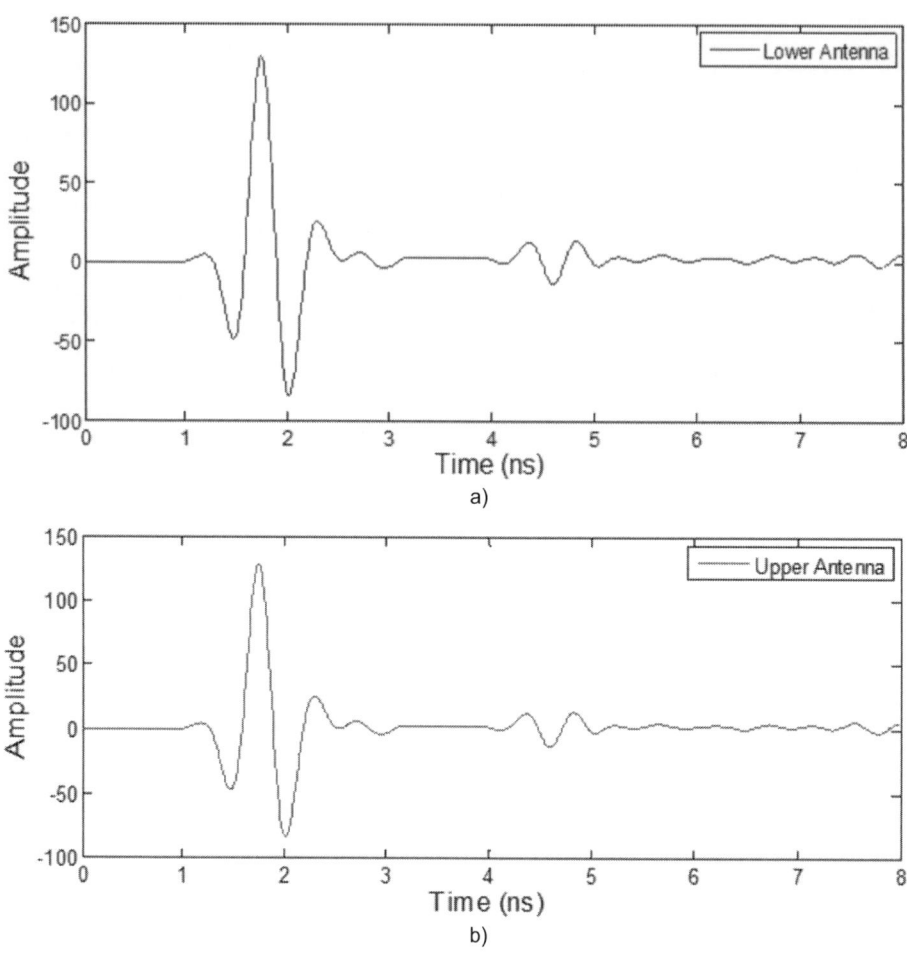

FIGURE 9.29

Received UHF signal in sound condition at: (A) Lower receiving antenna, and (B) Upper receiving antenna.

Then, a bulgy radial defect is created at (0, 350, 540) with dimensions of $20 \times 20 \times 40$ mm, shown in Fig. 9.30.

Fig. 9.31 displays the recorded signal by the receiving antennas in the defected condition.

The subtracted signal is calculated, as shown in Fig. 9.32, and the TD is converted to the distance difference according to Eq. (9.1).

$$x_1 - x_2 = c \times (t_1 - t_2) = 3 \times 10^8 \times (0.453 \times 10^{-9}) = 135mm \tag{9.38}$$

The hyperbolic parameters are achieved as $a = b = 133.9$. Now, a hyperbolic equation can be written, as follows:

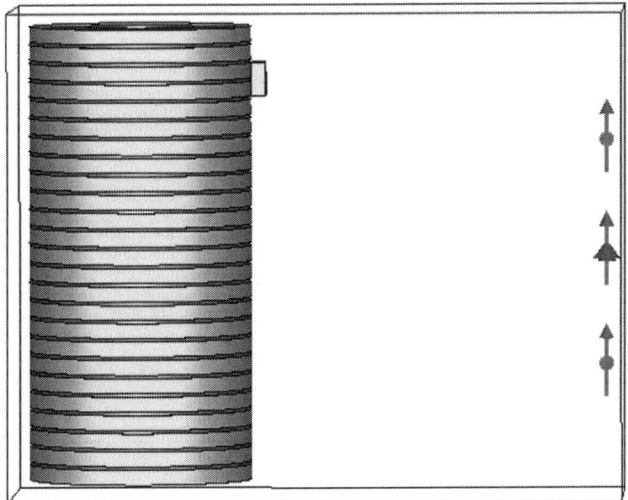

FIGURE 9.30

Y-Z view of model utilized with RD defect.

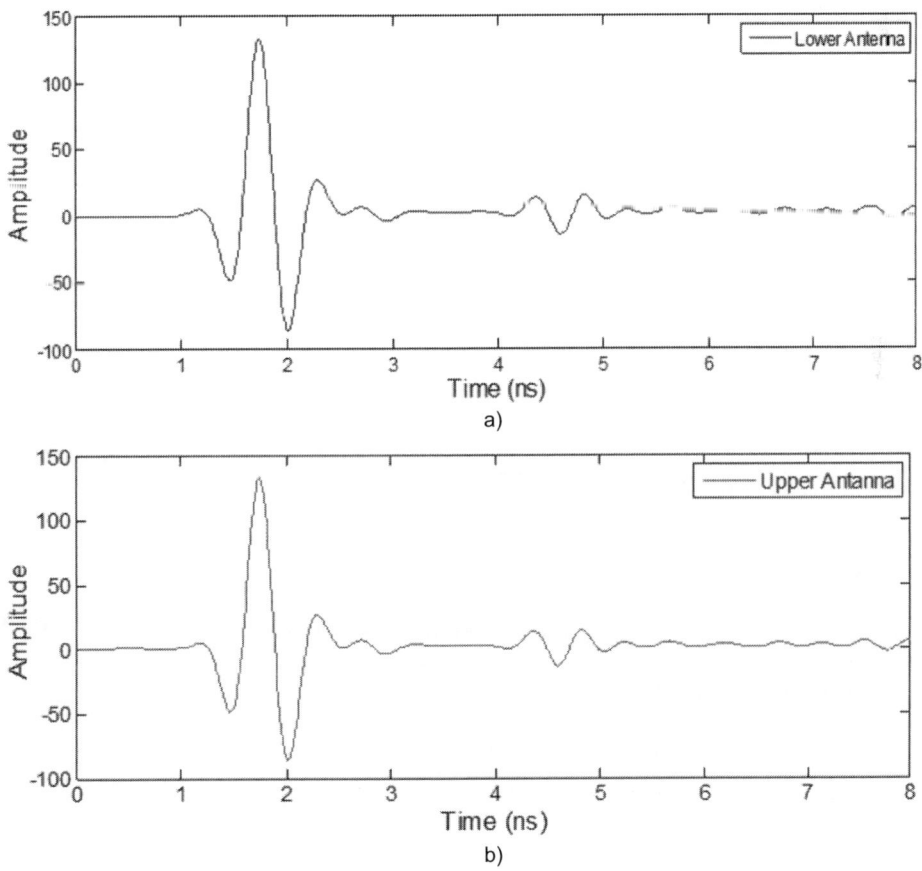

FIGURE 9.31

Received UHF signal in defected condition of winding at: (A) Lower receiving antenna, and (B) Upper receiving antenna.

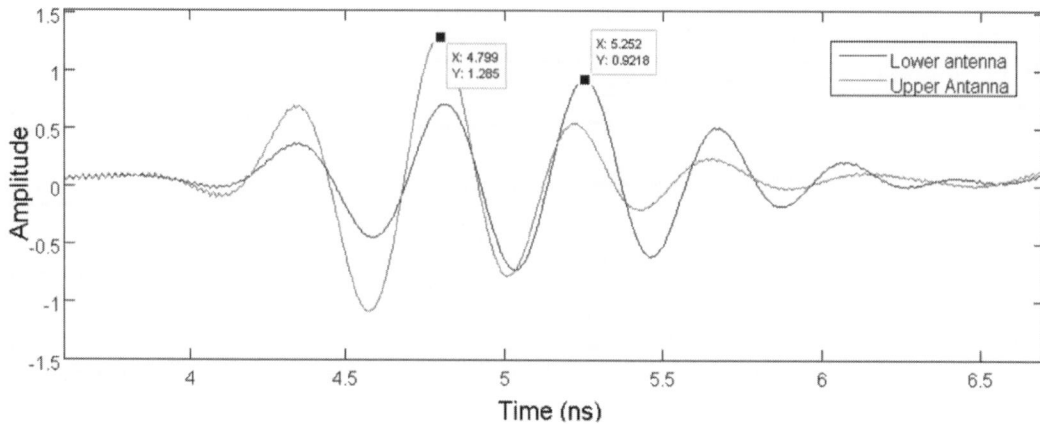

FIGURE 9.32

Substracted signal in UHF band for RD defect.

$$\frac{(z-300)^2}{4556} - \frac{x^2}{17929} - \frac{y^2}{17929} = 1 \tag{9.39}$$

The winding geometric location according to the simulated model of the winding is as follows:

$$\begin{cases} x = 0 \\ y = l = 485mm \\ 0 \le z \le h = 600mm \end{cases} \tag{9.40}$$

Two probable points of the RD location are determined by intersecting the hyperbolic and winding equations in 2D space at the heights of $z = 47$ mm and $z = 553$ mm. Based on Fig. 9.30, the upper antenna detects the defect effect earlier due to its proximity to the defect. So, the point with a height of $z = 553$ mm is considered acceptable between the two obtained points. The error of locating RD equals 2.1%.

9.4.2 Sensitivity of RD detection in UHF band

To assess the accuracy of the hyperbolic method in detecting RD in the UHF band, an RD with different extents is simulated, and the target section of the signal is recorded by the receiving antenna, which shows the effect of RD. Figs. 9.33 and 9.34 illustrate the target section of the received signal for the upper and lower antennas, respectively. The RD is investigated at the height of 540 mm with a size from 0 mm (SCALE = 20) to 12 mm (SCALE = 8).

As observed, increasing defect size raises the amplitude changes in the received signals. The amplitude changes in the upper receiving antenna are greater than the ones received by the lower one due to the location of the RD defect at the top of the winding, which results in receiving the defect effect by the upper receiving antenna more intensely.

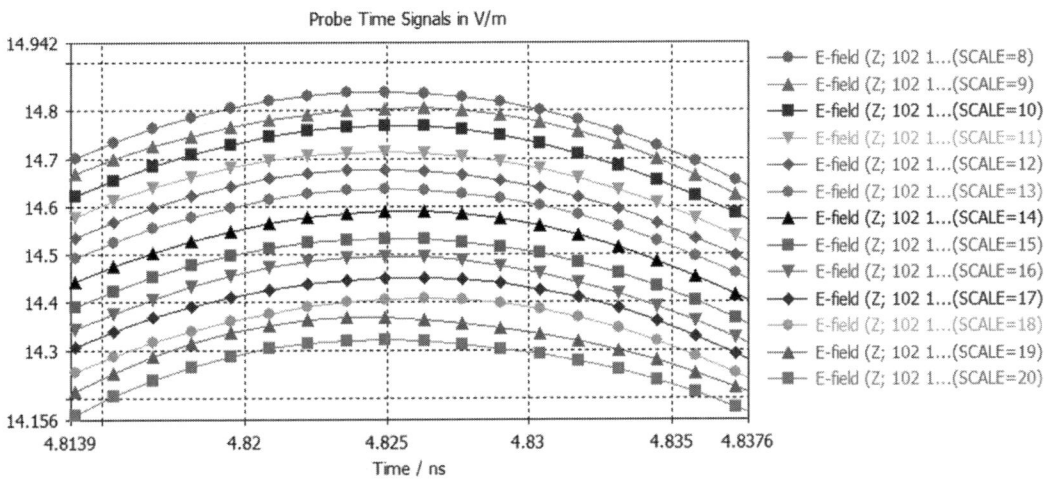

FIGURE 9.33

Target section of received signal for upper antenna with size of 0–12 mm.

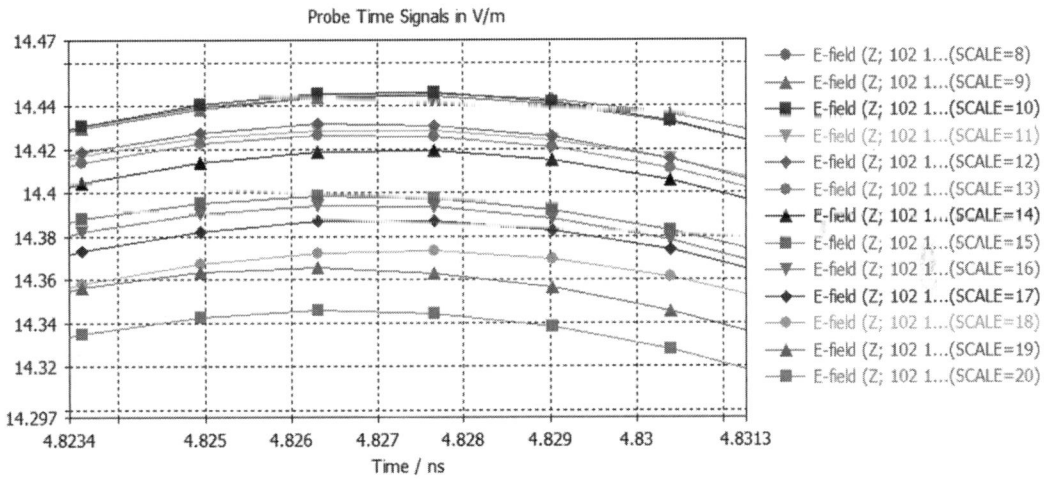

FIGURE 9.34

Target section of received signal for lower antenna with size of 0–12 mm.

The *TD* between the signals received by the upper and lower receiving antennas indicates the occurrence of an RD defect. In addition, the peak time in the received signals is regarded as fixed and is not changed with increasing the size of the defect, indicating the change in the size of the defect, not its location.

9.4.3 PD simulation in CST software

A Gaussian current is selected to model the PD. To this aim, a discrete port is used in CST software. The amplitude of the current is 5 mA [65]. The simulated model and receiving probes are shown in Fig. 9.28. The discrete ports delegating PD defect in CST software are shown in Fig. 9.35. A sample received signal transmitted from PD and recorded by the receiving antenna is shown in Fig. 9.36.

9.4.4 Effect of PD on hyperbolic method in UHF band

As explained, the difference between the target part of the received signal and the reference signal indicates the occurrence of an RD defect in the transformer HV winding. However, the combination of the PD signals and RD defect detection leads to false detection (Karami et al., 2018). For instance, the signal recorded by the receiving antennas is greatly distorted during monitoring a sound transformer winding when a PD defect occurs while transmitting and receiving Gaussian signals of the hyperbolic method. As illustrated in Fig. 9.37, the difference in the peak of the subtracted signal indicates the RD occurrence, while this detection is considered as incorrect and only a PD defect occurs during monitoring of the transformer winding in the sound mode. Thus, the PD defect should be detected during sending and receiving the signals in the hyperbolic method, and its effect should be eliminated to prevent false detection.

Similar to the previous chapter, applying the stepped-frequency approach and generalized likelihood ratio test (GLRT) can help to overcome mentioned issue. Fig. 9.38 demonstrates the PD detection algorithm while the signals of the hyperbolic method are transmitted and received. Next, the results of applying the method on an actual three-phase transformer will be described.

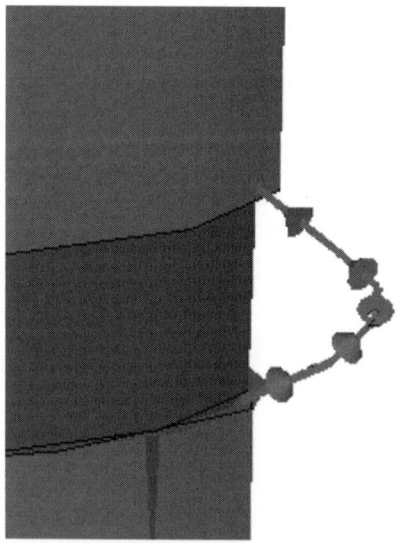

FIGURE 9.35

Discrete ports for modeling PD.

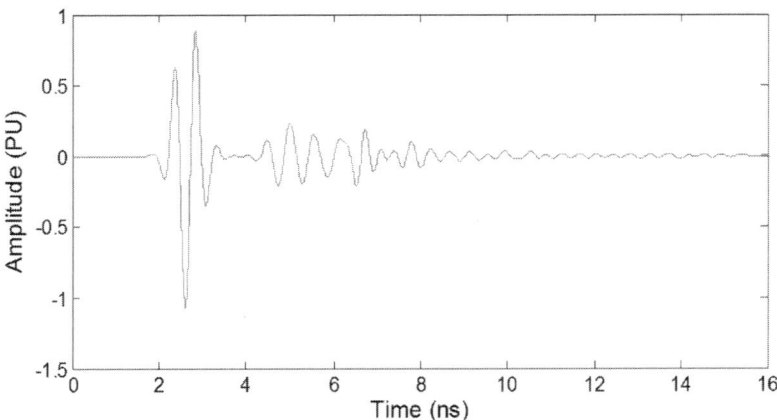

FIGURE 9.36

Sample received PD signal.

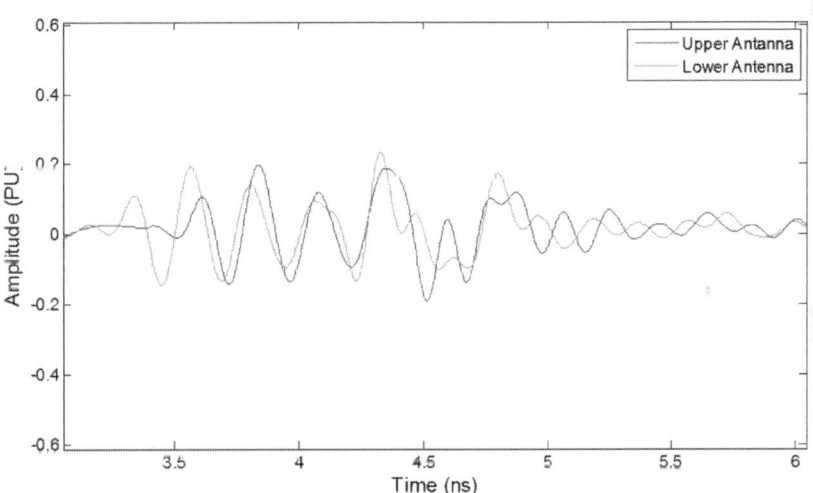

FIGURE 9.37

Effect of PD signal on subtracted signal while there is no RD.

9.4.5 PD detection during applying hyperbolic method in actual transformer

In this section, the AUTPDMD system, as described in the previous chapter, is used for applying the method. The monitoring system and proposed method should be tested on an actual transformer. A three-phase transformer is selected as the transformer under test for this aim. As displayed in Fig. 9.39, a metallic object is used to model the RD in the defective mode of the winding. Many tests can be performed for different locations of the RD defect because the winding is not damaged in this type of defect modeling, which is regarded as a reversible operation.

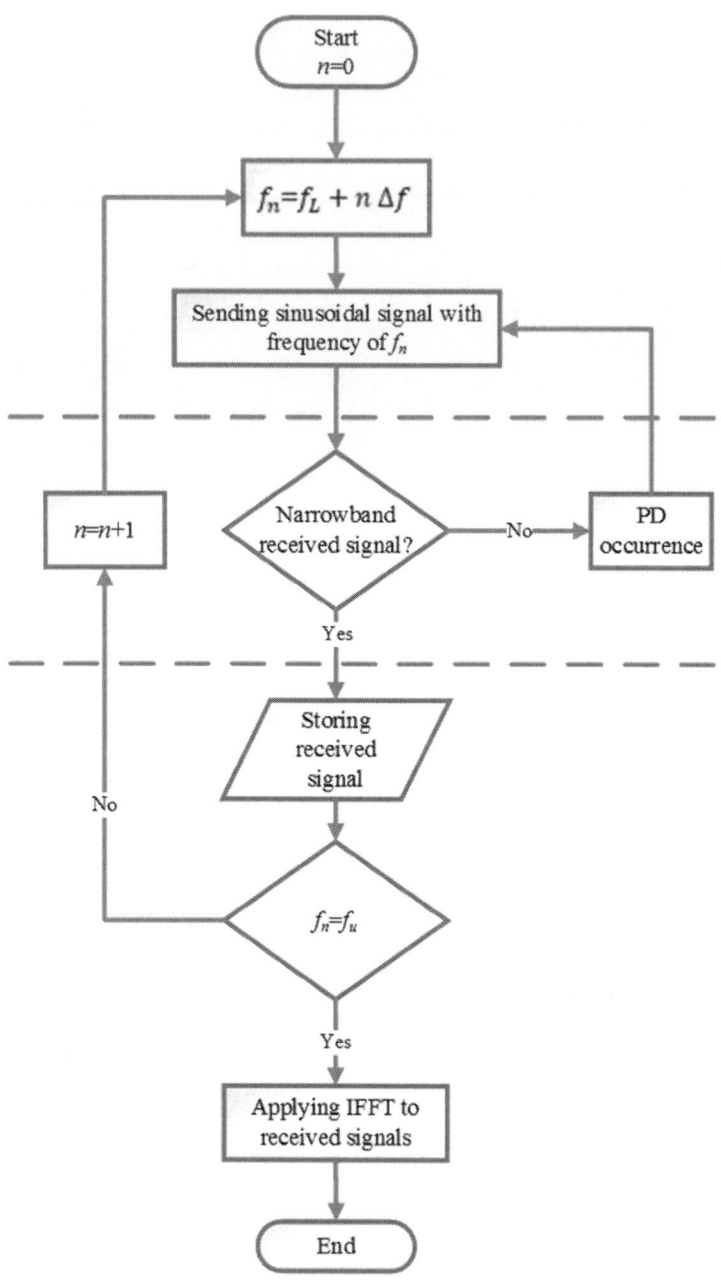

FIGURE 9.38

PD detection algorithm during applying hyperbolic method.

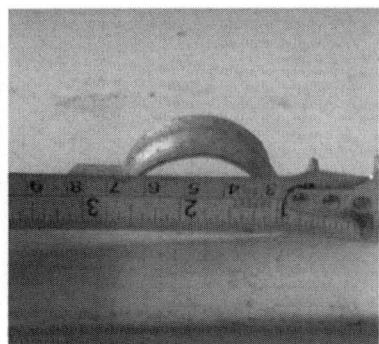

FIGURE 9.39

Metallic object for RD creation on winding.

9.4.5.1 Partial deformation defect modeling

To prevent damage to the windings in the PD test, two electrodes are installed inside the tank and near the winding to perform the discharge test and apply a PD at the appropriate voltage by connecting to the voltage source, as shown in Fig. 9.40.

FIGURE 9.40

Installed electrode for PD test.

9.4.5.2 Preparing transformer for testing

As illustrated in Fig. 9.41, the windings are placed in the transformer tank and the tank is filled with NYNAS oil.

A three-stage cascade transformer is utilized to generate voltage, as shown in Fig. 9.42. One stage of such transformer suffices for testing since each stage can generate 100 kV voltages. However, the second stage of such a power supply, which can generate voltage up to 200 kV, is applied for more assurance.

9.4.5.3 Partial deformation detection

The voltage values are increased step by step by starting the test. A PD occurs at a voltage of about 19 kV. The utilized oil creates a PD at this voltage due to its properties, such as aging and pollution. The AUTPDMD device receives signals in the frequency range of 100 MHz—3 GHz when it is in passive mode. The transmitted signals change from 100 MHz to 3 GHz in 10 MHz increments. The device can capture signal information at each frequency with a bandwidth of 20 MHz and perform sampling operations by reducing its frequency to the sampling one on the sampling board.

Fig. 9.43 shows the received noise signal in the passive mode of AUTPDMD when no PD occurred. Figs. 9.44 and 9.45 illustrate the received PD signal and it is magnified in the monitoring system while the AUTPDMD is in passive mode. The signal of both antennas is demonstrated since both are connected to the device. As observed, by occurring a PD, the amplitude of the received signal significantly differs from the noise, so that it can be detected easily by the AUTPDMD device.

9.4.5.4 Radial deformation detection

As indicated, a reference signal is needed to compare the signals in the sound and defective conditions in order to detect the RD defect. The reference signal includes a set of sinusoidal signals from 300 MHz to 3 GHz. The height of the transformer winding and the distance of the antennas from the

FIGURE 9.41

Filling transformer tank with oil and preparing it for practical testing.

FIGURE 9.42

View of cascade transformer used to create PD.

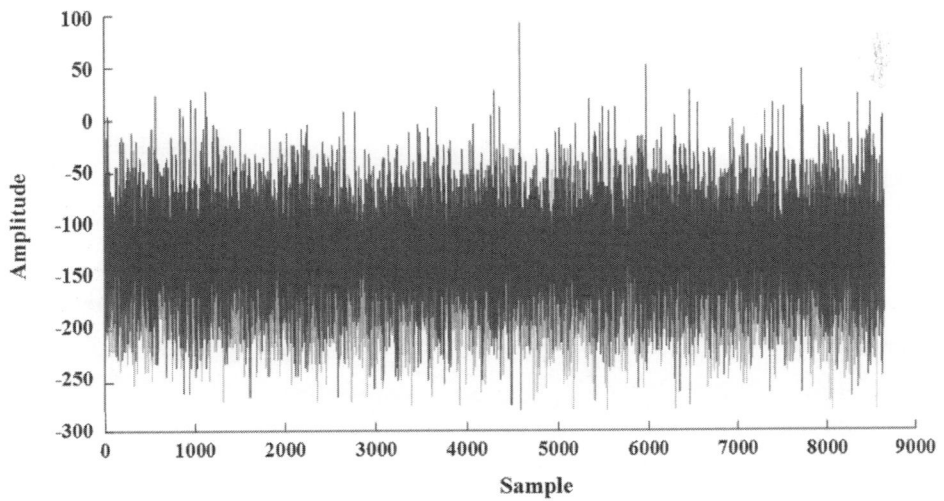

FIGURE 9.43

Noise recorded by AUTPDMD.

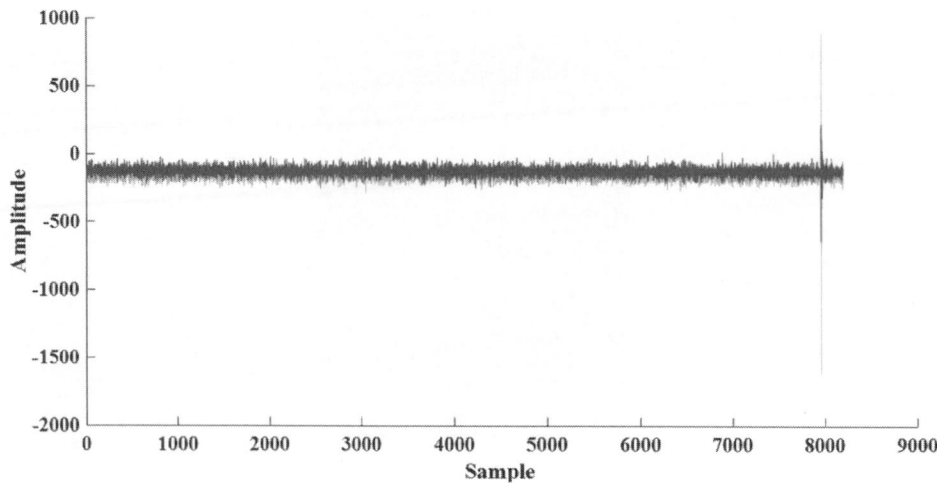

FIGURE 9.44

Received PD signal while AUTPDMD is used in passive mode.

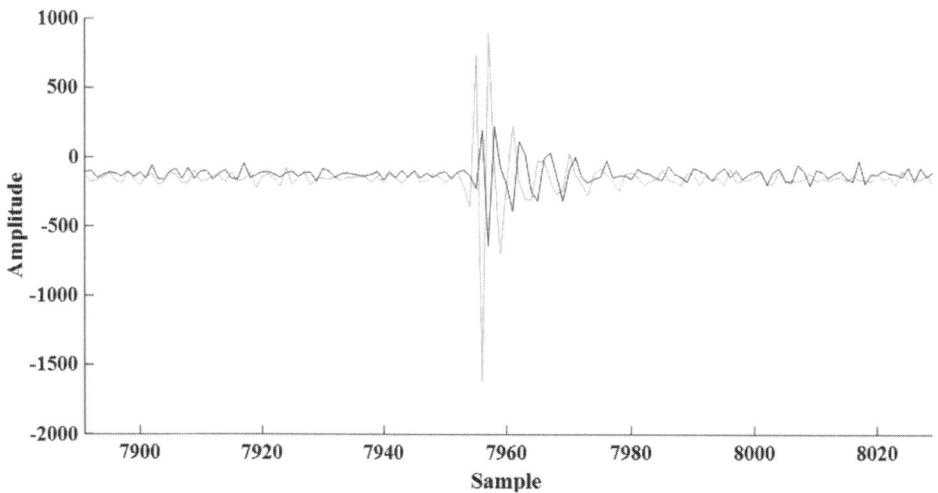

FIGURE 9.45

Magnified received PD signal.

winding surface equal 45 cm and 28 cm, respectively. In addition, the transmitting antenna is located at a height of 225 mm from the bottom of the winding, while the receiving antennas are placed at a distance of 100 mm from the transmitting antenna. Fig. 9.46 illustrates the transformer and test layout to detect and locate the RD of the three-phase transformer winding.

FIGURE 9.46

Test layout for RD detection of three-phase transformer.

First, sinusoidal signals should be sent for the sound condition of the winding, and their reflection should be recorded by the receiver. Fig. 9.47 demonstrates the reference signal made from the received signals in the sound condition for the upper and lower receiving antennas.

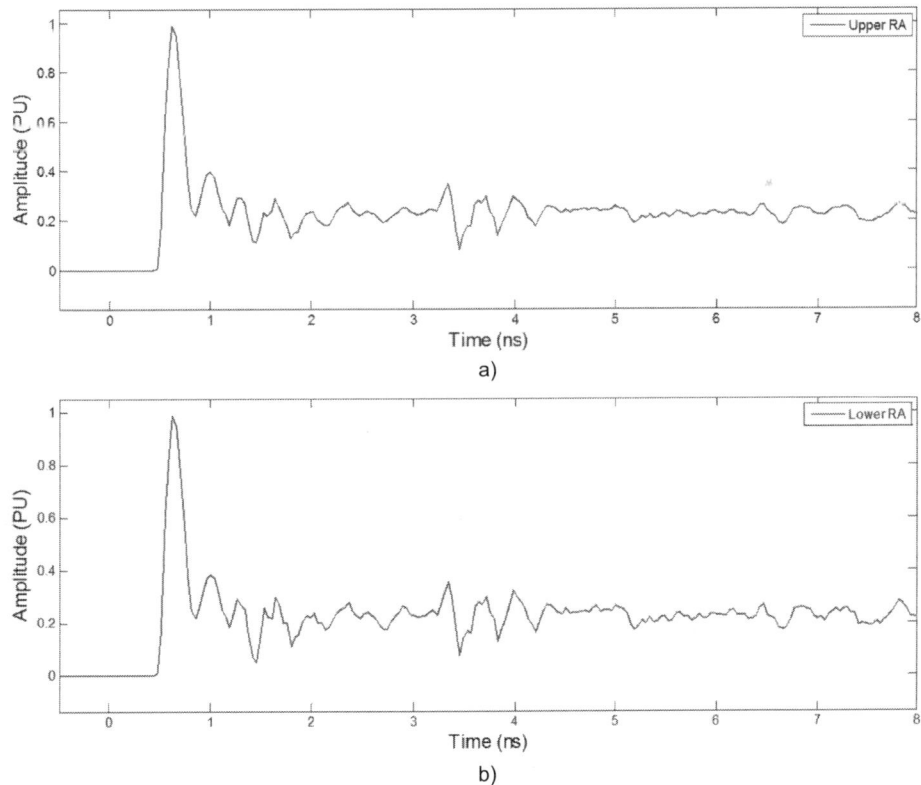

FIGURE 9.47

Reference signal in sound condition of winding at: (A) Upper and (B) Lower receiving antennas.

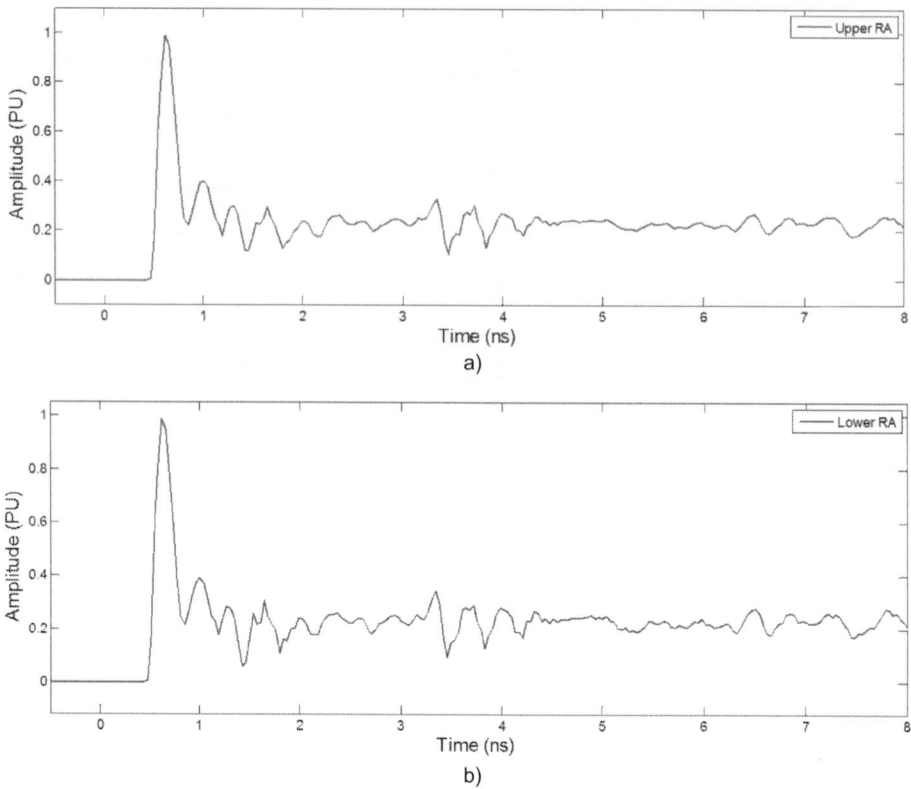

FIGURE 9.48

Received signal in defect condition of winding at: (A) Upper and (B) Lower receiving antennas.

As indicated, a metal with a length of 40 mm, width of 250 mm, and height (bulgy) of 20 mm is placed at a height of 305 mm from the bottom of the winding. The sinusoidal signals are sent, and their reflections are recorded by the receiver. Fig. 9.48 displays the reference signals made from the recorded signals in defect condition for the upper and lower receiving antennas.

The subtracted signal is created by considering the appropriate time window for each antenna. Fig. 9.49 shows the subtracted signals for the upper and lower receiving antennas.

The RD locating is proceeded by calculating the *TD* of the subtracted signals and finding the hyperbolic equation parameters as follows:

$$x_1 - x_2 = c \times (t_1 - t_2) = 3 \times 10^8 \times \left(0.15 \times 10^{-9}\right) = 45mm \tag{9.41}$$

$$b = c = \sqrt{d^2 - a^2} = 97.4 \tag{9.42}$$

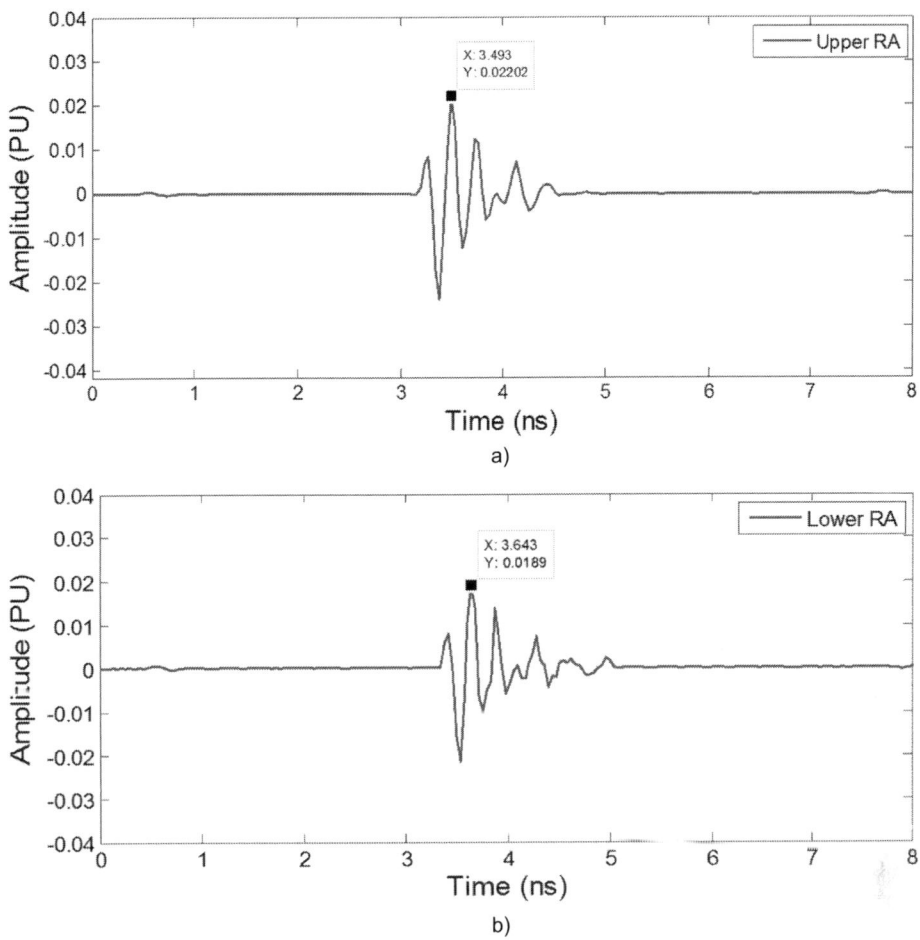

FIGURE 9.49

Subtracted signals at: (A) Upper and (B) Lower receiving antennas.

Now, the hyperbolic and the winding equations are created, as follows:

$$\frac{(z-225)^2}{506} - \frac{y^2}{9486} - \frac{x^2}{9486} = 1 \tag{9.43}$$

$$\begin{cases} x = 0 \\ y = l = 280mm \\ 0 \le z \le h = 450mm \end{cases} \tag{9.44}$$

In fact, two possible RD points at the height of $z = 293.4$ mm and $z = 156.6$ mm are determined by intersecting the hyperbolic and winding equations. The upper antenna detects the effect of the defect earlier due to its proximity to the defect. Thus, the point with a height of $z = 293.4$ mm is considered acceptable between the two points obtained, meaning that the RD location is estimated. The error value in detecting the RD equals 2.53%. Therefore, the RD can be detected and located with the described approach. Detecting the PD during sending and receiving signals in the hyperbolic method is not repeated here due to its similarity to what is already explained in Section 8.5.2.3.

References

Azadifar, M., Karami, H., Wang, Z., Rubinstein, M., Rachidi, F., Karami, H., Ghasemi, A., & Gharehpetian, G. B. (2020). Partial discharge localization using electromagnetic time reversal: A performance analysis. *IEEE Access, 8,* 147507–147515.

Golsorkhi, M. S., Hejazi, M. S. A., Gharehpetian, G. B., & Dehmollaian, M. (2012). A feasibility study on the application of radar imaging for the detection of transformer winding radial deformation. *IEEE Transactions on Power Delivery, 27*(4), 2113–2121.

Karami, H., Hejazi, M. S. A., Naderi, M. S., Gharehpetian, G. B., & Mortazavian, S. (2012). Three-dimensional simulation of PD source allocation through TDOA method. In *The 4th conference on thermal power plants* (pp. 1–4).

Karami, H., Gharehpetian, G. B., Norouzi, Y., & Hejazi, M. A. (2016). GLRT-based mitigation of partial discharge effect on detection of radial deformation of transformer HV winding using SAR imaging method. *IEEE Sensors Journal, 16*(19), 7234–7241.

Karami, H., Gharehpetian, G. B., Norouzi, Y., & Hejazi, M. A. (2018). Experimental study on elimination of partial discharge effect on detection of radial deformation of high voltage transformer winding using electromagnetic waves. In *2018 IEEE international conference on environment and electrical engineering and 2018 IEEE industrial and commercial power systems europe (EEEIC/I&CPS europe)* (pp. 1–5).

Karami, H., Tabarsa, H., Gharehpetian, G. B., Norouzi, Y., & Hejazi, M. A. (2019). Feasibility study on simultaneous detection of partial discharge and axial displacement of HV transformer winding using electromagnetic waves. *IEEE Transactions on Industrial Informatics, 16*(1), 67–76.

Karami, H., Azadifar, M., Mostajabi, A., Rubinstein, M., Karami, H., Gharehpetian, G. B., & Rachidi, F. (2020). Partial discharge localization using time reversal: Application to power transformers. *Sensors, 20*(5), 1419.

Karami, H., Gharehpetian, G. B., Hejazi, M. A. A., & Norouzi, Y. (2020). *Detection of radial deformations of transformers.* Google Patents.

Karami, H., Gharehpetian, G. B., Norouzi, Y., & Akhavan-Hejazi, M. (2020). Simultaneous radial deformation and partial discharge detection of high-voltage winding of power transformer. *IET Electric Power Applications, 14*(3), 383–390.

Karami, H., Mostajabi, A., Rachidi-Haeri, F., Azadifar, M., & Rubinstein, M. (2021). *Partial discharge localization using time reversal: Application to power transformers and gas-insulated substations.*

Mortazavian, S., Gharehpetian, G. B., Hejazi, M. A., Golsorkhi, M. S., & Karami, H. (2012). A simultaneous method for detection of radial deformation and axial displacement in transformer winding using UWB SAR imaging. In *The 4th conference on thermal power plants* (pp. 1–6).

Rahbarimagham, H., Porzani, H. K., Hejazi, M. S. A., Naderi, M. S., & Gharehpetian, G. B. (2015). Determination of transformer winding radial deformation using UWB system and hyperboloid method. *IEEE Sensors Journal, 15*(8), 4194–4202.

Rahbarimagham, H., Karami, H., Esmaeili, S., & Gharehpetian, G. B. (2019). Determination of transformer HV winding axial displacement extent using hyperbolic method—a feasibility study. *IET Electric Power Applications, 13*(7), 1004—1013.

Tabarsa, H., Hejazi, M. A., & Gharehpetian, G. B. (2019). Detection of HV winding radial deformation and PD in power transformer using stepped-frequency hyperboloid method. *IEEE Transactions on Instrumentation and Measurement, 68*(8), 2934—2942. https://doi.org/10.1109/TIM.2018.2868491

Partial discharge monitoring using EMWs

10

Gevork B. Gharehpetian[1], Hossein Karami[2] and Seyed-Alireza Ahmadi[3]

[1]*Amirkabir University of Technology (AUT), Electrical Engineering Department, Tehran, Iran;* [2]*Niroo Research Institute (NRI), High Voltage Studies Research Department, Tehran, Iran;* [3]*School of Electrical and Computer Engineering, College of Engineering, University of Tehran (UT), Tehran, Iran*

10.1 Partial discharge defect introduction

The partial discharge (PD) occurs when the electrical strength is not strong enough in a part of the insulating material due to a nonuniform electric field. Therefore, the energy collected in this part can be released in a short moment of time in the form of a spark. The PD emits light and heat, which can be detected depending on the type of the insulation.

The mechanism of electrical discharge in solid and liquid insulators is not yet fully understood. However, the discharge is due to mainly nonuniformity of a solid body in a solid insulation, which leads to lack of stability in the electrical strength of the solid body and its change due to minor factors. In addition, the voids and pores play a significant role in the nonuniformity of solid insulation, resulting in creating the PD inside the insulation. In liquid insulators, discharge occurs mostly around the sharp electrodes, and also in the voids and gas bubbles in the liquid insulator.

To model the PD phenomenon, an insulator is assumed to be placed between two electrodes. The electric field between these two electrodes takes a special shape by applying a high-voltage (HV) to the electrodes. The electric field is considered as linear when the electrodes are regarded as two relatively large planes and the insulators are assumed to perfectly be uniform and homogeneous. Fig. 10.1 shows the voltage distribution along with the insulation. The failure occurs and an electric arc is established

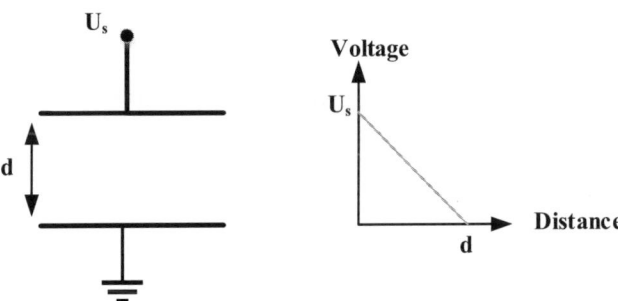

FIGURE 10.1

Electric field in insulation and its voltage drop.

Power Transformer Online Monitoring Using Electromagnetic Waves. https://doi.org/10.1016/B978-0-12-822801-2.00008-2

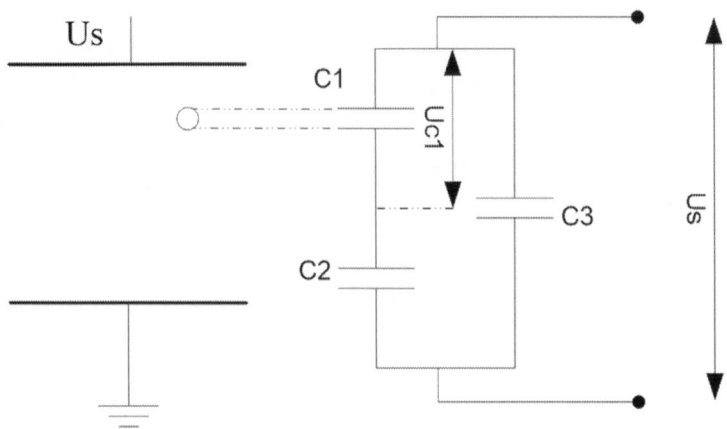

FIGURE 10.2

Insulation circuit model along with void.

between the two electrodes when the applied voltage to the electrodes is higher than the insulation strength. Such a phenomenon that destroys the insulation is called complete discharge (Coenen, 2012).

The voids are the main reason for PD in solid and liquid insulations. Fig. 10.2 illustrates the insulation and its void, along with a simple model, in which $C_3 >> C_1 > C_2$.

The capacitor values before and after discharge, i.e., C_A and C_B, respectively, are as follows:

$$C_A = C_3 + \frac{C_1 \times C_2}{C_1 + C_2} \tag{10.1}$$

$$C_B = C_3 + C_2 \tag{10.2}$$

Fig. 10.3 demonstrates the changes in the capacitor voltage and current passing through the electrodes. As observed, the voltage of C_1 is calculated by (Eq. 10.3) when voltage U is applied to the electrodes.

$$U_{C_1} = U_S \times \frac{C_2}{C_1 + C_2} \tag{10.3}$$

The discharge occurs inside the void modeled with C_1 when the voltage U_{C1} approaches the required level to start a PD. In addition, U_{C1} decreases to U_L in which the discharge is interrupted. Different authors have proposed various methods to calculate the PD inception voltage (PDIV). The following equation is presented in (Judd et al., 2001, 2004) to calculate the root mean square (rms) of PDIV in kV:

$$U_L = PDIV = 26.5 \times p \times t + 0.55 \tag{10.4}$$

where, p and t indicate the pressure inside the void in the atmosphere and the thickness of the void in centimeters, respectively. Then, the voltage U_{C1} increases, and the discharge happens again when it approaches the U_L limit. Such a process is repeated during the negative period. It is worth noting that the U_L value is often extremely small and negligible.

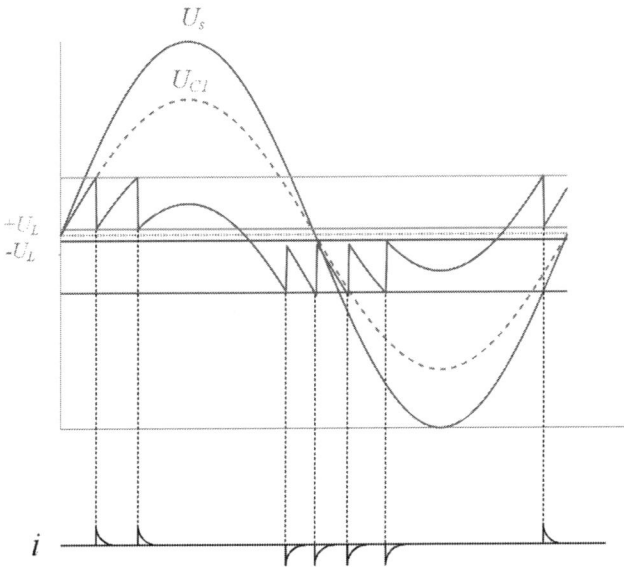

FIGURE 10.3

Electric discharge curve in void inside insulation.

10.2 Partial discharge monitoring methods

PD detection methods are divided into two general categories, including electrical and nonelectrical (chemical, light, and sound) detection methods. The most important ones are described in the following subsections.

10.2.1 Nonelectrical methods

Nonelectrical methods are used to detect the occurrence of electrical discharges, but most of them cannot be utilized to measure the amplitude of discharges. Nonelectrical phenomena used to detect discharges are as follows:

- Gas pressure
- Heat
- Light
- Chemical changes
- Acoustic
- Electromagnetic

The first two are less useful in power transformers. Other cases are briefly explained in the following.

10.2.1.1 Optical method

Utilizing the optical techniques is limited to electrical discharges in transparent insulators. Optical detection is applied for surface and corona discharges. In fact, such a method can hardly be used in transformers due to the opacity of mineral oil.

10.2.1.2 Chemical detection

Some PD detection methods are based on chemical reactions and can be utilized to determine the occurrence of discharge. The PD decomposes the oil into various hydrocarbon gases, such as acetylene. Other gases such as ethane, methane, ethylene, and similar gases are produced lower and hydrogen to a relatively greater extent. The PD in the oil can create a type of wax called X or unknown wax, which can damage the paper and explode the transformer. The presence of CO or CO_2 in the oil indicates the involvement of paper in the PD, the loss of which may damage the transformer irreversibly. Analyzing the oil-soluble gases can be applied as one of the PD detection methods (Ahmadi & Sanaye-Pasand, 2021). Despite advances, the chemical method cannot determine the PD location, nature, and intensity. Therefore it has the following disadvantages:

- It cannot be used to locate PD.
- It is regarded as time-consuming.

However, the chemical method is utilized when time constraints are insignificant.

10.2.1.3 Acoustic detection

The acoustic method is based on receiving and recording acoustic signals created by the PD, and it has the ability to detect and locate the PD with an acceptable accuracy, low cost, and high speed. The weakness of the acoustic signals produced by the PD is the complex nature of sound propagation. In addition, there are internal and external acoustic noises in the transformer environment, such as Barkhausen noise from the transformer core or that of sand or raindrops hitting the transformer tank. This method can be applied online with a nondestructive and nonoffensive nature, which is considered as appropriate for owners and manufacturers of power transformers (Lundgaard, 1992b). To this aim, various acoustic sensors can be used, the best type of which is a piezoelectric transducer. However, it has some limitations due to the complex propagation nature of acoustic waves hitting the inner wall of the transformer housing and the attenuation of acoustic waves. Such limitations and undeniable advantages paved the way for utilizing the internal detection systems, resulting in producing fiber-optic acoustic sensors to be installed inside the transformer. The complex propagation of acoustic waves hitting the inner wall of the transformer housing and the attenuation of acoustic waves are other disadvantages of the aforementioned method (Lundgaard, 1992a).

Different technologies have been applied to produce acoustic sensors. Piezoelectric transducers made from ferroelectric ceramic materials, such as lead zirconium titanate, have widely been utilized in external acoustic detection systems. The most commonly used are accelerometers and acoustic emission sensors.

An electrical signal is created in proportion to the acceleration of the surface connected to the sensor in the first category. Such sensors have a usable frequency range of up to 50 kHz. In contrast, acoustic emission sensors produce an electrical signal proportional to the surface velocity on which the sensor is placed. Acoustic emission sensors have a frequency range between 30 kHz and 1 MHz.

Optical sensors have been applied in recent years. Fiber optic sensors operate based on light modulation. The interaction of acoustic waves at audio and ultrasonic frequencies with optical fiber presses on the optical fiber. The acoustic pressure on the fiber in the ultrasonic frequency range is considered as uniform along the fiber and has axial symmetry, resulting in creating a uniform radial pressure on the fiber. The pressure sensitivity in the fibers is affected by the elastic and elasto-optic coefficients of the glass fiber and those of the fiber coating. Light is modulated in fiber in different methods. Interferometry technique is used in detecting PDs by fiber optic sensors among different methods. To this aim, the light emitted through the fiber is phase-modulated by the perturbation resulting from the ultrasonic pressure waves generated by the PD. The modulation coefficient depends on various parameters such as fiber refractive index, adaptive acoustic impedance of the material in contact with the fiber, and change in fiber length (Wild & Hinckley, 2008). Examining more powerful sensors is among the current studies.

10.2.2 Electrical methods

In these methods, the voltage and current of the transformer bushing are sampled and utilized to detect PD. On the one hand, the PD current pulses at the defect location have a high frequency, and their frequency spectrum has components up to several GHz. On the other hand, the measuring bandwidth of electrical devices for PD detectors is limited by their filtering system, detector impedance, and amplifier. Therefore each measuring device can only detect a part of the PD frequency spectrum in the range of its measured bandwidth (Standard, 2000). Measuring devices can be divided into narrowband, wideband, and ultra-wideband (UWB), depending on their frequency bandwidth. Based on the IEC standard, the bandwidth in the narrowband devices is in the range of 9930 kHz and their central frequency can be adjusted in the range of 50 kHz-1 MHz. In addition, the low cut-off frequency in wideband devices should be in the range of $30-100$ kHz, the high cut-off frequency below 50 kHz, and their bandwidth in the range of $100-400$ kHz. The apparent load quantities, occurrence phase in a PD pulse relative to the applied voltage, peak pulse, time of occurrence, and amount of voltage applied at the moment of occurrence for each PD pulse can be measured with the aforementioned devices.

The wideband measuring devices have better time resolution than narrowband ones, but they are more sensitive to noise. In addition, unlike narrowband ones, they can determine the polarity of PD pulses. New PD electrical measuring devices have a larger frequency bandwidth than conventional wideband ones. Thus such devices are called UWB devices. The frequency bandwidth of UWB devices is generally greater than 1 MHz. Unlike narrowband and wideband devices, UWB ones can detect more information on the waveform of PD single pulses in the time domain. The time resolution in wideband devices is extremely better than the latter two options with the lowest time width in response signal, despite having high sensitivity to noise. Narrowband and wideband devices apply the stability of the frequency spectrum at low frequencies and its proportionality to the detected signal amplitude to determine the apparent load of PD pulses, which is called frequency or quasi-integration. In contrast, the apparent load in UWB devices is obtained by integrating the sampled PD current signal in the time domain (IEC Standard, 2000).

The instantaneous voltage difference in successive PD pulses is closely related to changes in the electric charge transmitted at its location. Previous pulses and the loads left affect the electromagnetic conditions at the PD location, as the time required to recombine and neutralize that area with nonidentical charges is usually longer than the average time between two consecutive PD pulses due to the surface and spatial conductivity of the occurrence location. Therefore the occurrence of new PD pulses depends on the previous ones, resulting in achieving a new feature, called instantaneous voltage difference, for determining the type of defect using the instantaneous voltage subtraction of consecutive PD pulses.

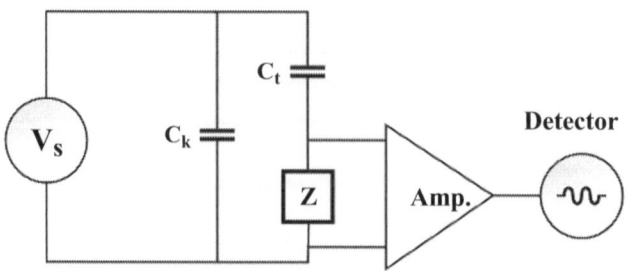

FIGURE 10.4

Standard PD detection circuit.

The patterns $\varphi - q - n$, $\varphi - n$, $\varphi - q_p$, $\varphi - q_a$, $\Delta u_i - \Delta u_{i-1}$, and $n - \Delta u$ can be obtained by electrical detection, in which the φ, n, q, q_p, q_a, and Δu indicate the phase of occurrence, number of occurrences, apparent load, apparent load peak value, the average value of the apparent load, and the difference in the values of the applied excitation voltage during the occurrence of two consecutive PD pulses, respectively (Mirzaei et al., 2008). Nowadays, significant information related to PD lies in the waveform of the PD signal. Only devices with a wide frequency bandwidth (several MHz) can measure sufficient information of the PD pulses waveform in the time domain.

The coupling capacitor (CC) is considered as the most significant component of the detector circuit, which allows for recording the high-frequency pulses. The CC may be used in several different forms. The types of capacitive couplings used are as follows (Paoletti & Golubev, 2001; Zeng, 1998):

- Utilizing an HV C_k capacitor connected to the detector impedance as the standard PD measurement circuit needs to be directly connected to HV. Fig. 10.4 illustrates the standard PD detection circuit. However, the CC is placed in series with the detector impedance in most cases.
- Applying a stray capacitor between an HV conductor and a metal electrode or a small capacitor connected to a detector circuit. This setup is influenced by different noise sources with relatively low circuit sensitivity.
- Receiving the signal through the bushing tap as a CC. The bushing tap is installed inside the bushing to uniform the voltage distribution. However, such a solution is widely used nowadays since no separate HV capacitor needs to be installed and noise sources affect the received signals much less than before. The last layer is connected to the ground, and the voltage is considered equal for all the layers. Thus disconnecting from the ground and using the connection as a measuring capacitor leads to a voltage distribution that increases the possibility of damage and electrical stress to the insulation, which is regarded as noticeable at HV levels.
- Applying a separate capacitive sensor that can be installed on the body of the transformer bushing. Using a capacitive sensor installed on the bushing insulation body as close to the transformer body as possible (first bushing step), which is detected by the last capacitive layer inside the bushing and the excess length of each layer, creates another capacitive layer that can be utilized as a CC due to lack of uniform distribution of the field in capacitive bushing in case of disconnection in the last capacitive layer and the absence of bushing tap in old transformers.

10.3 Comparing partial discharge detection methods

The most significant disadvantages of mentioned methods are as follows.

- The optical method can only be applied to transparent insulation materials and is not considered as appropriate for accurate locating.
- The chemical method is not capable of locating PD.
- The acoustic method has limitations, considering the complex propagation of acoustic waves hitting the inner wall in the transformer housing and their attenuation. Such a method is not considered as sensitive when the defect is detected inside the winding, leading to the attenuation of the acoustic waves.
- The possibility of error increases in the acoustic method since the acoustic noise inside the transformer is significant. In addition, the system is not completely isolated from outside events. Recent studies show that some PD defects create weak acoustic signals and cannot be detected by this method, despite having a high apparent load.
- The electrical method generates a lot of false alarms due to its high sensitivity since it receives a sample signal from the bushing head and is not regarded as isolated from outside events. In other words, the electrical method is significantly sensitive to external events and noise, and increases the possibility of faulty detection. In addition, such a method focuses on the ground connection of the transformer.
- The electromagnetic method in PD detection can be considered as nonelectrical detection method. Based on the experimental studies, a large number of PD defects produce current pulses with a wavelength of less than 1 ns, resulting in emitting emitted electromagnetic waves (EMWs) by the pulses with components in the range of a few GHz. The pulses related to PD defects have different time characteristics and frequency spectra. The frequency spectra in pulses related to several PD defects have been studied in (Raja & Floribert, 2002a, 2002b) with the help of laboratory results. The corona pulse in a gas-insulated substation environment with a wavefront of less than 40 ps is considered as the fastest PD pulse. It is noteworthy that the frequency spectra related to different PD defects differ from each other and depend on the type of defect, type of electrical discharge, and physics of the defect location.

10.4 EMW-based detection method

Measuring PD with the help of ultrahigh-frequency (UHF) EMWs is considered as a new method (Azadifar et al., 2020; Karami et al., 2016, 2018, 2019). In transformers, high-frequency PD EMWs in the UHF range are detected by antennas mounted directly inside the transformer tank or on its outer walls (Karami, Azadifar, et al., 2020; Karami et al., 2021; Meijer et al., 2006). Inserting the antenna into a tank with the help of oil drain valves is among the methods to install a UHF antenna in power transformers. According to (Tenbohlen et al., 2008), there is an appropriate signal to noise due to the isolation of the transformer tank from PDs occurring outside. The present study aims to prove that the types of PDs occurring in the transformer can be detected by EMWs.

In (Coenen et al., 2008), an example of a UHF antenna investigated, which mounted on the outer wall of a transformer tank using dielectric windows (DWs). An oil-compatible electrical insulator is installed on the hole created in the metal wall of the transformer tank. To this aim, epoxy resin or PTFE insulation can be utilized (Judd et al., 2005a). The DWs allow UHF waves to escape from inside the transformer and be detected by an antenna installed in its behind, in addition to preventing oil leakage and creating a shield against the electric field at electric frequency. In addition, the DWs realize the installation of the antenna on the transformer without the power outage and the measurement of the UHF signals generated by PD pulses inside the transformer.

The frequency characteristics of the antenna should be adjusted based on the frequency spectrum in the PD defects. The wide frequency spectrum in the PD reduces the corresponding wavelengths. Therefore smaller antennas can be applied, which plays a significant role in adapting antennas designed to the internal environment of a large transformer. In addition, high-frequency components with small wavelengths allow EMWs to be propagated more easily inside the transformer and improve the location determination accuracy by simplification of appropriate measurement.

Also, safety against external noise is among the advantages of PD detection using EMWs. The presence of a grounded metal tank, which acts as an electromagnetic shield against external noise, allows on-site measurements. The corona pulse has an extremely attenuated amplitude due to the shielding property of the tank. Thus the external defect EMWs, which enter the transformer through the bushing conductor, cannot interfere with the effective measurement of PD due to their highly attenuated amplitude (Judd et al., 2005a).

The EMWs resulting from a PD current pulse propagate as a spherical wave in the oil when they do not collide with other materials and tolerate refraction and reflection. The wave travels each meter in the oil in 5 ns due to its relative dielectric ratio ($\varepsilon_r = 2.2$). As indicated in (Convery & Judd, 2003), the propagation of EMWs in the UHF frequency band in transformer oil is not affected by factors such as aging, the presence of humidity, and its temperature changes under normal conditions.

10.5 Time difference of arrival method

Analyzing the signals received from the defect by the antennas using the Huygens-Fresnel principle is among the steps in locating PD in a power transformer. The time difference of arrival (TDOA) is the most common and widely used method in analyzing the received signals (Chan & Ho, 1994; Wenzhi et al., 2009). The TDOA method, widely used in telecommunications, was first utilized in GIS substations and turbine generators in power engineering (Kawada et al., 1998; Mizuno et al., 1996). Many studies have been conducted on the possibility of using such a method in oil over time (Sarathi et al., 2008). Applying the TDOA method for power transformers was assessed in (Karami et al., 2012; Karami, Gharehpetian, & Hejazi, 2013), leading to appropriate results.

The electromagnetic signal is scattered in the transformer environment when a current pulse is generated due to the PD at one point. Such diffusion follows the Maxwell equations. The emitted signals can be received by applying the antennas placed in the transformer tank. The propagation time of the signal from the PD source to the antennas cannot be directly determined. However, the time difference (TD) of the received signal helps a pair of antennas to detect the defect location with a hyperbolic equation based on the TDOA method. Three hyperbolic equations are formed with four antennas, which intersect at one point. Fig. 10.5 displays the TD related to one signal for four antennas.

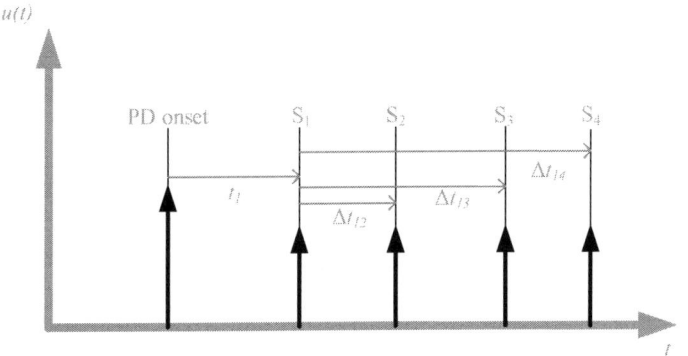

FIGURE 10.5

TDOA of PD signal from defect to antennas location.

As observed, the signal approaches the first antenna at time t_1. The hyperbolic equations for the TDOA method are as follows when the PD location is assumed to be at $P(x, y, z)$ and the antennas at $S_i(x_i, y_i, z_i)$, where $i = 1,2,3,4$ (Karami et al., 2012):

$$(c \times t_1)^2 = (x - x_1)^2 + (y - y_1)^2 + (z - z_1)^2 \tag{10.5}$$

$$(c \times (t_1 + \Delta t_{12}))^2 - (x - x_2)^2 + (y - y_2)^2 + (z - z_2)^2 \tag{10.6}$$

$$(c \times (t_1 + \Delta t_{13}))^2 = (x - x_3)^2 + (y - y_3)^2 + (z - z_3)^2 \tag{10.7}$$

$$(c \times (t_1 + \Delta t_{14}))^2 = (x - x_4)^2 + (y - y_4)^2 + (z - z_4)^2 \tag{10.8}$$

where c indicates the speed of light in the environment, and Δt_{12}, Δt_{13}, and Δt_{14} represent the TD between the signal received from the defect location to the first antenna and that received to the next antennas.

Finding the values of four unknowns t_1 and (x, y, z) from the above-mentioned equations can be conducted by knowing the variables c, Δt_{ij}, and the location of the antennas. The defect location is considered as a line when there are three antennas.

10.6 Simulation of PD in power transformer

CST software is used to analyze static problems in time and frequency domains, in addition to the motion behavior of charged particles (Karami et al., 2012; Karami, Gharehpetian, & Hejazi, 2013, 2020). The finite integration method is utilized for spatial discretization in the computer simulation technology (CST) software. To this aim, the problem structure should be divided into meshes. Three types of meshing can be applied in this software, six-dimensional (6D), four-dimensional (4D), and surface meshing. Almost all of the solution methods in the CST software use 6D meshing, and a small number of them can also utilize the other two types. All complex structures can be meshed, and the highest problem-solving speed is achieved by applying 6D meshing. First, the modeling of equipment such as transformers, oil, and pressboards will be evaluated. Then, the modeling of PD defects will be described in the following sections.

10.6.1 Transformer model

A background of modeling conducted for transformers in the literature is discussed to demonstrate the significance of the modeling performed in 3D in the CST software. The wave propagation in oil, considering the active part of the transformer, has been simulated in (Kang & Birtwhistle, 2001). The received signals have been studied in different places, indicating that the received signal by the sensors farther from the PD location is weaker.

The simulation performed in (Stone, 1991) in a 3D environment considered a single cylinder as the core. In addition, the windings were not fully regarded, and the yoke was not considered. The simulation results were used for PD locating in (Jeyabalan, 2011) after simulating the propagation of waves conducted in 2D without considering the insulation components. The model was simplified from 3D to 2D mode without any explanation. It is noteworthy that the PD was modeled in the oil during the performed simulations.

To simulate a transformer sample, its complete specifications, including material and insulation features, and the distance among components, are required. More real specifications give more reliable results due to the high accuracy of the CST software. Required specifications related to a real sample are extracted from (Karami et al., 2012; Sawhney, 1984), in which the low and high voltage winding type is helical and cross-over, respectively. Table 10.1 indicates the specifications related to one phase of the transformer. It is worth noting that a three-phase transformer cannot be simulated considering the actual sizes of active and passive parts due to hardware and software limitations. Therefore the model dimensions in the simulation are multiplied by 0.2 here. Fig. 10.6 demonstrates an overview of the transformer core. In addition, Fig. 10.7 displays the transformer simulated in the CST software from three different perspective views (Karami, Hejazi, & Gharehpetian, 2013).

10.6.2 Oil and pressboard models

In this subsection, the complex permittivity is utilized in this model.

10.6.2.1 Complex permittivity

Generally, a material cannot immediately show its polar behavior versus field changes (Kramers, 1927; Kronig, 1926). The loss and dielectric behavior of an insulator are a functions of temperature and frequency, like several other physical properties. Thus the temperature and frequency should be indicated when $\tan\delta$ and ε_r are declared for an insulator. The temperature and frequency are considered as 20°C and 5060 Hz, respectively, when they are not indicated.

The motion of ions and electrons or the rotation of dipoles in their places is called polarization. Therefore, moving ions and electrons by polarization takes time. Electronic polarization occurs quite fast, while ionic polarization, dipole rotation, and displacement happen slower due to the mass of the ions and dipoles, and the friction, which decelerates the motion. In other words, the force created by an electric field on a charged particle, such as an ion, is calculated as follows:

$$F = e \times E(t) \tag{10.9}$$

Table 10.1 Specifications related to understudy transformer.	
Number of phases	**3**
Tank body material	Steel
Core diameter	135 mm
Internal height of the window (H_w)	300 mm
Internal width of the window (W_w)	120 mm
External height of the window (H)	536 mm
External width of the window (W)	624 mm
Yoke diameter	118 mm
Tank height	750 mm
Tank width	840 mm
Tank length	350 mm
Insulation material between H.V. and L.V. and between the H.V. windings	Pressboard
Insulation width between H.V. and L.V.	1.5 mm
Duct width between H.V. and L.V.	5 mm
H.V. height	221 mm
H.V. outer diameter	239 mm
H.V. inner diameter	186.2 mm
Distance between the coils	5 mm
Insulation width between H.V. layers	0.3 mm
Number of H.V. coils	8
Number of H.V. layers	24
L.V. height	285 mm
L.V. outer diameter	156.2 mm
L.V. inner diameter	138 mm
Insulation material between L.V. layers	Pressboard
Insulating width between L.V. layers	0.5 mm
Insulation material between L.V. and core	Pressboard
Insulation width between L.V. and core	1.5 mm

where, e is regarded as the electric charge of the particle and E is considered as the intensity of the electric field at the particle location, which changes over time. Such force is neutralized by several others based on the principle of action and reaction. These are the forces of acceleration, friction, and tension between molecules known by $m\ddot{x}$, $b\dot{x}$, and ax, respectively. The latter force derives from the inhibition of adjacent molecules in the particle. In Eq. (10.10), a and b represent the coefficients of tension and friction, and m indicates the mass of the particle.

$$m\ddot{x} + b\dot{x} + ax = e \times E(t) \tag{10.10}$$

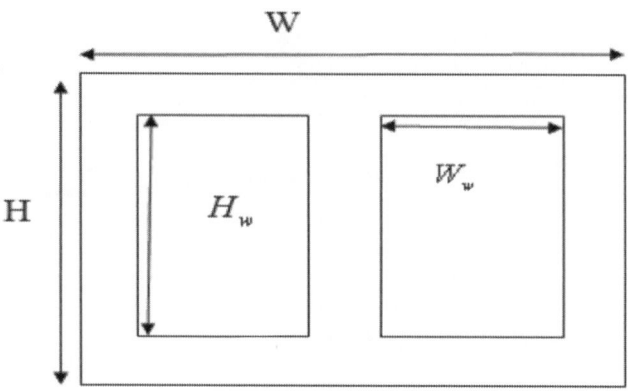

FIGURE 10.6

Overview of transformer core.

As observed, the aforementioned differential equation is related to an oscillating system when it is stimulated by the force $eE(t)$ and so, we have:

$$m\lambda^2 + b\lambda + a = 0 \tag{10.11}$$

The oscillation frequency can be calculated as Eq. (10.12).

$$f = \frac{1}{2\pi} \sqrt{\frac{a}{m} - \left(\frac{b}{2m}\right)^2} \tag{10.12}$$

The amount of attenuation equals the real component. However, $\left(\frac{b}{2m}\right)^2$ should be smaller than $\frac{a}{m}$ for the device to oscillate. In addition, resonance occurs when the driving force $e \times E(t)$ is regarded as an oscillating one as indicated in Eq. (10.13) and f_l is close to f.

$$e \times E(t) = F \sin(2\pi f_l t) \tag{10.13}$$

The general answer of the differential equation, i.e., Eq. (10.10), with the driving force according to Eq. (10.13) for a long time after connecting the oscillating device to the driving force can be written as follows:

$$x(t) = A \sin(2\pi f_l t) + B \cos(2\pi f_l t) \tag{10.14}$$

In addition, Eq. (10.15) can be written by assuming $\varphi = arc \tan\frac{B}{A}$ and $D = \sqrt{A^2 + B^2}$.

$$x(t) = D \cos(2\pi f_l t - \varphi) \tag{10.15}$$

The amplitude of the oscillation D and phase difference φ depend on the proximity of f_l to f. Full resonance occurs when f and f_l are considered equal. In addition, increasing the distance between f_l and f decreases the oscillation amplitude. Further, increasing the amplitude of the oscillation reduces the amount of energy converted to heat due to particle friction (insulation losses), resulting in raising the insulation loss coefficient concerning the resonance frequency.

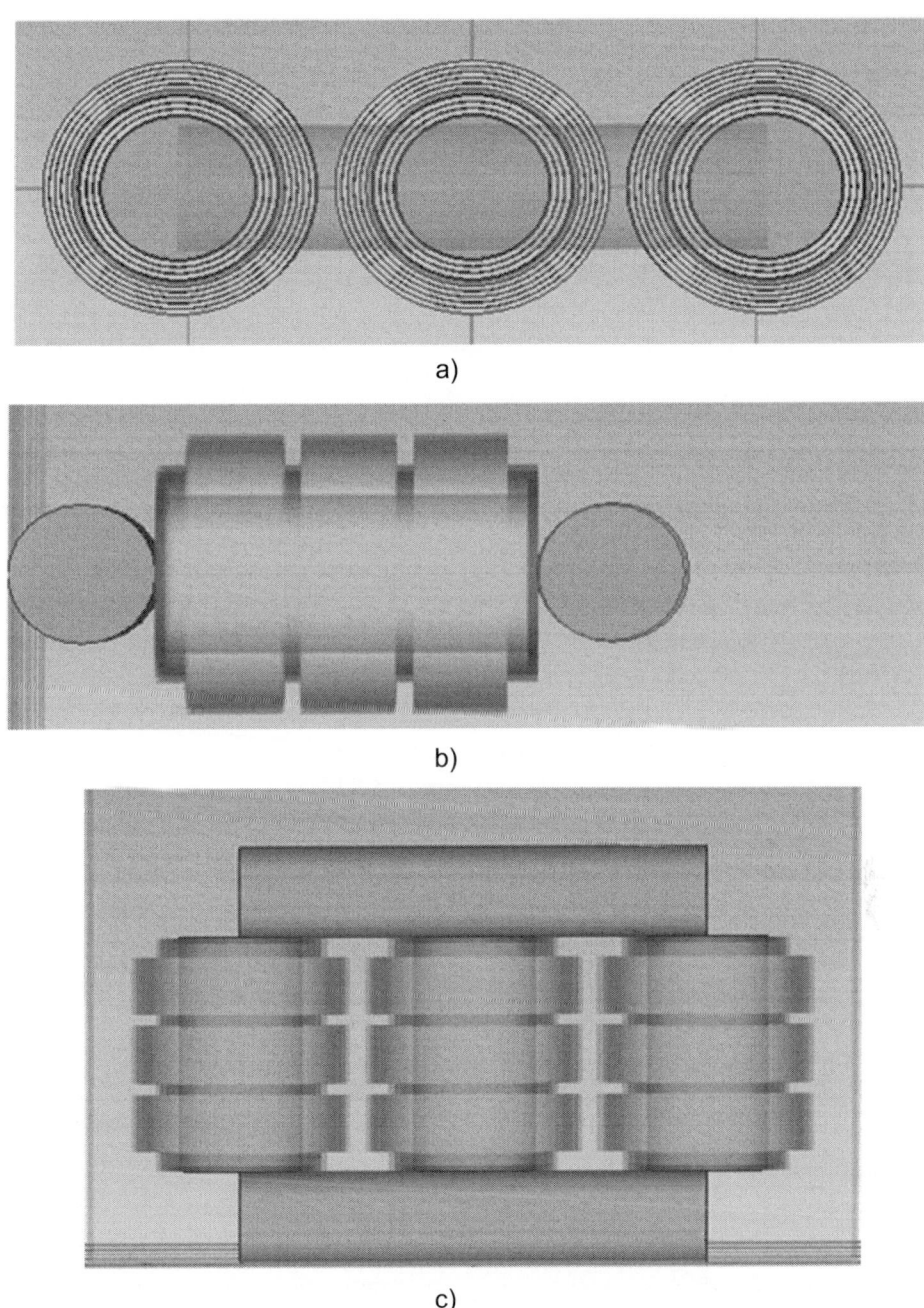

a)

b)

c)

FIGURE 10.7

Transformer simulated in CST software from different perspective views: (A) x−y, (B) y−z, and (C) x−z.

The frequency f depends on the coefficients a, b, and m. These coefficients differ for electronic, ionic, and rotational polarization. Specific coefficients are regarded as true for each material in an insulation, including a mixture of different materials. Thus sometimes the curve of changes in loss coefficient in an insulator in terms of frequency has several maximum and minimum points.

The system can oscillate with the excitation frequency at frequencies lower than the natural one. However, the amplitude of the oscillation increases significantly at the resonance frequency. The oscillation amplitude decreases as the frequency increases at frequencies higher than the natural one. In addition, the oscillation amplitude decreases more at frequencies that are significantly higher than the natural one since the principle of inertia prevents movement. Therefore polarization is not performed, and the dielectric value decreases for frequencies higher than the resonance one. The dielectric value of the material decreases when the frequency is significantly high. In addition, the dielectric value of water at low and optical frequencies is more than 80 and less than 2, respectively. The loss coefficient increases in resonance cases, and the dielectric value decreases as it passes through the resonance frequency.

As observed in Eq. (10.15), x and E indicate that the distances traveled by the charged particle in an alternating field and the intensity of the electric field are not in the same phase and have a phase difference of φ angle. Thus the amount of polarization has a phase difference with the field, meaning that the dielectric value for the alternating field is a complex one with a real and imaginary part.

Unlike vacuum response, the response of materials to external fields depends on the frequency of the applied field, indicating that the polarization of materials does not respond immediately to applied fields. Therefore the dielectric coefficient usually behaves as a complex function of the applied field frequency, meaning that $\varepsilon \rightarrow \widehat{\varepsilon}(\omega)$ (complex values can express phase in addition to size). The dielectric coefficient is defined as follows:

$$D_0 \exp(-i\omega t) = \widehat{\varepsilon}(\omega)E_0 \exp(-i\omega t) \tag{10.16}$$

The average response to static electric fields is expressed by the low-frequency range of the dielectric coefficient and is called the static dielectric coefficient ε_{DC} or ε_s.

$$\varepsilon_s = \lim_{\omega \to 0} \widehat{\varepsilon}(\omega) \tag{10.17}$$

The complex dielectric coefficient is usually displayed as ε_∞ in the high-frequency range. The dielectric acts as ideal materials at plasma frequencies and above. ε_{DC} is considered as an appropriate approximation for low-frequency alternating fields, and a measurable phase difference δ occurs between E and D when the frequency increases. The amount of frequency in which the fuzzy shift becomes significant depends on the temperature and details of the situation. D and E are regarded as proportional to each other for soft field strength (E_0), which can be written as follows:

$$\widehat{\varepsilon} = \frac{D_0}{E_0} = |\varepsilon|\exp(i\delta) \tag{10.18}$$

The real and imaginary parts of the dielectric coefficient are usually displayed separately and written in the following form conventionally when the response of materials to alternating fields is expressed by a complex dielectric coefficient:

$$\widehat{\varepsilon}(\omega) = \varepsilon'(\omega) + i\varepsilon''(\omega) = \frac{D_0}{E_0}(\cos\delta + i\sin\delta) \tag{10.19}$$

where, $\varepsilon'(\omega)$ and $\varepsilon''(\omega)$ indicate the real and imaginary parts of the dielectric coefficient, which are related to the stored energy and energy losses or scattering, respectively. In other words, the density of the current passing through the insulation is calculated as follows:

$$j(t) = i\varepsilon_0 \varepsilon_r \omega E(t) \tag{10.20}$$

The following equation can be written due to the sinusoidal changes in the intensity of the electric field, E, for alternating voltage and complex dielectric value, $\varepsilon' - i\varepsilon''$:

$$j(t) = \varepsilon_0 E_0 \omega(\varepsilon'' \cos \omega t - i\varepsilon' \sin \omega t) \tag{10.21}$$

In addition, the following equation can be written for the alternating current in the phasor domain:

$$J = i\varepsilon_0 \omega(\varepsilon' - i\varepsilon'')E = \varepsilon_0 \varepsilon'' \omega E + i\varepsilon_0 \varepsilon' \omega E \tag{10.22}$$

where J and E are considered as phasor. The loss coefficient or the ohmic to capacitive current ratio can be calculated as follows:

$$\tan \delta = \frac{\varepsilon''}{\varepsilon'} \tag{10.23}$$

10.6.2.2 Characteristics of model
Determining the correct characteristics related to oil and pressboard is among the effective factors in detecting PD defects utilizing EMWs. The oil characteristics were measured in the laboratory. Tables 10.2 and 10.3 show the characteristics of the oil and pressboard applied in the simulation based on the values obtained for different frequencies (Karami, Hejazl, & Gharchpetian, 2013).

10.6.3 Partial discharge model
10.6.3.1 Partial discharge current characteristics
Various practical works have been conducted over the years to find the characteristics of a PD defect in the transformer, dating back to 1968 (Wherry et al., 1968). The characteristics of the PD waveform in oil were calculated approximately in (Levy et al., 1996).

PD waveforms have extremely fast peak times of a few tenths to 10 ns and the Gaussian function is regarded as the most similar waveform to reality (Sarathi et al., 2008; Zhou & Li, 2005). The PD current function model can be demonstrated numerically as follows (Bojovschi et al., 2010; Judd et al., 2005b; Sellars et al., 1995; Zhou et al., 2005):

$$i(t) = I_0 \exp\left(-\frac{(t - t_0)^2}{2\sigma^2}\right) \tag{10.24}$$

where, I_0, t_0, and σ represent current amplitude, initial time, and the characteristic parameter of the waveform, respectively, and the latter determines the pulse width half maximum (PWHM). The PWHM parameter for the PD pulse equals 2.36σ, indicating the close relationship between the

Table 10.2 Dielectric characteristics of oil in simulation.

Frequency (GHz)	ε'	ε''	tan δ	Frequency (GHz)	ε'	ε''	tan δ
0.2	3.1585	1.4124	0.4471	6.14	2.0584	0.1614	0.07841
0.596	2.4957	0.2225	0.08915	6.536	2.0597	0.174	0.08448
0.992	2.3142	0.1934	0.08357	6.932	2.0425	0.1634	0.08
1.388	2.2973	0.1455	0.06333	7.328	2.0283	0.1666	0.08213
1.784	2.2622	0.1585	0.070065	7.724	2.0204	0.1521	0.07528
2.18	2.2415	0.1522	0.067901	8.12	1.9954	0.1524	0.07638
2.576	2.2136	0.1456	0.06577	8.516	1.9956	0.146	0.07316
2.972	2.1954	0.1457	0.06636	8.912	1.9836	0.1399	0.07053
3.368	2.1933	0.1673	0.07627	9.308	1.9793	0.1392	0.07033
3.764	2.164	0.1567	0.07241	9.704	1.9549	0.1261	0.064505
4.16	2.1493	0.1768	0.08226	10.1	1.955	0.1221	0.062455
4.556	2.1308	0.1765	0.08283	10.496	1.9445	0.1052	0.054101
4.952	2.1188	0.1665	0.07858	10.892	1.9322	0.1051	0.054394
5.348	2.1028	0.1856	0.08826	11.288	1.9322	0.0942	0.04873
5.744	2.0888	0.1586	0.07593				

Table 10.3 Dielectric characteristics of pressboard in simulation.

Frequency (GHz)	ε'	ε''	tan δ	Frequency (GHz)	ε'	ε''	tan δ
0.2	5.2461	2.3459	0.4471	6.14	3.4189	0.2681	0.7841
0.596	4.1452	0.3696	0.08915	6.536	3.4010	0.27	0.08448
0.992	3.8438	0.3212	0.08357	6.932	3.3925	0.2714	0.08
1.388	3.8157	0.2417	0.06333	7.328	3.3689	0.2767	0.08213
1.784	3.7574	0.2633	0.070065	7.724	3.3558	0.2526	0.07528
2.18	3.723	0.2528	0.067901	8.12	3.3142	0.2531	0.07638
2.576	3.6767	0.2418	0.06577	8.516	3.3	0.2425	0.07316
2.972	3.6464	0.242	0.06636	8.912	3.2946	0.2324	0.07053
3.368	3.6429	0.2779	0.07627	9.308	3.2875	0.2312	0.07033
3.764	3.5943	0.2603	0.07241	9.704	3.2470	0.2094	0.064505
4.16	3.5699	0.2937	0.08226	10.1	3.2471	0.2028	0.062455
4.556	3.5391	0.2932	0.08283	10.496	3.2297	0.1747	0.054101
4.952	3.5192	0.2765	0.07858	10.892	3.2093	0.1746	0.054394
5.348	3.4926	0.3083	0.08826	11.288	3.2109	0.1565	0.04873
5.744	3.4694	0.2634	0.07593				

insulation strength and the geometric shape of the PD gap. Generally, the small geometric diameter of the defect location increases the slope of the waveform and decreases the peak time, resulting in reducing the characteristic parameter σ (Jarman et al., 2008). The characteristic parameter in the current waveform helps to understand the PD situation created better.

The frequency of the PD current can be achieved by taking the Fourier transform from Eq. (10.24), as follows:

$$I(i\omega) = \sqrt{2\pi}I_0\sigma \, \exp\left(-\frac{\sigma^2\omega^2}{2}\right)\exp(-i\omega t_0) \tag{10.25}$$

where, ω is considered as the frequency angle. The PD pulse current can be written as an infinite sine series. The emitted electromagnetic signal can be modeled with a dipole antenna, the length (L) of which is determined by the geometric diameter of the insulation fault. The variable current with the corresponding time can be expressed as follows:

$$I = I(i\omega)\cos \omega t \tag{10.26}$$

Based on the theory of antennas, the amplitude of the field propagated in the far-field can be calculated using the following relation.

$$e_k \propto I_0\omega\sigma \, \exp\left(-\frac{\sigma^2\omega^2}{2}\right) \tag{10.27}$$

In addition, the frequency spectrum function of the transmitted signal is related to the PD current parameters as follows:

$$p(\omega) \propto \left(\frac{I_0\omega\sigma l}{c}\right)^2 \exp\left(-\sigma^2\omega^2\right) \tag{10.28}$$

Fig. 10.8 displays the frequency spectrum $p(\omega)$ of PD current in terms of ω. Based on Eq. (10.28) and Fig. 10.8, the highest value of $p(\omega)$ occurs at $\omega = \frac{1}{\sigma}$.

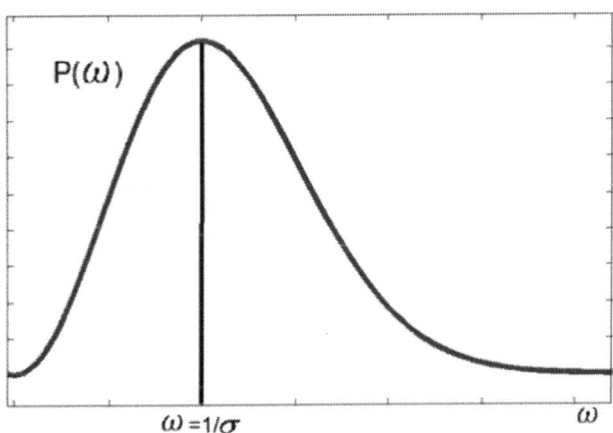

FIGURE 10.8

Frequency spectrum of PD signal.

10.6.3.2 Modeling in CST software

As indicated, a Gaussian current is selected to model the PD. To this aim, a discrete port is utilized in the CST software. The amplitude of the aforementioned current is mostly 5 mA in the simulation and experimental results (Zhou & Li, 2005). Changing the peak time of the PD current results in altering the frequencies in the signal emitted from the defect. Higher frequencies are received through the antenna in a short peak time. To change the peak time in the CST software, changes should be applied in the frequency range of the Gaussian signal. In other words, the low and high frequency of the signal should be 0 and 1.1 GHz, respectively, when the Gaussian signal with the characteristics indicated in the equations requires a peak time of 0.5 ns. In addition, the above-mentioned values should be 0 and 2.8 GHz, respectively, when the peak time is 0.2 ns. Figs. 10.9 and 10.10 show a Gaussian current modeled in the CST software and a discrete port, respectively.

The results obtained in the present section are related to the PD occurrence at the location (0,21.75,78.4) between the third and fourth turns of HV winding from the middle phase inside the pressboard insulation. In this section, it is assumed that the locations of the antennas are as $S_1(0,22.82,137.82)$, $S_2(0,-22.82,137.82)$, $S_3(74.8,0,137.82)$, and $S_4(-74.8,0,137.82)$, respectively. Fig. 10.11 illustrates the defect location in the simulation.

10.7 Simulation results of locating PD
10.7.1 Base case simulation

This case is regarded as the base and the others are compared with this. The peak time of the PD current is considered 0.1161 ns and the current peak amplitude is about 5 mA in open add space or normal air environment space. The maximum mesh length is 0.6317 mm with 55280640 meshes in the whole environment. There are 591, 245, and 385 meshes along the three axes X, Y, and Z, respectively. Fig. 10.12 displays the received signal in four antennas.

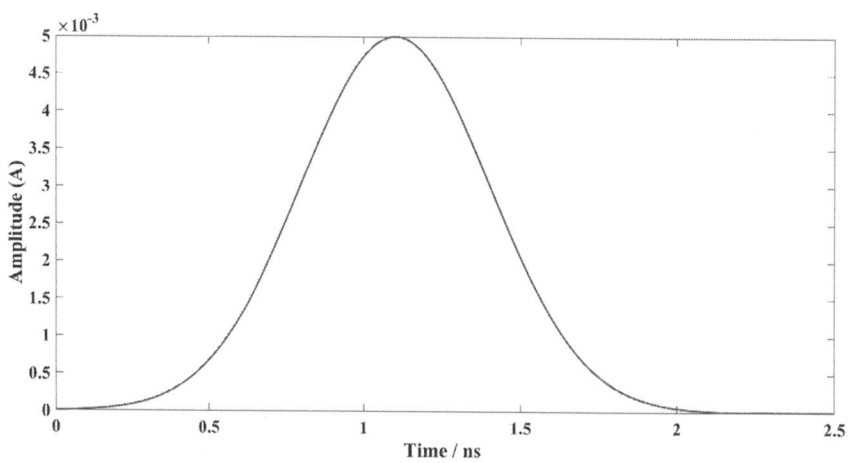

FIGURE 10.9

PD source current.

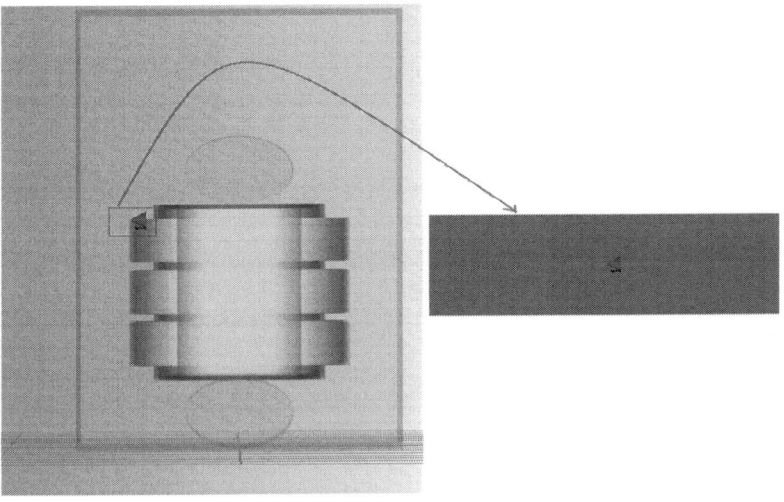

FIGURE 10.10

PD discrete ports.

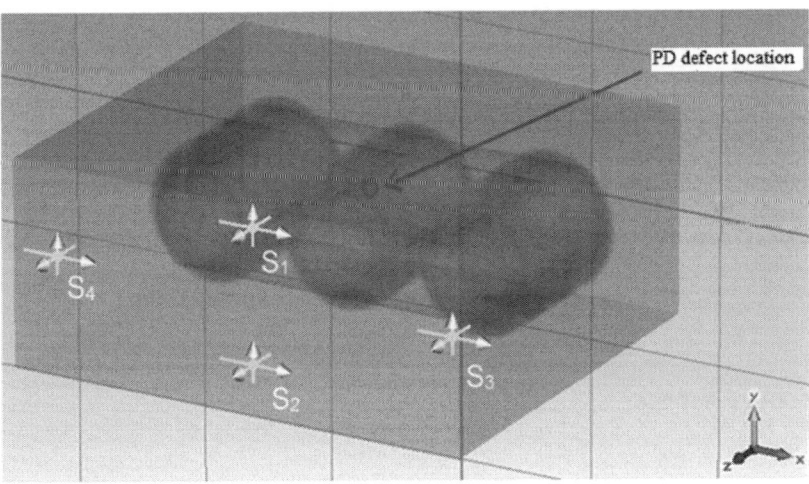

FIGURE 10.11

3D view of PD defect location.

As observed, the signals with the colors green and pink overlap because antennas S_3 and S_4 are symmetrical with respect to the defect location and receive the signal simultaneously. The TD between the signals approaching the antennas compared to the first antenna is 0.055, 0.22, and 0.22 ns, respectively, which result from the TD at the first peak in each of the signals received by the antennas. The error of estimating the defect relative to its actual location can be achieved from the equations

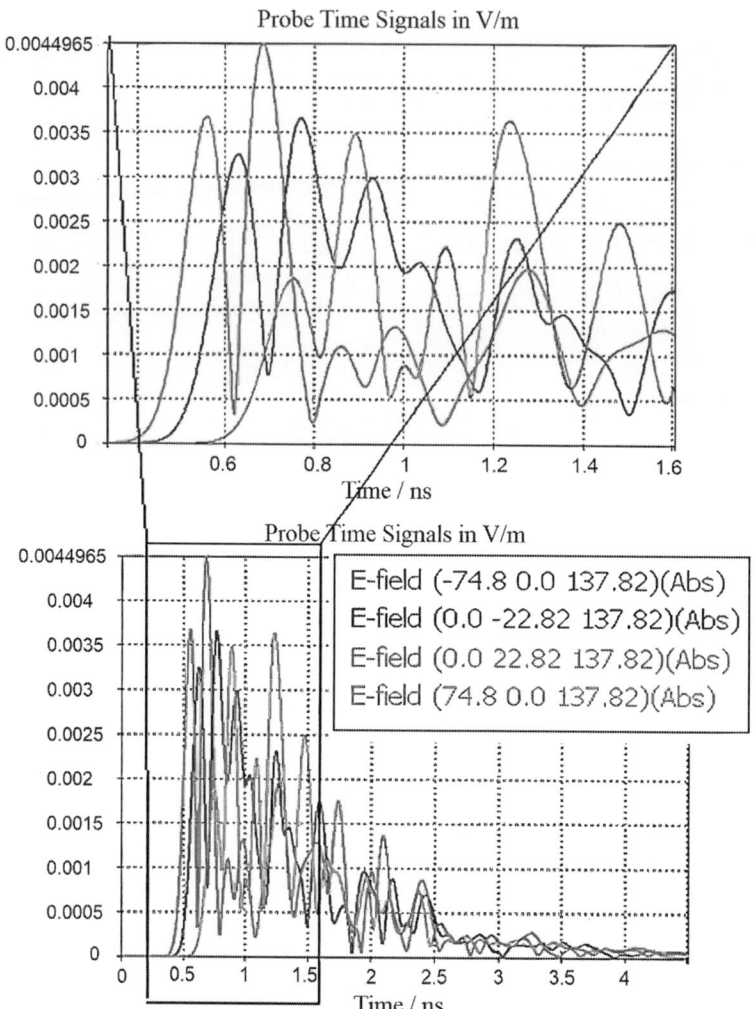

FIGURE 10.12

Signal received from four antennas in base case.

indicated in the TDOA method hypothesizing the speed of light in the environment to be $163.4 \frac{mm}{ns}$. It means that c, Δt_{ij}, and the location of the antennas are known and the location of PD, (x, y, z), and t_1 are unknown. Such nonlinear equations can be solved using the fsolve function in the MATLAB software. The estimated location is at $(0,12.58,77.88)$, which is 8.91 mm away from the actual one.

The error derives from the approximation taken at the constant speed of light in the environment since such speed differs from that in oil for two reasons with a constant dielectric coefficient of 2.33:

- The simulated oil has wave scatter and has different velocities for different frequencies. The transmitted signal has various frequencies, the velocity of which is calculated difficultly.
- The transmitted signal does not pass through the oil directly. Rather, the signal faces various objects, including the pressboard and different metals, in its path to the antenna.

Therefore the speed of light changes in the simulation and leads to localization error.

10.7.2 Effect of PD peak current value

The only difference between the present case and the previous one is the PD peak current value. Here, the effect of the PD current value on defect location is studied. The current peak is considered 100 mA in the current case. Like the previous case, the peak time of the PD current amplitude is considered 0.1161 ns and simulated in open add space. In addition, all of the mesh characteristics are the same as before. Fig. 10.13 shows the received signals on four antennas.

As observed, Figs. 10.2 and 10.13 differ only in terms of the peak value. In addition, the defect location is the same as before and does not differ in error. Further, all the parameters are the same as in the previous simulation, such as the TD between the signals. Thus, the location is placed at $(0, 12.85, 77.88)$, which is 8.91 mm away from the actual location as before.

10.7.3 Effect of wave scattering

The complex insulation value of the oil and pressboard was discussed in the oil model section. In the present section, the difference in the PD defect location is investigated by performing a simulation in which all of the characteristics other than the wave scattering effect are the same as the base sample simulation. The PD current and the open add space is the same as for the base case. Fig. 10.14 illustrates the magnitude of the received signal in the four antennas for the no-scatter wave mode.

The attenuation of the wave decreases in the absence of the wave scattering. As demonstrated in Fig. 10.14, the field strength is still high after 14 ns and is hundreds of times greater than the wave scattering condition, as shown in Fig. 10.12. The field strength is considered as negligible compared to the initial time values after 4 ns due to wave scattering.

The TD between the signals approaching the antennas and after the simulation compared to the first antenna are 0.0743, 0.2047, and 0.2047 ns, respectively. The error of estimating the defect relative to its actual location can be obtained from the equations indicated in the TDOA method, assuming the speed of light in the environment to be $186.2 \frac{mm}{ns}$. The achieved location is placed in the coordinates $(78.32, 20.14, 0)$, which is 1.614 mm away from the actual one. The error amount decreases while it increases when the number of meshes reduces. Comparing such error with the previous results indicates that the existence of the wave scattering in the oil and pressboard insulation considering complex values is among the reasons for an increase in error values due to the effect of the imaginary and real parts of the insulation dielectric value, described in the oil model section. The imaginary part deforms the waveform sent from the defect location by attenuating and dispersing energy, resulting in losing some location information. The signal emitted from the defect location has several frequencies,

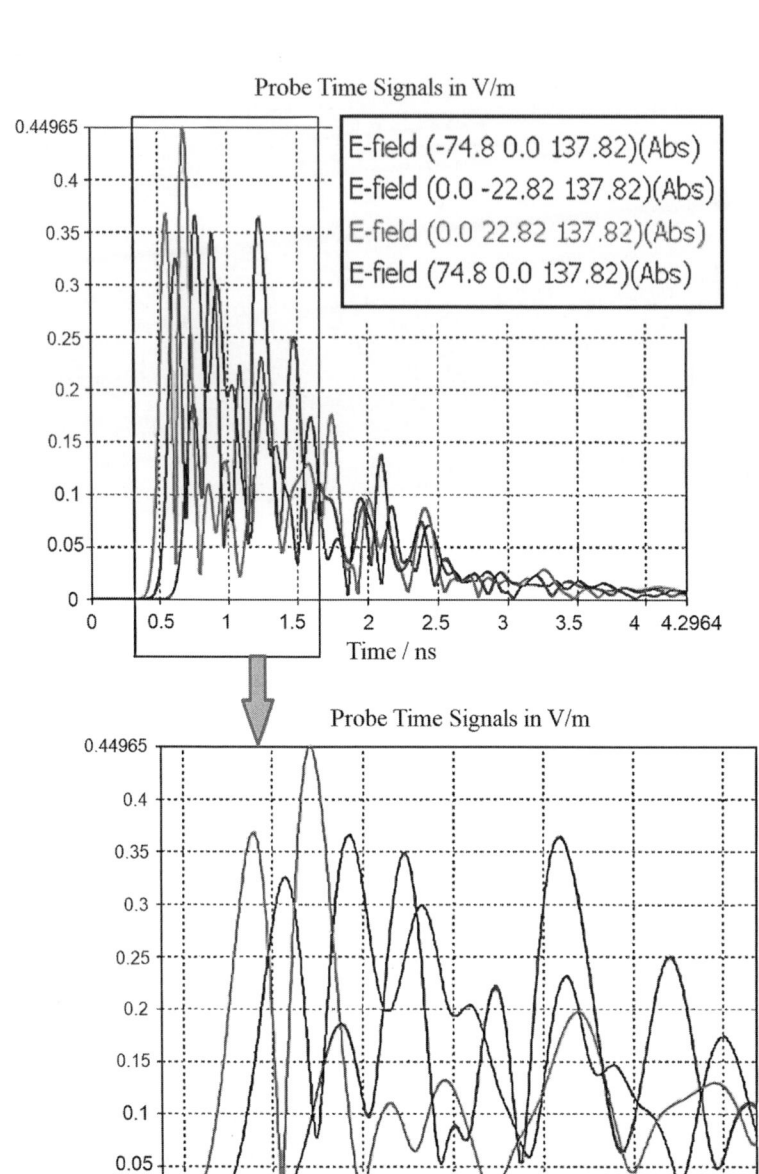

FIGURE 10.13

Signal received from four antennas considering effect of PD peak current value.

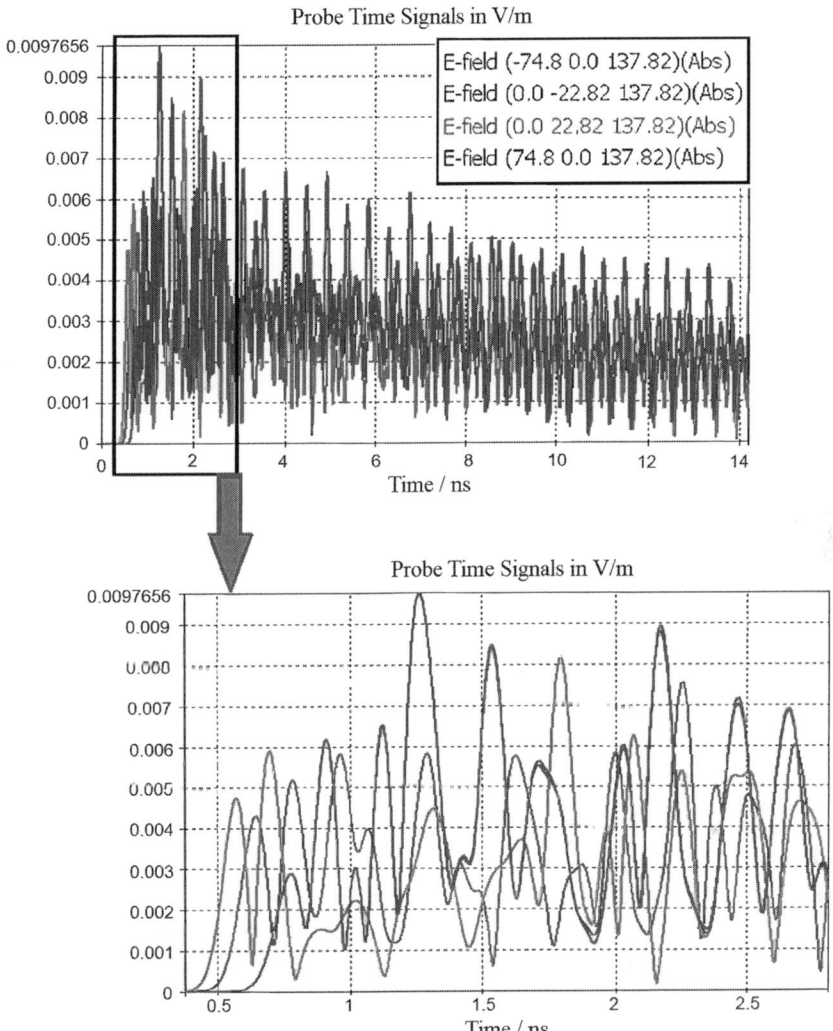

FIGURE 10.14

Signal received from four antennas without wave scatter.

and the variableness of the imaginary part in the insulation number for different frequencies makes the changes incompatible with such frequencies in the signal.

10.7.4 Effect of δ or PWHM

The PD model was discussed in previous sections and the PWHM or δ parameter is defined by the PD current waveform characteristics. This section analyzes the effect of increasing the PWHM parameter after its change. To this aim, the peak time in PD current increases to 0.3 ns from the amplitude range between 0.1−0.9 and is simulated in the open add space. The maximum mesh length is 0.6895 mm with 46105476 meshes in the whole environment. In addition, there are 552, 229, and 368 meshes along the three axes X, Y, and Z, respectively. Fig. 10.15 displays the received signals in four antennas and the S_1 signal in three different dimensions.

As observed, Fig. 7.16 differs from Fig. 7.13. The TDs are 0.0767, 0.1623, and 0.1623 ns. The error of estimating the defect location relative to its actual location can be achieved from the equations

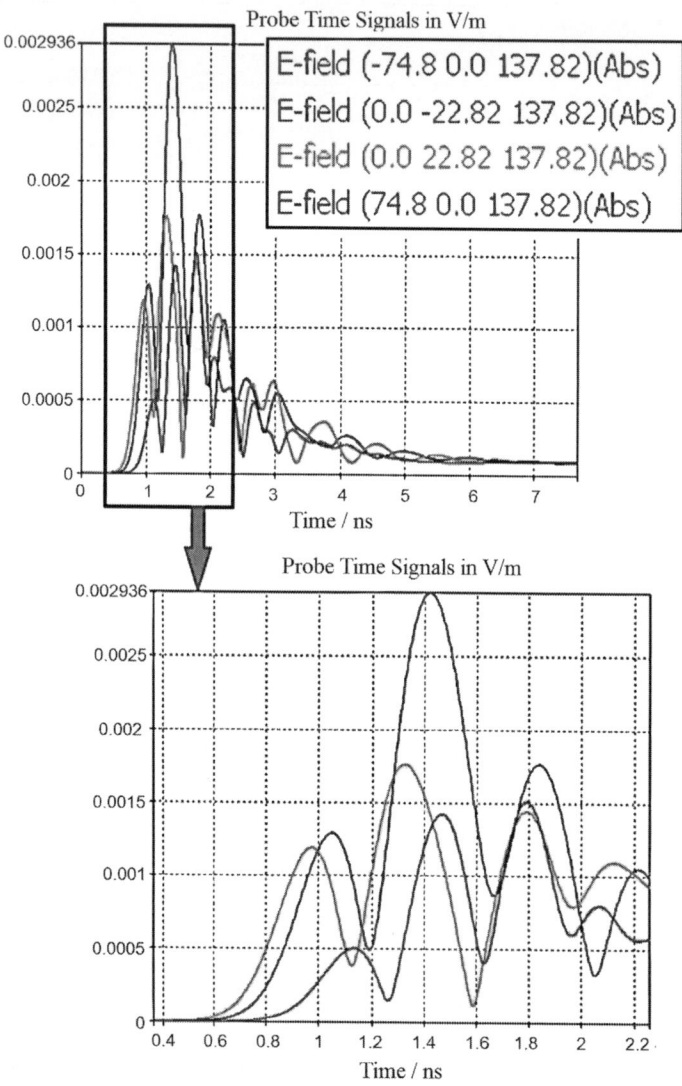

FIGURE 10.15

Signal received from four antennas with increasing PWHM.

indicated in the TDOA method, hypothesizing the speed of light in the environment to be $248 \frac{mm}{ns}$. The location obtained is placed at (79,28.573,0), which is 6.84 mm away from the actual one.

Based on the results, increasing the peak time in the PD current decreases the error in defect detection since the frequencies in the signal emitted from the defect location decrease with an increase in the PD current. The signal changes reduce with a decrease in the number of signal frequencies in the environment due to the variableness of insulation dielectric value, resulting in detecting the defect location more accurately.

References

Ahmadi, S.-A., & Sanaye-Pasand, M. (2021). A robust multi-layer framework for online condition assessment of power transformers. *IEEE Transactions on Power Delivery, 37*(2).

Azadifar, M., Karami, H., Wang, Z., Rubinstein, M., Rachidi, F., Karami, H., Ghasemi, A., & Gharehpetian, G. B. (2020). Partial discharge localization using electromagnetic time reversal: A performance analysis. *IEEE Access, 8*, 147507–147515.

Bojovschi, A., Rowe, W., & Wong, A. K. L. (2010). Electromagnetic field intensity generated by partial discharge in high voltage insulating materials. *Progress in Electromagnetics Research, 104*, 167–182.

Chan, Y. T., & Ho, K. C. (1994). A simple and efficient estimator for hyperbolic location. *IEEE Transactions on Signal Processing, 42*(8), 1905–1915.

Coenen, S. (2012). *Measurement of partial discharges in power transformers using electromagnetic signals.* BoD—Books on Demand.

Coenen, S., Tenbohlen, S., Markalous, S. M., & Strehl, T. (2008). Sensitivity of UHF PD measurements in power transformers. *IEEE Transactions on Dielectrics and Electrical Insulation, 15*(6), 1553–1558.

Convery, A. R., & Judd, M. D. (2003). Measurement of propagation characteristics for UHF signals in transformer insulation materials. In *2003 13th international symposium on high voltage engineering (ISH)*.

Jarman, P., Tenbohlen, S., Judd, M. D., Olof Stenestam, B., & Viereck, K. (2008). Recommendations for condition monitoring and condition assessment facilities for transformers. *Electra, 237*, 48–57.

Jeyabalan, V. (2011). Coherent phase detection technique for location of partial discharge in transformer windings. *IEEE Transactions on Power Delivery, 26*(4), 2885–2886.

Judd, M. D., Farish, O., Pearson, J. S., & Hampton, B. F. (2001). Dielectric windows for UHF partial discharge detection. *IEEE Transactions on Dielectrics and Electrical Insulation, 8*(6), 953–958.

Judd, M. D., Yang, L., Bennoch, C. J., & Hunter, I. (2004). UHF diagnostic monitoring techniques for power transformers. In *EPRI substation equipment diagnostics conference XII*.

Judd, M. D., Yang, L., & Hunter, I. B. (2005a). Partial Discharge Monitoring for Power Transformers Using UHF Sensors Part 2: Field Experience-The results of PD tests on power transformers provide sufficient evidence to justify making provision. *IEEE Electrical Insulation Magazine, 21*(3), 5–13.

Judd, M. D., Yang, L., & Hunter, I. B. (2005b). Partial discharge monitoring of power transformers using UHF sensors. Part I: Sensors and signal interpretation. *IEEE Electrical Insulation Magazine, 21*(2), 5–14.

Kang, P., & Birtwhistle, D. (2001). Condition monitoring of power transformer on-load tap-changers. Part 2: Detection of ageing from vibration signatures. *IEE Proceedings-Generation, Transmission and Distribution, 148*(4), 307–311.

Karami, H., Gharehpetian, G. B., & Hejazi, M. S. A. (2013). Oil permittivity effect on PD source allocation through three-dimensional simulation. In *International Power system conference, Tehran-Iran* (pp. 1–5).

Karami, H., Hejazi, M. S. A., & Gharehpetian, G. B. (2013). Simulation of transformer oil effect on PD source allocation. In *, PDC'13. 4th conference on partial discharge in electrical apparatus* (pp. 26–27).

Karami, H., Gharehpetian, G. B., Norouzi, Y., & Hejazi, M. A. (2016). GLRT-based mitigation of partial discharge effect on detection of radial deformation of transformer HV winding using SAR imaging method. *IEEE Sensors Journal, 16*(19), 7234–7241.

Karami, H., Gharehpetian, G. B., Norouzi, Y., & Hejazi, M. A. (2018). Experimental study on elimination of partial discharge effect on detection of radial deformation of high voltage transformer winding using electromagnetic waves. In *2018 IEEE international conference on environment and electrical engineering and 2018 IEEE industrial and commercial power systems Europe* (pp. 1–5). EEEIC/I&CPS Europe).

Karami, H., Hejazi, M. S. A., Naderi, M. S., Gharehpetian, G. B., & Mortazavian, S. (2012). Three-dimensional simulation of PD source allocation through TDOA method. In , *1–4. The 4th conference on thermal power plants.*

Karami, H., Azadifar, M., Mostajabi, A., Rubinstein, M., Karami, H., Gharehpetian, G. B., & Rachidi, F. (2020). Partial discharge localization using time reversal: Application to power transformers. *Sensors, 20*(5), 1419.

Karami, H., Gharehpetian, G. B., Norouzi, Y., & Akhavan-Hejazi, M. (2020). Simultaneous radial deformation and partial discharge detection of high-voltage winding of power transformer. *IET Electric Power Applications, 14*(3), 383–390.

Karami, H., Mostajabi, A., Rachidi-Haeri, F., Azadifar, M., & Rubinstein, M. (2021). *Partial discharge localization using time reversal: Application to power transformers and gas-insulated substations.*

Karami, H., Tabarsa, H., Gharehpetian, G. B., Norouzi, Y., & Hejazi, M. A. (2019). Feasibility study on simultaneous detection of partial discharge and axial displacement of HV transformer winding using electromagnetic waves. *IEEE Transactions on Industrial Informatics, 16*(1), 67–76.

Kawada, M., Kawasaki, Z., Matsuura, K., Kuroki, S., Osawa, T., & Tanaka, H. (1998). Detection of partial discharge in operating turbine generator using GHz-band spatial phase difference method. In *Proceedings of 1998 international symposium on electrical insulating materials. 1998 Asian international conference on dielectrics and electrical insulation. 30th symposium on electrical insulating ma* (pp. 71–74).

Kramers, H. A. (1927). La diffusion de la lumiere par les atomes. *Atti Congresso Internazionale di Fisica, 2,* 545–557.

Kronig, R. de L. (1926). On the theory of dispersion of x-rays. *Josa, 12*(6), 547–557.

Levy, A. F., Oliveira Filho, O. B., & Olivieri, M. M. (1996). Ultra wideband measurements of partial discharges in air and insulating oil. In *ICDL'96. 12th international conference on conduction and breakdown in dielectric liquids* (pp. 218–221).

Lundgaard, L. E. (1992a). Partial discharge. XIII. Acoustic partial discharge detection-fundamental considerations. *IEEE Electrical Insulation Magazine, 8*(4), 25–31.

Lundgaard, L. E. (1992b). Partial discharge. XIV. Acoustic partial discharge detection-practical application. *IEEE Electrical Insulation Magazine, 8*(5), 34–43.

Meijer, S., Agoris, P. D., Smit, J. J., Judd, M. D., & Yang, L. (2006). Application of UHF diagnostics to detect PD during power transformer acceptance tests. In *Conference record of the 2006 IEEE international symposium on electrical insulation* (pp. 416–419).

Mirzaei, H. R., Akbari, A., Kharezi, M., & Jafari, A. M. (2008). Experimental investigation of PD diagnosis based on consecutive pulses data. In *2008 international conference on condition monitoring and diagnosis* (pp. 1262–1265).

Mizuno, K., Ogawa, A., Nojima, K., Murase, H., Koyama, H., Wakabayashi, S., & Sakakibara, T. (1996). Investigation of PD pulse propagation characteristics in GIS. In *Proceedings of 1996 transmission and distribution conference and Exposition* (pp. 204–212).

Paoletti, G. J., & Golubev, A. (2001). Partial discharge theory and technologies related to medium-voltage electrical equipment. *IEEE Transactions on Industry Applications, 37*(1), 90–103.

Raja, K., & Floribert, T. (2002a). Comparative investigations on UHF and acoustic PD detection sensitivity in transformers. In *Conference record of the the 2002 IEEE international symposium on electrical insulation (cat. No. 02CH37316)* (pp. 150—153).

Raja, K., & Floribert, T. (2002b). Source characterization of discharges in transformers using UHF PD signatures. In *2002 IEEE power engineering society winter meeting. Conference proceedings (cat. No. 02CH37309), 2* (pp. 1383—1388).

Sarathi, R., Giridhar, A. V., Mani, A., & Sethupathi, K. (2008). Investigation of partial discharge activity of conducting particles in liquid nitrogen under DC voltages using UHF technique. *IEEE Transactions on Dielectrics and Electrical Insulation, 15*(3), 655—662.

Sawhney, A. (1984). *Principles of electrical machines design* (5th ed.). Published by JC.

Sellars, A. G., MacGregor, S. J., & Farish, O. (1995). Calibrating the UHF technique of partial discharge detection using a PD simulator. *IEEE Transactions on Dielectrics and Electrical Insulation, 2*(1), 46—53.

Standard, I. (2000). High-voltage test techniques: Partial discharge measurements. *IEC*, 13—31.

Stone, G. C. (1991). Practical techniques for measuring partial discharges in operating equipment. In *[1991] proceedings of the 3rd international conference on properties and Applications of dielectric materials* (pp. 12—17).

Tenbohlen, S., Denissov, D., Hoek, S. M., & Markalous, S. M. (2008). Partial discharge measurement in the ultra high frequency (UHF) range. *IEEE Transactions on Dielectrics and Electrical Insulation, 15*(6), 1544—1552.

Wenzhi, C., Zhiguo, T., Chengrong, L., Hao, W., Yuan, Y., & Yimin, J. (2009). A novel sensor for UWB RF PD location and experimental investigation on a 220kV transformer. In *2009 IEEE conference on electrical insulation and dielectric phenomena* (pp. 413—416).

Wherry, F. E., Toothman, L. R., Yakov, S., & Preston, L. L. (1968). The significance of corona measurements on transformers. *IEEE Transactions on Power Apparatus and Systems, 11*, 1889—1898.

Wild, G., & Hinckley, S. (2008). Acousto-ultrasonic optical fiber sensors: Overview and state-of-the-art. *IEEE Sensors Journal, 8*(7), 1184—1193.

Zeng, D. (1998). An improved method of measuring C1 power factor of RG (resistance-graded) bushings. *IEEE Power Engineering Review, 18*(10), 59.

Zhou, L., & Li, W. (2005). Characteristic estimation of partial discharge from its radiating signal. In *2005 5th international conference on information communications & signal processing* (pp. 757—760).

Zhou, L., Li, W., & Su, S. (2005). The deduction of partial discharge pulse current from its radiating UHF signal. In *2005 international power engineering conference* (pp. 1—193).

Index

CPI Antony Rowe
Eastbourne, UK
February 21, 2023